T4-AKS-542

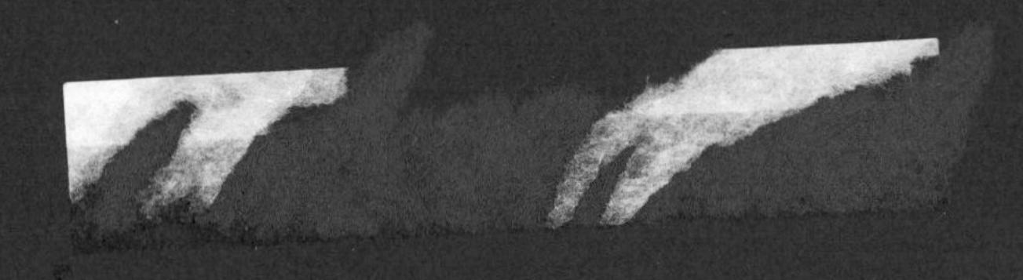

Principles of
Industrial Measurement
for Control Applications

Principles of **Industrial Measurement** *for Control Applications*

Ernest Smith
IBM United Kingdom Limited

INSTRUMENT SOCIETY OF AMERICA

Principles of Industrial Measurement for Control Applications

Printed in the United States of America

This work is an expansion of the author's earlier work entitled *Sensor Based Engineering*, published by International Business Machines Corporation © 1976. The republication of portions of *Sensor Based Engineering* has been licensed to Instrument Society of America which expresses its appreciation for making this enlarged work available.

In preparing this work the author and publisher have not investigated nor considered patents which may apply to the subject matter hereof. It is the responsibility of the readers and users of the subject matter to protect themselves against liability for infringement of patents. The information contained herein is of a general educational nature. Accordingly, the author and publisher assume no responsibility and disclaim all liability of any kind, however arising, as a result of using the subject matter of this work.

The equipment referenced in this work has been selected by the author as examples of the technology. No endorsement of any product is intended by the author or publisher. In all instances, the manufacturer's procedures should prevail regarding the use of specific equipment. No representation, expressed or implied, is made with regard to the availability of any equipment, process, formula, or other procedures contained herein.

Instrument Society of America
67 Alexander Drive
P.O. Box 12277
Research Triangle Park, NC 27709

The publishers extend their thanks to Gadi Kaplan, D.Sc. E.E., for his assistance as technical consultant.

Library of Congress Cataloging in Publication Data

Smith, Ernest, 1927–
Principles of industrial measurement for control applications.

Includes index.
1. Engineering instruments—Handbooks, manuals, etc.
2. Mensuration—Handbooks, manuals, etc. I. Title.
TA165.S58 1984 620'.044 84-12992
ISBN 0-87664-465-5

Design and book production
by Publishers Creative Services Inc. New York

PREFACE

Most books are written as entities; this one, like little Topsy, "just growed." Its origins lie in an accumulation of notes dating from 1951, when I started my career as an instrumentation engineer, with the then Bristol Aeroplane Company. In this, my new career, it did not take me long to discover that information pertinent to my particular technical concerns was sparse. To a considerable extent it resided either in the experience of my colleagues, or was hidden in a mountain of largely irrelevant literature. As a consequence, I formed the habit of making notes of any material I could glean that I thought might someday be useful. Over the years this collection became voluminous and would have remained incomprehensible to anyone but myself had it not been for the accident of a change in my career.

I joined IBM in 1969 as an instrument specialist in one of the company's Data Acquisition and Control Centers where I became involved with engineers responsible for the installation of computers in industrial operations. At a very early state in this involvement, I was to receive two surprises. First, I was very surprised to discover that many of these engineers had no formal training in the application of instrumentation. Even more surprising, however, was the discovery that there was an apparent lack of any general literature in a form convenient to their requirements. In considering ways in which this situation might be alleviated, it occurred to me that I might make good use of my notes, and so I set about assembling and editing them. The outcome of this work was a manual entitled *Sensor Based System Engineering*, published by IBM in 1976.

The manual was eventually noticed by the Publication Department of ISA. On review, it was considered that it would form an ideal basis for a more comprehensive work, whose style and content would complement existing ISA publications. I was subsequently approached by ISA as to whether I might be interested in undertaking such a task. I must admit that for some time I hesitated—mainly because of my misgivings about the age of some of the material I had used in the manual. My first thoughts were that any book based on these materials would be out of date, but on reflection I realized this was far from the truth. Many of the principles used by instrumentation engineers were discovered several hundred years ago. While scientific devices exploiting these principles have been improved, their basic form remains substantially unchanged. Any book, therefore, that follows the format that I had adopted in the manual must be open ended and in a sense timeless. It is dated only by the absence of new developments and not by the supersession of the old. With this realization, I decided that the effort to expand the work would be worthwhile.

The final stage of growth has been long and tortuous. At the outset the decision to produce an ISA book based on an IBM manual created the need for extremely complex negotiations between the two organizations. The task was undertaken by ISA and its successful conclusion speaks well for ISA's tenacity and persuasiveness. That the negotiations were successful also reflects the generosity of IBM, which allowed large sections of the original manual and many of the original drawings to be reproduced.

From my own point of view, I do not find writing the easiest of occupations and even under ideal circumstances the task would have been difficult. In retrospect, had I realized the difficulties I might have hesitated considerably longer before accepting the invitation. It is incredible how difficulties are magnified when the author and publisher are separated by the Atlantic Ocean and the nuances of what is supposed to be a common language. Simple problems that could have been solved by a few minutes of personal contact became the subjects of extensive correspondance and frequent misunderstanding. I shall always remember with affection the fortitude with which Sidney Solomon and his staff at Publishers Creative Services bore this burden and I thank them for their forebearance and cooperation.

If friendship is defined as the extent of effort that one person will expend for another, then I have found a very good friend in one of my colleagues from Boca Raton, Florida. Thomas J. Harrison spent many hours reading the text and offered good advice that I would have been foolish to ignore. In addition, Dr. Harrison took on the mammoth job of correcting those differences in spelling and word usage that we have come to describe as "Britishisms." The amount of red ink he expended on the manuscripts is an indication of my naivete in believing that the English and American people speak the same language.

Throughout the book there are many illustrations and quotations from proprietary material. It is a pleasure to express my thanks to the many firms involved, not only for allowing me to publish this material, but also for the assistance I received in ensuring that I was supplied with the latest and most complete information available. Without exception my requests were met with prompt and courteous reponses.

Finally, I wish to thank a large number of people who are now no longer part of my recollection. When I began to collect my notes, more than 30 years ago, it never occurred to me that I might one day wish to publish them. Consequently I made no effort to record my sources of information in a meticulous manner. It is to my profound regret that I should now wish to recognize so many colleagues, firms, representatives, and teachers who have all contributed indirectly to this book and be unable to do so. I can only reflect that by including the knowledge they imparted to me I am paying them the highest tribute that is in my power to offer.

Ernest Smith

List of symbols

A	Area
a	Area
α	Temperature coefficient of resistance (alpha)
β	Magnetic flux density (beta)
C	Capacitance
c	Coefficient of damping
c_c	Coefficient of critical damping
d	Diameter
E	Electric field intensity
ϵ	Permittivity
ϵ_r	Relative permittivity or dielectric constant
ϵ_0	Permittivity of free space
ϵ	Young's modulus
F	Force
f	Frequency
ϕ	Magnetic flux (phi)
G	Modulus of rigidity
G	Amplifier gain factor
G_F	Gage factor
g	Acceleration constant due to gravity
γ	Damping ratio (gamma)
H	Intensity of magnetic field
h	Height

I Second moment of area
i Current $\sqrt{-1}$
j Mathematical operator
k Spring constant
L Inductance
L Load
l Length
λ Coefficient of linear expansion (lambda)
M Bending moment
m Mass
μ Magnetic permeability (mu)
μ_0 Magnetic permeability of free space
μ_r Relative permeability
N Number of turns in a coil
ν Mechanical stress (nu)
p Pressure
q Electric flux
q Shear strain
R Resistance
r Radius
$\mathcal{R}$ Gas constant
ρ Resistivity (rho)
ρ Liquid density (rho)
σ Poisson ratio (sigma)
σ Electric field (sigma)
T Absolute temperature
T Torque
t Temperature
t Thickness
t Time
τ Time constant (tau)
τ Period of pendulum (tau)
V Volume
V Voltage
v Voltage
v Velocity
W Watts
X_c Capacitive reactance
X_L Inductive reactance
Z Electrical impedance
ω Angular velocity (omega)

Contents

Principles of **Industrial Measurement** *for Control Applications*

CHAPTER 1

THE DEVELOPMENT OF INDUSTRIAL INSTRUMENTATION

INTRODUCTION

Making measurements is a means to an end and not an end in itself. Whenever people undertake a project they very quickly reach the point where a measurement must be made. If instruments to make that measurement are not available, then they must be invented. Necessity is the catalyst that has led to the development of modern instrumentation, and it is reasonable to assume that it has always existed. It follows, therefore, that people must have developed some sort of crude measurement almost from the time that they began to use tools. If this is so, then those who make measurements today must be participating in the oldest science. This chapter does not purport to be a history of measurement techniques nor does it claim to be an accurate chronology of events. It is rather a reflection on what appears to be an inseparable relationship between the development of society to its present state and the corresponding development in the science of measurement. It is also, by its nature, a tribute to a host of unknown people whose efforts from the beginning of time have created the profession we call instrumentation engineering.

The early history of measurement is a fascinating subject for idle contemplation. So much has been lost in the obscurity of time that now we may only look at the wrecks of many dead civilizations and guess how their builders managed to work to such accuracy. From time to time archeologists find instruments in these heaps of rubble, and deduce their use. In these instances it is often possible to obtain glimpses of different solutions to a common problem developed by people completely isolated from each other by time and distance. The measurement of time is a classic example, and the history of horology demonstrates how many individual solutions have been applied to that problem.

Although our knowledge of prehistoric instrumentation is nonexistent, it is possible to imagine some of the problems which early man faced and guess at possible solutions. In more recent years the problems associated with making measurements in an environment of rapidly growing complexity are better understood and are documented in considerable detail. While it is not possible to examine these problems in depth, even a superficial consideration of their progressive solution makes possible an appreciation of the pressures that forced the development of the instrumentation industry as we know it today.

IN THE BEGINNING

If we regard the application of continuous error correction to estimations of distance and speed as a form of measurement, then it follows that the ability to make measurements is not the sole prerogative of the human species. An animal stalking a moving prey is making these measurements, and also subconciously making complex calculations, which, if correct, will lead to a meal, and if in error, can lead to the predator's extinction. Before the discovery of tools the primitive human must have possessed these abilities, possibly in a more highly developed form than other animals. Where humans separated themselves from the animal world was in the way that they began to use artificial aids to enhance their own abilities. The first tools were probably nothing more or less than naturally occurring materials, whose shape and composition they came to recognize as an aid to performing a task more easily. Could this be the origin of the first measurement?

A problem which prehistoric people must frequently have faced was that of crossing a fast-flowing, murky river, with an attendant danger of stepping in over their depth. At some time someone must have realized that it was possible to test the depth of the water ahead with a stick. In the process that person invented the first dipstick.

Considering such a trivial measurement from the end of a million years of evolution it is easy to overlook the level of intelligence required to grasp the concept. This primitive person was not only learning to make a measurement with an instrument but was learning to compare the measurement with his or her own stature and from it to estimate the chances of survival. There must have been times when the choice between facing the dangers of drowning or turning to face the danger from which the person was escaping must have been a fine one.

Writers of prehistory appear at times to underestimate the problems of making a measurement without the benefit of precedence, and in the process they unwittingly attribute a degree of intelligence to a primitive person which they would not normally allow. We are told, for example, that people became farmers at an early stage of development, farming crops as well as animals. To farm crops a knowledge of the seasons is essential, and it has been suggested that to satisfy this need these early people invented a shadow stick. It is true that this is a device sufficently simple for a primitive person to erect, and a relatively primitive person may in time recognize the movement of the shadows to be cyclic from day to day. To recognize that the change in position of the shadow at sunrise is also cyclic from year to year, and to interpret the position in terms of the seasons, however, requires the application of intelligence of a different order. It is reasonable, therefore, to assume that either the primitive farmer had a much higher intelligence than is credited, or alternatively, that the shadow stick did not come into use until much more recent times.

INSTRUMENTATION IN EARLY CIVILIZATIONS

The process by which measurement techniques developed in prehistoric times can never be more than conjecture, and even in the civilizations which existed within the last 4500 years our knowledge is sparse. Best-known measuring instruments of these civilizations are those used for building and those used for measuring time, probably because they are the most easily recognized. Enough evidence exists to show that ingenious devices were produced to make less common measurements, and it is possible that some of these, entitled "Ritual Objects," adorn the shelves of museums. Even so, some instruments have been recognized, and from them it is possible to obtain a tantalizing glimpse of the state of excellence to which instrumentation had developed. In the Far East, for example, it is known that nomadic tribes navigated their chariots round the Mongolian desert with the aid of a sun compass. The designer's name may have been lost in antiquity, but the ingenuity of the device must be a source of continuing admiration. Also, in the Far East about 130 A.D. a Chinese astronomer by the name of Chang Hang invented a device for recording the occurrence of earthquakes and indicating their approximate direction. Seventeen hundred years later, the principle was reinvented by a European scientist by the name of Niccolo Cacciatore.

As civilization spread westward the Middle East also contributed to the development of instrumentation. One significant discovery recovered from the Mediterranean was part of a device identified as a planetarium. Also we know from the writings of Cicero that Archimedes had built such a planetarium some 150 years previous to the one recovered from the sea of Antikythera. Another name which receives much attention in Greek and Egyptian literature is that of Strato who appears to have been the father of experimental science. Very little is known of his work except for a treatise on pneumatics originally attributed to Hero of Alexandria, but now claimed to be mainly the work of Strato. It is unfortunate that only remnants of Strato's writing exist, for it could have provided much information about the techniques of measurement at the time.

One of the more spectacular monuments of prehistory is the megalith Stonehenge, whose construction began about 4000 years ago and took approximately three centuries to complete. The prodigious effort of moving massive stones over a distance of 200 miles in itself is impressive, as is their erection into a monument capable of surviving so many years. If Professor Gerald Hawkins's thesis in his book *Stonehenge Decoded* is correct, however, the considerations of building the monument are negligible compared with the accuracy with which the stones were raised. The ability to erect massive sighting structures capable of predicting sun and moon phenomena and movement accurately over many centuries and with errors of less than one degree should excite the imagination of anyone who makes a measurement.

MODERN INDUSTRIAL INSTRUMENTATION

About the end of the seventeenth century a few talented and dedicated men began to examine natural phenomena in a logical and rational manner and to then test their theories by experiment. As a consequence, they were forced to develop techniques of measurement, many of which have been adapted to modern requirements.

In parallel with the rapid accumulation of scientific knowledge, the development of steam engines also had a profound effect on the development of measuring instruments. Steam engines provided unprecedented sources of power which allowed the concentration of industrial plants into what was, by the standards of the time, large operational units. To operate such industrial complexes safely and economically, a degree of monitoring and control was essential, which could only be supplied by a specialist industry. During the nineteenth century the output of this industry was largely confined to instruments for measuring basic parameters. Most instruments were of the pointer and dial display type, and if they were not mounted adjacent to the measuring points, they were piped short distances and grouped on small, local display panels. Such an arrangement was satisfactory for comparatively small industrial units but as operations became more complex, it was often necessary to concentrate displays from points widely separated in distance. This required that new types of instruments be evolved to meet these situations.

The advent of continuous process plant compounded the problems of the instrumentation industry. Not only did the distances between the measurement point and the display become very long, but the presence of explosive gases and corrosive liquids in some plants caused much concern. One method developed to overcome the problems of fire and explosion hazard used instruments with remote sensing heads. These sensing heads converted the measured signals into pressure changes in a pipe connecting the head to the indicator. Indicators were, therefore, pressure gages calibrated in terms of the parameter being measured. Pressure-operated transmission systems offered a high degree of safety and were developed to their ultimate reliability and accuracy. Under these circumstances it is not surprising that they are still in widespread use and likely to be for many years.

PROCESS CONTROL

Monitoring instrumentation signals is only part of the problem faced by the operators of continuous process plants; there is also the problem of control. It is a feature of such plants that they include large valves requiring periodic adjustment. Often these valves are located in hazardous and inaccessible places. Early process-control engineers used compressed air as their power source, developing "air motors" which were very reliable and positive in their action. Basically, air motors contain flexible diaphragms connected to the valves by spring-loaded actuator rods. The position taken up by the valve depends on the pressure applied to the diaphragm which operates against the spring force. In the event of an air failure, the valve is driven to its safe position by the spring. It is not very practical, or economic, to pipe high-pressure line air from a remote control room to an air motor, and thus servo valves were developed which operated from low-pressure, low-flow air supplies.

The concentration of instrument displays at a central point and the ability to concentrate control valves established the principle of the control room, from which a few people could monitor the operation of an entire plant and make any adjustments necessary for the optimization of the process.

AUTOMATIC CONTROL

Superficially the adjustment of a valve appears to be a simple operation; in fact, operators are performing a series of complex mental calculations. If they are experienced operators, they will deliberately change a control more than is required so that the initial movement of the valve is quite rapid. Change in the process is monitored by an indicator, and as the pointer approaches a *desired* setting they partially close the control so that movement of the valve is arrested at an anticipated point.

With practice it is possible to adjust a single valve with little apparent mental effort, but a much more difficult situation exists when an operator is required to adjust a process whose control is interdependent on two valves. More than two valves creates a task which is for all practical purposes beyond the abilities of one person.

So long as control functions remained simple, an operator's task was taxing but not impossible, but as control became more complex, it became necessary to provide mechanical assistance. One way of easing the task uses the signals from a process sensing instrument to adjust the valve automatically, and to meet this requirement the instrumentation industry designed the *automatic controller*.

For its operation an automatic controller requires two signals, a *set-point* signal, and a *feedback* signal. The set-point signal is adjusted by the operator to a required value. The controller is provided with a comparator which compares the sensor signal measuring the process with the set point. From the difference between the signals an adjustment to the valve motor is calculated causing the valve to move to a new position. The comparison between the set-point signal and the feedback signal from the measuring point is continuous, and when the two are equal, the valve is brought to rest. This type of controller is known as a *proportional controller*. It is comparatively easy to design and works directly from the air-pressure signals available in the manually controlled plant. Unfortunately, proportional control is often not adequate for many situations, and controllers providing much more complex functions are required.

ELECTRICAL INSTRUMENTATION

To give the impression that all instruments in early process plants were air operated would be wrong. Industrial electrical instruments were in use at the beginning of the twentieth century, but their insensitivity and their high current consumption restricted them to heavy electrical engineering functions. Electrical instruments, however, offered many attractive features which stimulated the development of types more suited to control applications. But with the introduction of electrical instruments came the problem of safety. The hazards of using electricity in the presence of inflammable gases is very real and much research has been required to overcome the problems. To some extent the problems of explosion hazard were reduced by the development of the pressure-to-current converter. These devices allowed the use of measuring instruments with air transmission in hazardous areas and provided a conversion to electrical signals outside. In the reverse process it was possible to use electrical controllers and convert electrical signals to pressure signals by current-to-pressure converters.

The use of electrical controllers potentially provided an ideal solution to the problem of providing complex responses to plant operational disturbances. Complex control functions are much easier to simulate electrically than mechanically and much easier to modify if the necessity should arise. Early electronic circuits, however, were unstable, and components were large. As a result the full advantage of electrical control was not realized until the introduction of the transistor.

CHART RECORDERS

The use of visual displays presented process operators with a number of problems, especially when used in large numbers. There is a limit to the number of displays that can be viewed by an operator and still remain meaningful. This is particularly true when it is necessary to monitor long-term trends, and when the trends from several displays are interactive, the operator's problems become extreme. Another problem with a visual display is that it indicates only the present state of a signal. If the operator does not have the display under constant surveillance, it is possible for transient problems to occur without his or her knowledge.

To overcome these problems chart recorders were developed that allowed signals to be recorded as continuous traces on a moving chart. Early recorders used adaptations of the visual display, in which the pointer was replaced by a pen that drew a line on a clockwork-driven chart. Although pen recorders solved the problems of trend recognition and transient signal monitoring, they did so at the sacrifice of accuracy, and they increased the cost of the display. Some improvement was made in economy by sharing a chart between a number of pens, but this did not improve the accuracy.

The development of the electrical-servo-driven pen recorder eliminated many of the shortcomings of the early pen recorders. Servo pens can be used with large deflections that still retain good accuracy. The operation of the servo pen is somewhat similar to that of an automatic controller. An input sensor signal is applied to an electronic comparator which drives a servo motor. A feedback sensor connected to the motor shaft produces a signal proportional to the number of rotations of the motor, and this is compared with the input signal. When the two are equal the motor is stopped. A pen drive mechanism connected to the shaft is used to move a pen across a chart, and with suitable gearing this movement may span charts 10 to 12 inches (250 to 300 mm) wide. The disadvantage of the servo-driven recorder is its cost. The cost of servo pens is comparatively high, and in many cases economy is restored by switching a group of signals sequentially to the pen to provide a multichannel display. This technique of sharing one display among a number of signals is known as *multiplexing*, and the switch is known as a *multiplexer*.

The electrical automatic controller and the servo pen recorder were probably the last significant innovations in process control instrumentation prior to the introduction of the computer and this period established a standard of control room design which is still used in many plants.

THE PROCESS CONTROL COMPUTER

The improvements in electronics and the introduction of the transistor stimulated the development of the *digital computer* for process control. Many process-instrument signals are of a type that vary in amplitude proportionally to the change producing them. Such signals are described as *analog*. Since the first electrical process controllers operated directly on these signals and produced an analog output that was a function of the input, the controllers were, effectively, *analog computers*.

About the time that electronic analog controllers were being introduced into the process industry another type of computer was being developed. Primarily intended for scientific and mathematical calculations, the *digital computer* was soon recognized as an ideal basis for the design of a central process controller. In this type of computer, numbers are represented as codes made up of groups of on-and-off signals described as *binary numbers*, and a special arithmetic called binary arithmetic was used for the calculations. In addition to being able to calculate, digital computers are able to store large quantities of binary information and make predefined logical decisions on both the stored and calculated information. In theory the introduction of a central digital computer into a process control system could provide an excellence of control not previously possible and at the same time remove many potential sources of human error. In practice, many problems required solving before the application and the tool could be reconciled.

If a computer is to be used as a process controller it is necessary for it to accept an analog signal from the process, calculate a response, and output a control signal. The first problem to be overcome was one of signal compatibility. Digital computers operate on binary representations of an analog and not on the analog signal itself. The method of converting analog signals to binary representations is described as *analog-to-digital conversion*, and the device that produces this conversion is described as an *analog-to-digital converter*. The converse conversion is known as *digital-to-analog* conversion, and the device known as a *digital-to-analog converter*. At the time under discussion both devices were expensive, and if the system was to be made economic, multiplexing techniques both for input and output were required.

Another problem that designers of the process-control computer faced was that of *real-time operation*. Malfunction in a process may be sudden, and if not corrected quickly it can become catastrophic. With manual operation it is normal to use sensing instruments in critical areas to operate off-limit alarms. When an alarm is operated, the process operator interrupts the current task and gives immediate attention to correcting the situation. When the situation has been normalized the operator returns to the original task. If two alarms should operate simultaneously, it is necessary to register both and decide which should be given priority.

With the computer, designers were faced with the same problem and invented interrupt systems that would suspend the normal sequential work of the computer and operate on a special "alarm program" before returning it to its routine work.

Early process-control computers were usually part of a large central system. Economics would have forced this situation, even if the central control room had not already existed. The concept of the central controller had several disadvantages, the worst being the effect of computer failure. Fall-back systems ameliorated the worst effects of failure, but usually it was only possible to operate the plant in some degraded mode while the computer was inoperative.

The ideal solution to the failure problem is to divide the plant into a number of local areas, each controlled by its own computer, with each control computer monitored by a central system. With such configuration, failure of any component control will cause only the minimum of plant disturbance. Until recently the cost of such a system would have been prohibitive, but now costs have fallen to the point where it is possible to install a computer in each control circuit if the engineer considers this desirable.

THE PRESENT

It would be a gross oversimplification to suggest that the process industry has been the only stimulus to the development of instrumentation, or to attempt to conceal the fact that many of the solutions to problems were acquired from other industries. Nevertheless, process industries are closely associated with the perfection of many measurements, and the debt that instrumentation engineering owes to the genius of early process engineers cannot be overstated. It is appropriate, therefore, that this industry should feature in a discussion designed to illustrate some of the solutions which were devised to a progression of problems experienced in all industries of growing complexity.

Today's industries are advancing the frontiers of technology in many directions, but whether their particular project is the conquest of space or the dissection of the atom, their problems of measurement are as critical as those of the prehistoric man crossing the river. The essential difference between the two ends of history is that early man was his own instrumentation engineer; now the problems lie firmly at the feet of professionals who have chosen to make the science of measurement a way of life.

A further difference we are experiencing is the speed of development. It took man a million years to develop the mechanical clock; it took him about 700 years to take it off the church steeple and put it into his pocket. Similarly with the computer. It took him until the 1940s to build a practical digital computer that would do his scientific calculations. It only took the next 30 years to take it out of a room full of equipment racks and put it into a matchbox.

As time passes, the pace of development appears to be accelerating, and since no one can accumulate all knowledge, there is a growing tendency for instrumentation engineers to specialize early in their careers. Under these circumstances it is becoming increasingly important that students obtain as wide a base of knowledge as possible and that specialists have access to material which allows them easy reference into areas of the discipline with which they are not normally associated. This book has been written with these factors in mind, and it is hoped that it will go some way to meet these needs.

CHAPTER 2

INTRODUCTION TO SENSOR FUNDAMENTALS

SENSORS AND TRANSDUCERS

A transducer is a device that converts one form of energy into another. This conversion may be pressure to movement, an electric current to a pressure, a liquid level to a twisting movement on a shaft, or any number of other combinations. One group has special significance; transducers used to drive electrical recorders invariably have an electrical output. These are often described as *electrical sensors.*

Although the final output of a sensor may be electrical, there may also be one or more intermediate transducing steps. In a pressure sensor, for example, a diaphragm may be used to convert changes in pressure into the movement of push rod. The push rod may be used to apply mechanical strain to resistance wires and the corresponding change in electrical resistance of the wires, when strained, used to produce the electrical output. In another type of sensor, the push rod may be used to operate a simple switching action to give warning when a pressure falls below a predetermined value.

Therefore, there are two basic types of sensor. One that produces an output proportional to a change in a parameter is described as an *analog* device; one that produces an on/off type of output is described as a *digital* device. To complicate matters, there are sensors that provide digital outputs proportional to changes in the parameter; these are regarded as digital sensors.

If one were to catalog sensors by the parameters they are required to measure, the task would be almost limitless. The list is already incredibly long and grows as instrumentation engineering continues to break new ground. Fortunately, a very large proportion of the sensors in common use are required for a limited number of types of measurement. For the rest, there is a limited number of transducing techniques, and it is the ingenuity of the engineer in applying these techniques that makes possible the wide range of sensor measurements in use.

In any introductory study of instrumentation engineering it is reasonable to restrict the number of types of measurement discussed to those in common use and to place the emphasis on an understanding of transducing fundamentals.

TERMINOLOGY

As with every other subject, instrument engineering has its own terminology. Some of the terms have subtle meanings, and a misunderstanding can lead to a completely wrong impression of the performance of a system. The following definitions are intended as an introduction to the use of the terminology. Readers requiring more complete and precise definitions should refer to ISA Standard S51.1.

Range

Every sensor is designed to work over a specified *range*. While an electrical output may be variable and adjusted to suit the application, this is not usually practical with mechanical transducing elements. The design ranges of these mechanisms are usually fixed, and if exceeded, result in permanent damage to or destruction of a sensor.

It is customary to use transducing elements over only the part of their range where they provide predictable performance and often enhanced linearity. This practice also provides a greater safety factor against overloading. To compensate for the accompanying loss of signal (or sensitivity) the electrical output is often amplified and adjusted to a convenient size. Unfortunately, there is a limit to which amplification can be applied and still retain the enhanced performance of the transducer, as the enhancement is offset by amplifier errors, and *noise*. Noise is an interference signal superimposed on the measurement signal. It may be "picked up" from external sources, or be caused by equipment instability.

Zero

When making a measurement it is necessary to start at a known datum, and it is often convenient to adjust the output of the instrument to zero at the datum. The output of a centrigrade thermometer is zero at the freezing point of water; the output of a pressure gage may be zero at atmospheric pressure. Zero, therefore, is a value ascribed to some defined point in the measured range.

Zero Drift

One of the problems experienced with sensors occurs when the signal level varies from its set zero value. This introduces an error into the measurement equal to the amount of variation, or *drift* as it is usually termed.

Zero drift may occur for a number of reasons. Changes in ambient temperature, causing physical changes in the sensor, or electrical changes in the amplifier, are common causes, but aging of components or mechanical damage may also be responsible. It is a particularly troublesome problem when the sensor is used for long-term measurements and may not be returned to its zero position for periodic checking.

All sensors are affected by drift to some extent and it is sometimes specified in terms of *short-term* and *long-term drift*. Short-term drift is usually associated with changes in temperature or electronics stabilizing. Long-term drift is usually associated with aging of the transducer or electronic components.

Sensitivity

Sensitivity of a sensor is defined as the change in output of the sensor per unit change in the parameter being measured. The factor may be a constant over the range of the sensor, or it may vary. Such devices are described as having a linear or a nonlinear output, respectively.

Sensitivity depends on a number of factors which are variable. The mechanical properties of a transducer may vary with temperature, causing a variation in sensitivity, but often it is the electrical part of the sensor which is responsible for the greatest changes. An amplifier may change its gain because of temperature effects on components or variations in power supplies or even faulty operation. In addition, many sensors must be provided with a voltage from a power source before they give an output. Often the sensor output is directly proportional to this excitation voltage, and any changes in it during operation will introduce corresponding errors. It may be argued that the worst effects of sensitivity changes do not occur when the sensor fails completely, but when the sensitivity changes sufficiently to cause an error, without the error being easily recognizable.

Resolution

Resolution is defined as the smallest change that can be detected by a sensor. It is not a universally accepted term, some people claiming that it is a parameter with no absolute value and no precise definition. Sensors that use wire-wound potentiometers or digital techniques to provide their electrical output have finite resolution, which without doubt is meaningful and serves to describe the quality of the instrument. Frequently, however, manufacturers describe their sensors as having infinite resolution. There is no such device. Apart from restrictions incurred by electrical instability, the mechanical construction of a sensor imposes its own restriction, below which changes in output are meaningless.

Response

The time taken by a sensor to approach its true output when subjected to a step input is sometimes referred to as its *response time*. It is more usual, however, to quote a sensor as having a flat response between specified limits of frequency. This is known as the *frequency response*, and it indicates that if the sensor is subjected to sinusoidally oscillating input of constant amplitude, the output will faithfully reproduce a signal proportional to the input. Beyond the limiting frequencies the output may rise or fall, depending on the construction of the device.

Linearity

The most convenient sensor to use is one with a linear transfer function. That is an output that is directly proportional to input over its entire range, so that the slope of a graph of output versus input describes a straight line. This allows a single conversion factor to be applied over the range. In practice, this is never quite achieved, although most transducers exhibit only small changes of slope over their working range. To these curves, a *best* straight line is fitted whose error is usually well within the tolerance of the measurement. Some sensors, particularly those using inductive transducing principles, demonstrate considerable changes in the slope of their output versus input graph and may even reach a point where, regardless of change of input, there is no change of output. The working range of such a sensor is restricted and must be limited to where the graph is relatively linear, or alternatively a different factor must be applied to each reading.

Hysteresis

Hysteresis becomes apparent when the input to a sensor is applied in a cyclic manner. If the input is increased incrementally to the sensor's maximum and returned to its zero datum in a similar manner, the calibration may be seen to describe two curves that meet at the maximum. In returning to zero input, the calibration has not returned to its original datum. If now the calibration is continued in the negative direction of input, two further curves will be produced which are a mirror image of the previous ones. Further cycling will eventually link these two halves into one complete loop, which will then be repeatable with every cycle.

This loop is normally referred to as the *hysteresis loop* of the sensor, although it contains any of the other nonlinearity effects which may be present. Consequently, it is usual when specifying a sensor to quote nonlinearity and hysteresis as one parameter.

Calibration

If a meaningful measurement is to be made, it is necessary to measure the output of a sensor in response to an accurately known input. This process is known as *calibration*, and the devices that produce the inputs are described as *calibration standards*.

It is usual to provide measurements at a number of points of the working range of the sensor, so that a ratio of output to input may be determined. This ratio may be obtained by plotting a graph and calculating its slope, or it may be determined from the measured points by calculation. Such a ratio is described as a *calibration factor*.

The ratio output to input is not always a constant over the range of a sensor, and the calibration graph describes a curve. In these instances a *best straight line* may be fitted through the points and the errors accepted, or a different calibration factor must be provided for every measurement.

Accuracy

The *accuracy* of a sensor is a term causing some confusion, possibly because it is made up of a number of independent quantities, each of which is quoted separately, and possibly because it is usual to quote the accuracy of a measurement as a percentage of *full-scale output* (% FS).

The strict meaning of accuracy when quoted as a percentage of full-scale output is that of a *value of uncertainty* which is applied to converted sensor outputs throughout the entire range of measurement. For example, a measurement with an accuracy of ±1% FS and with a range of 0-100 units has a value of uncertainty of ±1 part in 100 or ±1 unit which applies to every measurement. A measurement of 50 units would be made with a value of uncertainty of ±1 unit and an uncertainty of ±2% of the value. Although the word accuracy is still widely used the word uncertainty is becoming commonplace.

Manufacturers do not normally quote an overall figure of accuracy for a sensor, since some of the variations may not be relevant to certain measurements. A sensor being used at constant temperature would not be affected by errors associated with temperature. Its uncertainty would be largely influenced by its linearity and hysteresis. So that users may have some control in estimating the performance of their sensors under their own operating conditions, it is usual to provide a breakdown of parameters, which may include:

- Linearity and hysteresis ± %FS
- Residual output at zero ± %FS

- Zero drift with temperature ± %°C
- Repeatability ± %FS
- Sensitivity change with temperature %/°C

This list is not exhaustive, nor is it standard, and this can make comparison of specifications difficult.

If a calibration over the operating limits of a sensor is not practical, then the overall uncertainty of the sensor signal can only be determined by summing up the manufacturer's specifications. Absolute uncertainty is the algebraic sum of the uncertainties of several parameters relevant to a particular mode of operation of the sensor. It is very unlikely, however, that in practice all the variations would be additive, and the *root-sum-square* method of summation is commonly used to determine a practical figure. This method is derived from probability theory and is summarized in the formula:

$$a = \pm \sqrt{a_1^2 + a_2^2 + a_3^2 + \ldots}$$

where a represents the overall uncertainty to a probability of about 70%, and a_1, a_2, and a_3 represent the parameter uncertainties expressed as a percentage of full scale.

In making a measurement, the accuracy or uncertainty of power supplies, amplifiers, and recorders also contributes to the overall value. Some instrumentation engineers treat all these quantities as a "measuring chain" and do not attempt to break them down, arguing that the accuracy of the measurement can only be the accuracy of the chain.

Another way of estimating the overall uncertainty of a measuring chain is to take the algebraic sum of the component values and divide this figure by the square root of the number of components. Although this method is valid, according to statistical theory, and approximates the root-sum-square answer, it is not so widely used.

SENSORS IN THE INDUSTRIAL ENVIRONMENT

There are many practical differences between a measurement made in a laboratory and one made in a typical engineering environment. In the laboratory the atmosphere is relatively free from contaminants, the temperature is relatively stable, the area is free from vibration, and personnel are specialized in handling the equipment they are using. The industrial measurement is often made under reverse conditions, and the sensor often positioned where conditions are worst. Not only must the sensor survive these adverse conditions, but the signal it produces must be capable of transmission to recording equipment which may be a considerable distance away. The wiring between the two may of necessity be routed close to heavy electric machinery capable of inducing electrical noise or routed through hazardous areas where there is a possibility of explosion.

It is not surprising under these circumstances that traditional sensors, which have demonstrated their reliability, maintain their place, where measurements could be made by a more accurate but less proven device. Before choosing a sensor for an industrial application the following questions should be considered:

Is the sensor satisfactory for its working environment? The choice of a sensor is governed by its ability to survive the environment where it has been installed. A thermometer, for example, may have to survive conditions of high vibration, or a vibration-measuring device conditions of high temperature.

Has the sensor adequate range? The choice of range can be a difficult compromise. It is not always sufficient to choose a sensor with adequate resolution for an application, for it must also cope with the extremes of the parameter it is measuring. The thermometer with a range of 50° may provide the required resolution for normal operation, but if periodically the temperature of fluid being measured can rise to 100°, it may not survive.

Has the sensor adequate resolution? If it is necessary to sacrifice resolution to meet the range requirements, there must be a point at which the measurement is unsatisfactory. At this point, a decision must be made, either to change the type of sensor to one more suited to the environment, or alternatively to find a different way of making the measurement.

Is the instrument accurate enough for the measurement? The absolute value of a reading can only be guaranteed if the instrument has been calibrated against adequate standards.

Is the measurement too good? The object when making a measurement is to make an adequate one. The more accurate the measurement, the more costly it will be. It is of little use to make a measurement of very high accuracy if the process being controlled by that measurement requires only nominal adjustment. To obtain measurements of high accuracy usually requires expensive sensors and calibration standards that are also expensive. It is only after all other methods of making the measurement have been explored and rejected that high accuracy measurements should be allowed.

CHAPTER 3

BASIC ELECTRICAL THEORY

Although the electrical theory involved in transducer and interface design can become extremely involved, for the most part the instrumentation engineer requires only a sound knowledge of basic electrical principles. This chapter covers those principles most frequently required. It is intended as a review aid, not as a primer. Any reader requiring a fuller or more basic explanation is recommended to consult one of the standard textbooks.

OHM'S LAW

Basic to the development of all electrical theory is Ohm's law. This states that when a current of i amperes flows through a resistance of R ohms, the voltage V required to maintain the current will be: $V = iR$. See Figure 3.1.

RESISTANCES IN SERIES

When a number of resistances are connected in series, the total resistance R of the circuit will be $R_1 + R_2 + R_3 \ldots$, where R is the total resistance of the circuit and R_1, R_2, R_3, . . ., are the component resistances connected in series. See Figure 3.2.

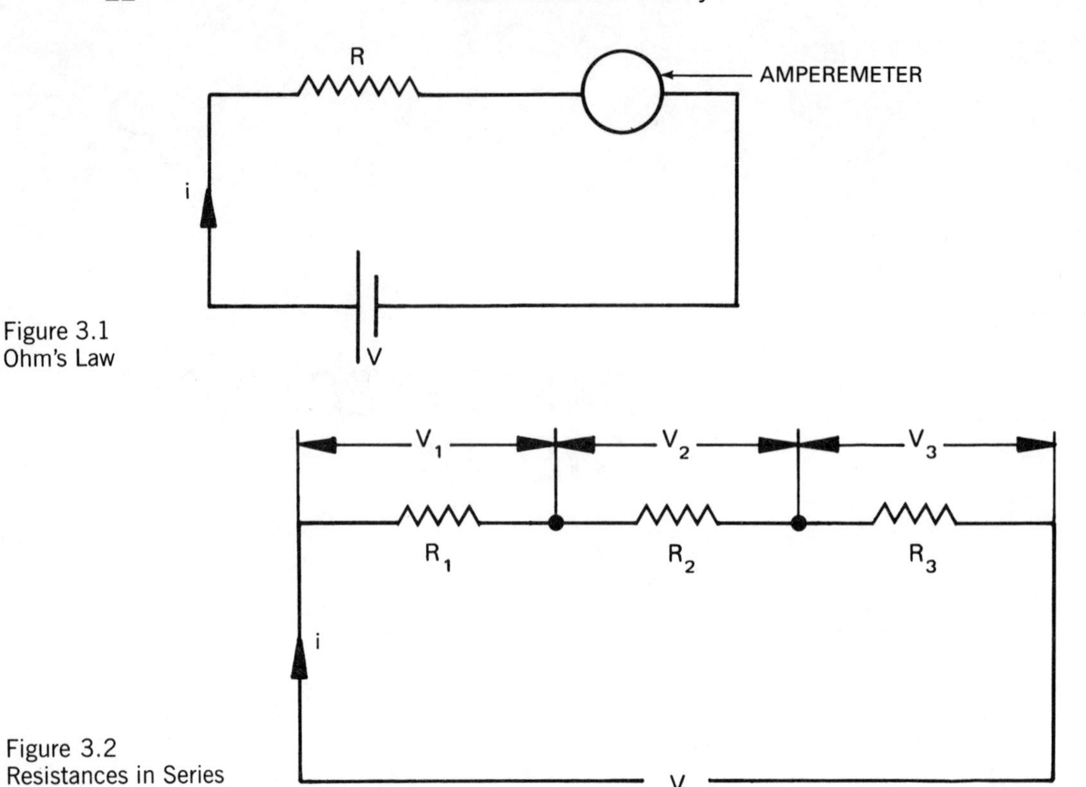

Figure 3.1
Ohm's Law

Figure 3.2
Resistances in Series

RESISTANCES IN PARALLEL

When a number of resistances are connected in parallel, the effective resistance of the group will be given by the expression:

$$\frac{1}{R} = \frac{1}{R_1} + \frac{1}{R_2} + \frac{1}{R_3} \cdots$$

where R is the effective resistance of the parallel group and R_1, R_2, R_3, . . ., are the component resistances connected in parallel. See Figure 3.3.

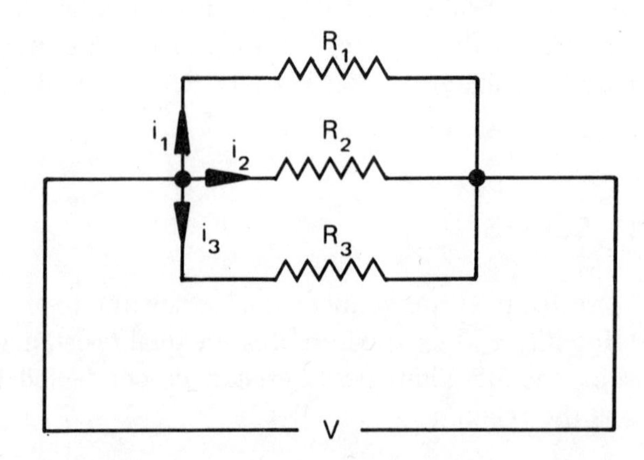

Figure 3.3
Resistances in Parallel

POWER IN A CIRCUIT

The power consumed by a circuit is defined as the product of the current flowing in the circuit and the voltage producing it. When a voltage of 1 V produces a current of 1 A the power consumed in the circuit is 1 watt.

$$\begin{aligned} \text{POWER} &= iV \\ &= i^2R \\ &= V^2/R \end{aligned}$$

RESISTIVITY

The resistance of a conductor with a uniform cross section is directly proportional to its length and inversely proportional to its cross-sectional area, such that:

$$R = \frac{\rho l}{a}$$

where

R = the resistance of the conductor,

l = its length,

a = its cross-sectional area,

ρ = a material constant known as resistivity.

EFFECT OF TEMPERATURE ON RESISTANCE

Most conductors change their resistance when subjected to changes of temperature, in the relation which approximates:

$$R = R_0 (1 + \alpha t)$$

where

R = the resistance at temperature, t,

R_0 = its resistance at 0°C,

t = the change in temperature from 0°C,

α = a material constant known as the temperature coefficient of resistance.

This is only true for a limited temperature range; beyond the limits extra factors must be taken into account. Platinum is probably one of the more important materials and it has the relationship:

$-200°C$ to $0°C$ $R = R_0\,[1 + \alpha t + \beta t^2 + C(t - 100t^3)]$

$0°C$ to $100°C$ $R = R_0\,(1 + \alpha t)$

$100°C$ to $600°C$ $R = R_0\,(1 + \alpha t + \beta t^2)$

where α, β, and C are constants determined experimentally.

THE POTENTIOMETER

Potentiometers are frequently used as transducers in electrical sensors and are, therefore, worth considering in some detail. Consider a resistor which is tapped at some point P so that the two parts have values of R_1 and R_2 (Figure 3.4a).

If a voltage V is applied across the resistor the current flowing will be:

$$i = \frac{V}{R_1 + R_2} \qquad (1)$$

The voltage developed across R_2 will be:

$$iR_2 = \frac{VR_2}{R_1 + R_2}$$

Such a circuit is known as a voltage divider or potentiometer, and if the point P is made variable, an adjustable voltage can be obtained across PB.

Suppose the tapping point at P is moved so that R_2 increases by δR and R_1 decreases by δR. The new voltage at P will be:

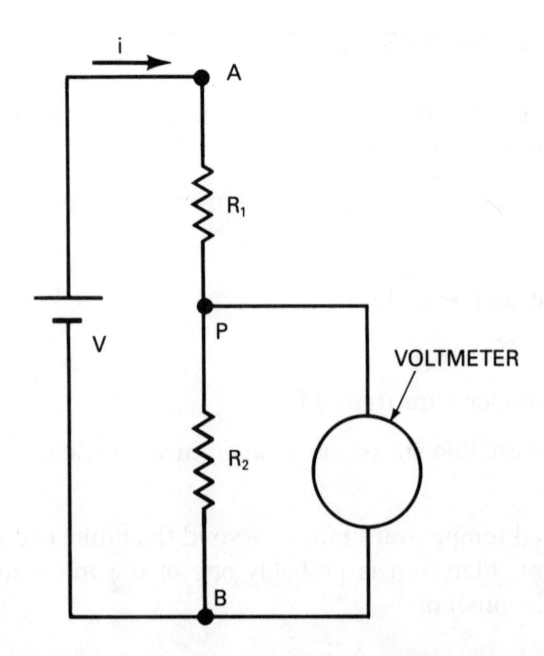

Figure 3.4a
Voltage Divider

$$\frac{V(R_2 + \delta R)}{(R_1 - \delta R) + (R_2 + \delta R)} = \frac{VR_2 + V\delta R}{R_1 + R_2}$$

or

$$\frac{VR_2}{R_1 + R_2} + \frac{V\delta R}{R_1 + R_2}$$

But

$$\frac{V}{R_1 + R_2} = i \text{ (from 1)}$$

and

$$\frac{VR_2}{R_1 + R_2} = \text{original voltage at } P.$$

Therefore,

Voltage at the new point = Voltage at the old point + $i\delta R$

or

Voltage at the new point = Voltage at the old point
+ Voltage change across δR. (2)

Consider the potentiometer shown in Figure 3.4b.

The voltage applied across the winding is V,

The voltage drops across individual turns are V_1, V_2, V_3, . . .,

The resistances of individual turns are R_1, R_2, R_3, . . .,

The distances between consecutive turns are l_1, l_2, l_3, . . .

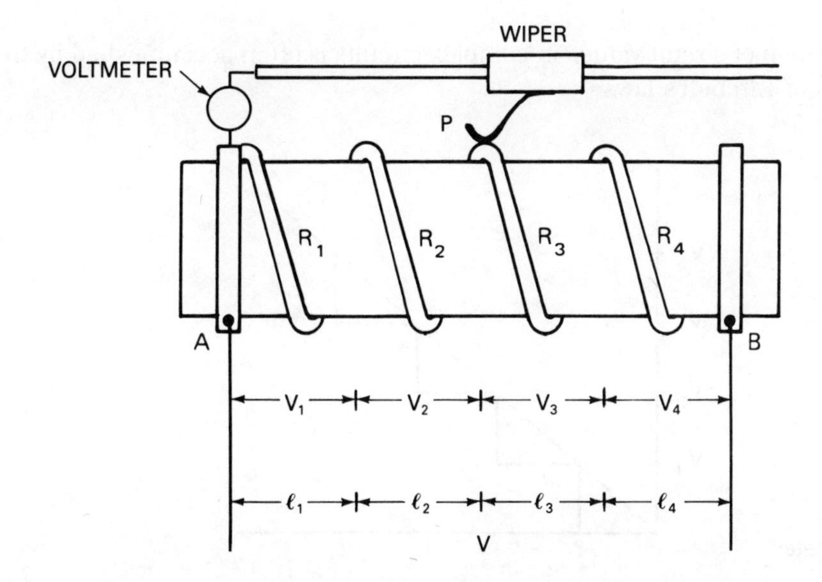

Figure 3.4b
The Potentiometer

It follows from (2) that the output of the potentiometer between the wiper and point *A* is:

$V_1 + V_2 \ldots$

Providing that the winding is made of a uniform-diameter wire and wound on a form of constant diameter,

and that $l_1 = l_2 = l_3 = l_4$

then $R_1 = R_2 = R_3 = R_4$

and $V_1 = V_2 = V_3 = V_4$

The output voltage of this potentiometer, therefore, is proportional to the wiper's distance from *A*. However, this voltage does not appear as a continuous change, but as a series of steps (Figure 3.5).

The size of these steps as the wiper moves from one turn to the next is a measure of the resolution of the potentiometer and it is every manufacturer's aim to keep the steps as small as possible. Close winding is one way; using the smallest possible diameter wire another but but there is an obvious practical limit to which resolution may be increased and still retain a reasonable life.

Wheatstone Bridges

Wheatstone bridges are used extensively to measure small changes in resistive-type sensors such as strain gages. An understanding of the fundamental principles of bridge operation is, therefore, an important aspect of an instrumentation engineer's training.

Kirchoff's Laws

The calculation of circuit values in complex circuits is often accomplished by the application of Kirchoff's laws:

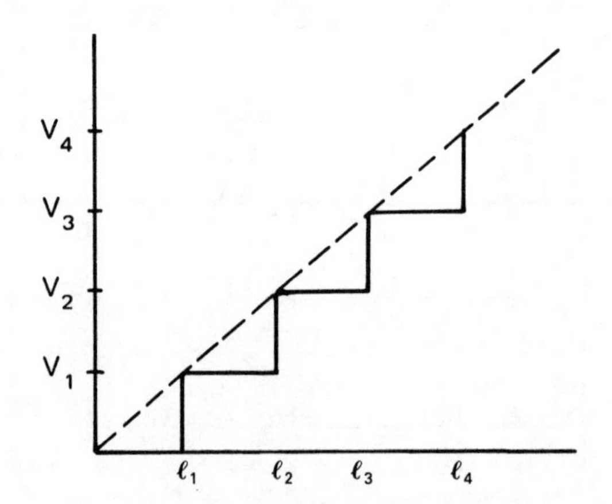

Figure 3.5
Output of a Potentiometer

1. The algebraic sum of currents at any junction in a circuit is zero.
2. In any closed circuit the algebraic sum of the iR drops across the elementary parts of the circuit is equal to the electromotive force acting round the circuit.

If Kirchoff's laws are applied to a Wheatstone bridge as shown in Figure 3.6, the circuit may be presented by three equations:

$$i_1R_1 + (i_1 - i_3)R_2 + i_4R_5 = E$$

$$i_1R_1 + i_3R_g - (i_4 - i_1)R_3 = 0$$

$$(i_1 - i_3)R_2 + (i_1 - i_3 - i_4)R_4 - i_3R_g = 0$$

with resistances R_1, R_2, R_3 and R_4 forming the bridge's arms and Rg representing the galvano-meter's resistance. Given sufficient circuit constants, it is possible to determine the unknowns from these equations.

Conditions of Balance for a DC Bridge

Consider the bridge in Figure 3.7 with a supply E across BD and a meter of very high resistance across AC. The voltage across R_1 will be i_1R_1, and the voltage across R_3 will be i_2R_3. When the bridge is balanced, the potentials at A and C are identical and $V_g = 0$. By Kirchoff's Law:

$$i_1R_1 - i_2R_3 = 0$$

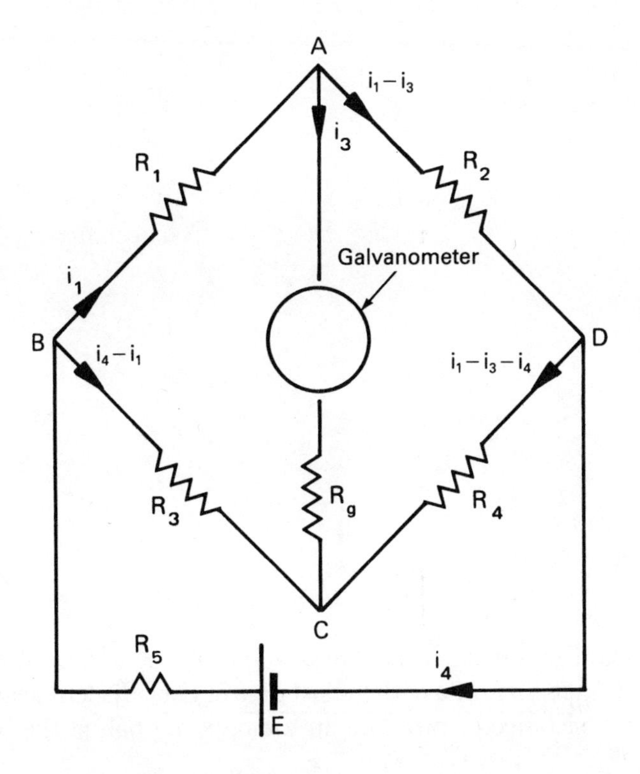

Figure 3.6
Wheatstone Bridge—Analysis

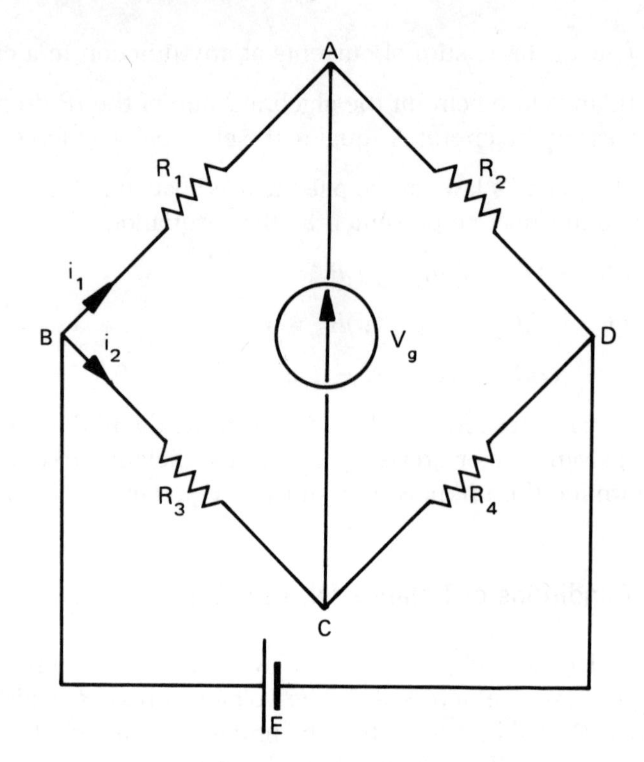

Figure 3.7
Wheatstone Bridge—Conditions of Balance

Therefore:

$$i_1R_1 = i_2R_3 \tag{1}$$

Similarly:

$$i_1R_2 = i_2R_4 \tag{2}$$

From (2):

$$i_1 = \frac{i_2R_4}{R_2} \tag{3}$$

and substituting (3) in (1):

$$\frac{i_2R_4}{R_2}R_1 = i_2R_3$$

Transposing and canceling:

$$\frac{R_4}{R_2} = \frac{R_3}{R_1}$$

From this expression it can be seen that to achieve a condition of balance across a Wheatstone bridge, the ratio of the resistance of two adjacent arms must equal the ratio of the resistance of two arms in the opposite half of the bridge.

Changes in Resistance of Bridge Arms

Consider the bridge in Figure 3.8 where for convenience all the arms have an equal value R. With the bridge in balance, the voltage across AC is zero, and the potential at A and C is $-V/2$ with respect to B.

Suppose now arm AB is changed by δR; the potential drop across AB is:

$$i_1 (R + \delta R) \qquad (1)$$

But,

$$i_1 = \frac{V}{2R + \delta R} \qquad (2)$$

Substituting in (1) the potential drop is:

$$\frac{V(R + \delta R)}{2R + \delta R}$$

The voltage V_g considering the loop ABC:

$$= \frac{V(R + \delta R)}{2R + \delta R} - \frac{V}{2}$$

$$= \frac{V(2R + 2\delta R - 2R - \delta R)}{2(2R + \delta R)}$$

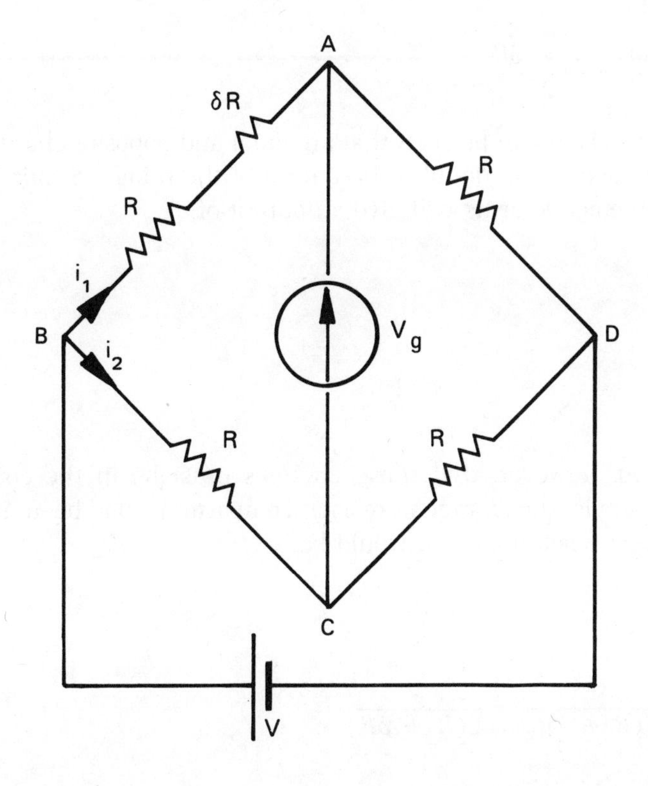

Figure 3.8
Wheatstone Bridge—Resistance Change in One Arm

Therefore:

$$V_g = \frac{V\delta R}{2(2R + \delta R)} \tag{3}$$

If δR is small compared with R:

$$V_g = \frac{V\delta R}{4R} \tag{4}$$

Now suppose the arm AD is also changed in the opposite sense by an amount δR. The voltage drop across AB will be:

$$i_1(R + \delta R)$$

as before, but now, assuming the current through the voltmeter between A and C is negligible, i_1 becomes:

$$\frac{V}{(R + \delta R) + (R - \delta R)} = \frac{V}{2R} \tag{5}$$

and the potential at B with respect to A becomes:

$$\frac{V(R + \delta R)}{2R} \tag{6}$$

The potential at C relative to B remains unchanged and equals $-V/2$, and V_g becomes:

$$\frac{V(R + \delta R)}{2R} - \frac{V}{2}$$

$$= \frac{V(R + \delta R - R)}{2R} = \frac{V \times \delta R}{2R} \tag{7}$$

If this is compared with (1), it can be seen that an equal and opposite change in the resistance of an adjacent arm doubles the output of the bridge. Similarly, it can be shown that three active arms will give an output of:

$$V_g = \frac{3V \times \delta R}{4R} \tag{8}$$

and four active arms:

$$V_g = \frac{V \times \delta R}{R} \tag{9}$$

It must be emphasized, however, that these changes must be in the correct sense. Suppose, for example, the change in resistance in arm AD had been in the same direction as AB. The potential at A would be:

$$i_1\ (R + \delta R)$$

and

$$i_1 = \frac{V}{(R + \delta R) + (R + \delta R)} = \frac{V}{2(R + \delta R)}$$

and

$$V_g = \frac{V(R + \delta R)}{2(R + \delta R)} - \frac{V}{2}$$

$$\frac{V(R + \delta R - R - \delta R)}{2(R + \delta R)} = 0 \qquad (10)$$

If the change in two adjacent arms is equal and in the same sense, there is no change in the bridge balance. This concept is important, because it provides a means of compensating bridges for changes in balance caused by temperature change in the component arms.

FARADAY'S LAW

When a magnetic flux varies with time an EMF is induced in an associated coil whose value is given by the expression:

$$V = -N\frac{d\phi}{dt}$$

where:

V is the induced EMF

$\frac{d\phi}{dt}$ is the rate of change of magnetic flux

N is the number of terms in the coil

The sign indicates that the induced voltage is in the opposite sense to the change producing it.

It is possible to consider the induced EMF from another standpoint. Suppose a conductor of length l moves a distance ds in a uniform magnetic field of flux density β. Let the area swept by the conductor be dA and the change in flux be $d\phi$.

The area swept equals

$$dA = lds \qquad (1)$$

and the change in flux through the circuit equals

$$d\phi = -\beta dA \qquad (2)$$

From (1) and (2)

$$d\phi = -\beta ld$$

and dividing both sides by dt

$$-\frac{d\phi}{dt} = \beta l\frac{ds}{dt} = \beta lv$$

where v is the velocity of the conductor in the magnetic field and $d\phi/dt$ is the voltage induced between the ends of the conductor.

The equation $V = \beta lv$ is sometimes described as the generator equation because it forms the basis for the calculation of voltages generated by equipment where conductors move in magnetic fields.

INDUCTANCE

The unit of inductance is the *henry* and is defined as the inductance of a circuit in which an EMF of 1 V is induced when a current varies at the rate of 1 A per second, or:

$$-L\frac{di}{dt} \tag{1}$$

From Faraday's law:

$$-N\frac{d\phi}{dt} \tag{2}$$

and equating (1) with (2)

$$N\frac{d\phi}{dt} = L\frac{di}{dt}$$

or

$$Nd\phi = Ldi$$

Integrating both sides,

$$N\phi = Li + k$$

But $\phi = 0$ when $i = 0$. Therefore $k = 0$ and $N\phi = Li$.

$$L = \frac{N\phi}{i} \tag{3}$$

ELECTROMAGNETISM

When a current flows through a conductor a magnetic field is produced around the conductor whose strength is inversely proportional to the distance from the center.

Magnetic Field of a Single-Turn Coil

If a current flows around a single-turn coil, a magnetic field is set up whose direction is at right angles to the plane of the conductor at its center line. The density of this field is given by the expression:

$$\frac{\varphi}{a} = \beta = \frac{\mu_0 i}{2r}$$

where

φ is the magnetic flux, in webers

β is the magnetic flux density, in webers per square meter

i is the current flowing in the conductor

r is the radius of the coil, in meters

a is the cross-sectional area of the turn

μ_0 is a constant known as the *permeability of empty space*

Magnetic Field of a Long Solenoid

When a current i flows through a coil, a magnetic field is created along its length whose axis at the center of the coil is parallel with the length of the coil. The density of this field is given by the expression:

$$\beta = \frac{\mu_0 N i}{l}$$

where N is the number of turns in the coil and l is the length of the solenoid. This relationship is only true for a point within a long solenoid.

Permeability

In the relationships defined in the previous sections a constant of proportionality has been assumed with the symbol μ_o. This is known as *the permeability of empty space*. This constant has a numerical value of 12.57×10^{-7} webers per ampere meter, and is determined experimentally.

If magnetic materials are present in the solenoid, the flux density will be much higher and governed by the relationship:

$$\beta = \frac{\mu N i}{l}$$

In this case μ is the permeability of the magnetic material. And since

$$\frac{Ni}{l} = H$$

$$\beta = \mu H \qquad (1)$$

It is often convenient to express μ in terms of μ_o and the ratio $\mu/\mu_o = \mu_r$, or the *relative permeability*. The expression (1) therefore becomes:

$$\beta = \mu_o \times \mu_r \times H$$

Magnetic Circuits

In many applications it is necessary to calculate the number of turns in a coil, and the current through that coil to produce a given magnetic flux in a magnetic circuit. Often this flux is required to produce a force to attract an armature such as a relay, but it could equally be used to drive a galvanometer. In such applications the magnetic circuit contains an air gap and it is usually required to calculate a working flux in that air gap.

Consider Figure 3.9 where a magnetic circuit consists of a movable armature hinged to a core member which carries the energizing coil.

Suppose the length of the magnetic path in the core is l_1, the length of the magnetic path in the armature is l_2, the length of the magnetic path in the air gap is l_g, the cross-sectional area of the core is a_1, the cross-sectional area of the armature is a_z, and the cross-sectional area of the air gap is a_g. In the general case:

$$\beta = \mu_o \mu_r H \quad (1)$$

where:

β is the flux density in the element under consideration

H is the magnetic intensity

μ_r is the relative permeability of the material

μ_o is the permeability of free space

Also

$$\beta = \frac{\phi}{a} \quad (2)$$

where ϕ is the working magnetic flux in the circuit and a is the cross-sectional area of the magnetic circuit element and

$$H = \frac{Ni}{l} \quad (3)$$

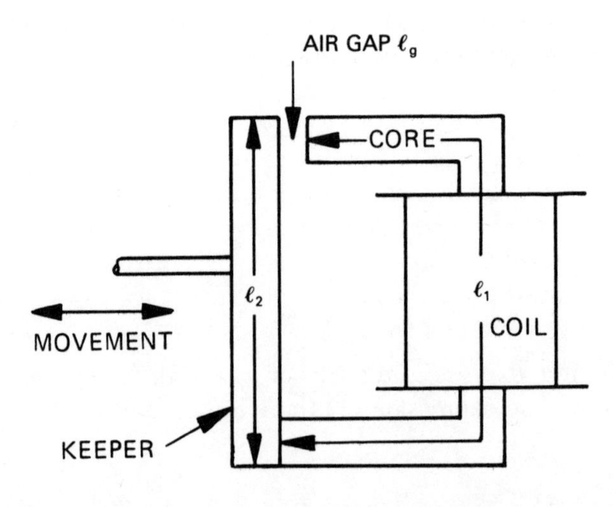

Figure 3.9
A Magnetic Circuit

where l is the length of the magnetic path in the element and Ni is the product of the current and the number of turns, usually referred to as ampere turns. From (1) and (3):

$$H = \frac{\beta}{\mu_0 \mu_r} = \frac{Ni}{l}$$

or

$$Ni = \frac{\beta l}{\mu_0 \mu_r}$$

Substituting from (2), the ampere turns required to produce the working flux in an element of the magnetic circuit is:

$$\text{Ampere turns} = \frac{\phi l}{\mu_0 \mu_r a}$$

Considering the circuit shown in Figure 3.9, the total ampere turns required to produce a working flux in the air gap is the sum of the ampere turns to produce the working flux in each element of the circuit

$$\text{Ampere turns for the air gap} = \frac{\phi l_g}{a_g} \text{ since } \mu_0 \mu_r \text{ for air} \simeq 1$$

$$\text{Ampere turns for the core} = \frac{\phi l_1}{\mu_0 \mu_r a_1}$$

$$\text{Ampere turns for the armature} = \frac{\phi l_2}{\mu_0 \mu_r a_2}$$

$$\text{Total ampere turns} = \phi \left[\frac{l_g}{a_g} + \frac{l_1}{\mu_0 \mu_r a_1} + \frac{l_2}{\mu_0 \mu_r a_1} \right]$$

This calculation does not take into account magnetic leakage across the circuit or fringe effects in the air gap. To allow for these losses the working flux is usually increased by a factor of about 1.2.

Force Exerted by a Magnetic Field

Consider two poles of an electromagnet with a flux density in the air gap of β. It can be shown that the energy density produced in the gap by this flux density has a value of $\beta/2$.

Suppose the poles of the magnet illustrated in Figure 3.10 each has an area a, and the force of attraction between the poles is P. Suppose now one pole is moved a small distance δl against the force P, and the current is increased to maintain the energy density. The work done in moving the pole is $P\delta l$ and the energy stored in the field is increased by

$$\frac{\beta^2}{2} \cdot adl$$

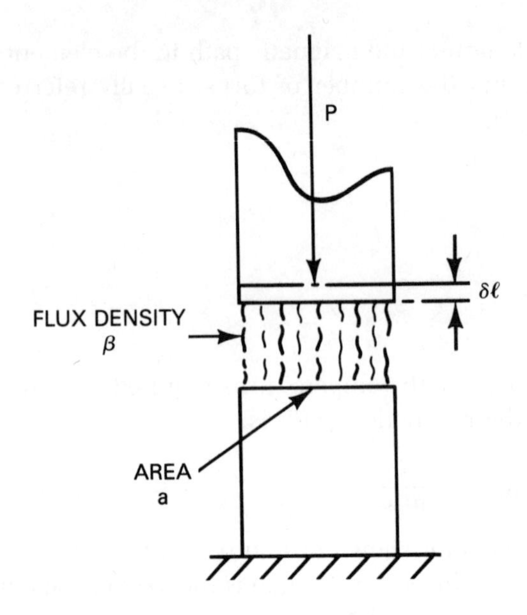

Figure 3.10
Force Exerted by Magnetic Field

These two are equal. Therefore:

$$P\delta l = \beta^2 adl/2$$

or

$$P = \beta^2 a/2$$

INDUCTANCE IN AC CIRCUITS

Consider a magnetic circuit with N turns and a current i flowing in them, producing a flux ϕ. The circuits' inductance L is given by the equation

$$L = \frac{N\phi}{i} \tag{1}$$

Suppose an alternating current with a peak value i_m flows in the circuit and suppose the peak magnetic flux produced is ϕ_m: Substituting in (1):

$$L = \frac{\phi_m N}{i_m}$$

The instantaneous current in the coil produced by the alternating current will be:

$$i = i_m \sin i\omega t \tag{2}$$

where $\omega = 2\pi f$ and f is the frequency of the alternating current.

The rate of change of current at any instant will be:

$$\frac{di}{dt} = i_m\ \omega \cos \omega t$$

But the instantaneous voltage drop across an inductor is

$$L\frac{di}{dt}$$

or

$$E = L\omega i_m \cos\omega t \qquad (3)$$

If E_m is the maximum value of E:

$$E_m = \omega L i_m$$

or

$$i_m = \frac{E_m}{\omega L}$$

The similarity to Ohm's law is easily recognized. The value ωL is called the reactance of the circuit, is expressed in ohms, and is given the symbol X_L. If (2) and (3) are compared, it is apparent that there is a 90° phase difference between the current and the voltage. To take this phase difference into account, equation (3) may be written as a phasor relationship:

$$i = \frac{E}{j\omega L}$$

between the current phasor i and voltage phasor E.

Circuits Containing Inductance and Resistance

In practice, no coil can be purely inductive, for it must also have resistance. If a coil that has both resistance and inductance is considered, it can be regarded as a circuit made up of two separate components in series, as shown in Figure 3.11.

The voltage E_1 across R is:

$$iR$$

The voltage E_2 across L is:

$$ji\omega L$$

where j indicates that the voltage leads the current by 90°.

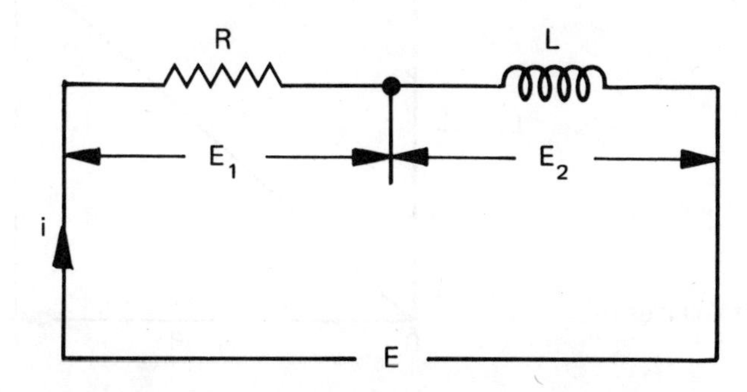

Figure 3.11
R and L in Series

If a vector shown in Figure 3.12. OI is drawn representing the current in the resistive circuit, E_1 will lie along this line, and E_2 will be a vector at right angles. Summing these two vectorially:

$$E = \sqrt{E_1^2 + E_2^2}$$
$$= \sqrt{i^2R^2 + i^2\omega^2L^2}$$
$$= i\sqrt{R^2 + (\omega L)^2}$$

If the effective ac resistance of the circuit is called Z, then

$$E = iZ$$

or

$$Z = \sqrt{R^2 + X_L^2}$$

where Z is the impedance of the circuit and X_L is the inductive reactance.

LENZ'S LAW

When the current in a coil changes, an electromotive force (EMF) is induced in the coil such as opposes the cause producing it. Such a coil is described as an *inductance* and the circuit is said to be *inductive.*

Rise in Current in an Inductive Circuit

When a voltage step function E is applied to a resistive circuit, the rise in current is instantaneous, but where a resistance coil has inductance, the applied voltage is opposed by a self-induced EMF.

The voltage drop across the inductor at any instant is:

$$L\frac{di}{dt}$$

where i is the instantaneous current.

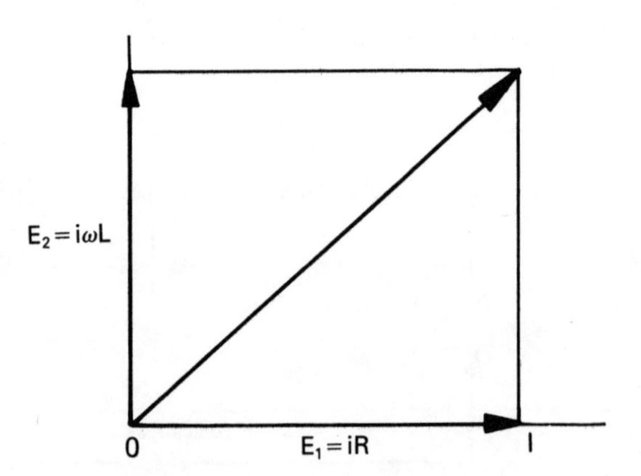

Figure 3.12
Vector Representation of Voltages in R and L Circuit

In a circuit containing resistance R in series with an inductance L, at any instant:

$$e = iR + L\frac{di}{dt}$$

where e is the instantaneous voltage. The solution of this differential equation is:

$$i = \frac{E}{R}(1 - e^{-Rt/L}) \qquad (1)$$

$$= I(1 - e^{-Rt/L}) \qquad (2)$$

where I is the final value of the current and L/R is the time constant.

The rise in current in an inductive circuit to which a step function voltage has been applied is illustrated in Figure 3.13a.

Decay of Current in an Inductive Circuit

Consider a coil that contains a resistance R and an inductance L and that carries a steady current I. Suppose that the coil is short-circuited and simultaneously the EMF producing the current is disconnected. The EMF equation becomes:

$$0 = iR + L\frac{di}{dt}$$

the solution to which is:

$$i = \frac{Ee^{-Rt/L}}{R} = Ie^{-Rt/L}$$

This decay is illustrated in Figure 3.13b.

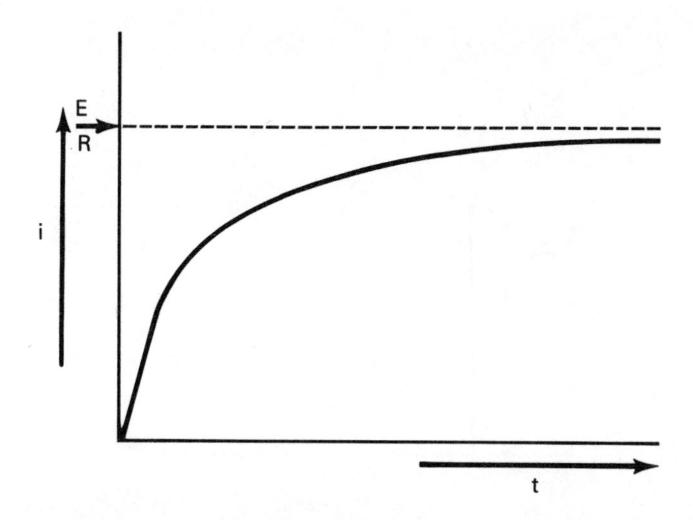

Figure 3.13a
Rise of Current in an Inductive Circuit

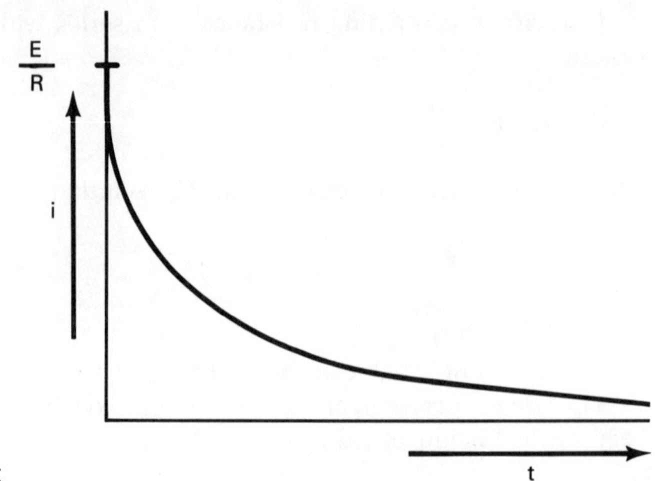

Figure 3.13b
Decay of Current in an Inductive Circuit

AC BRIDGES

The conditions of balance for an ac bridge are somewhat more complex than the dc equivalent. Consider the bridge in Figure 3.14 which is excited by an alternating current.

The impedance, or ac resistance, of the *AB* and *BC* are Z_1 and Z_2 and of *AD* and *CD*, Z_3 and Z_4, respectively.

The conditions of balance will be:

$$\frac{Z_1}{Z_2} = \frac{Z_3}{Z_4} \quad (1)$$

In its complex notation:

$$Z = R + jX \quad (2)$$

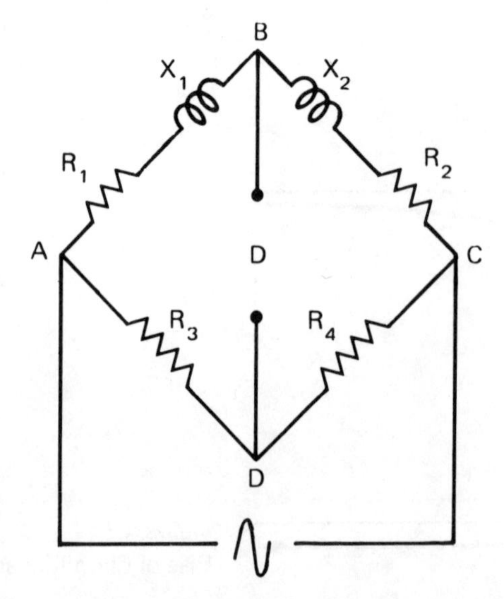

Figure 3.14
The AC Bridge

$$\frac{R_1 + jX_1}{R_2 + jX_2} = \frac{R_3 + j0}{R_4 + j0}$$

$$R_1R_4 + jR_4X_1 = R_2R_3 + jX_2R_3 \tag{3}$$

Equating the real parts:

$$\frac{R_1}{R_2} = \frac{R_3}{R_4} \tag{4}$$

as for the dc bridge, and equating the imaginary parts:

$$\frac{R_3}{R_4} = \frac{X_1}{X_2}$$

Hence:

$$\frac{R_1}{R_2} = \frac{R_3}{R_4} = \frac{X_1}{X_2}$$

The balance of the bridge, however, is usually complicated by the fact that the arms in addition to resistance and inductive reactance also contain stray capacitive reactance.

Output of an AC Bridge

The bridge shown in Figure 3.15 is fed by an alternating current. The output points of the bridge are shown connected to one channel of an oscilloscope and the ac supply connected to a second channel. At balance, the signal trace on the

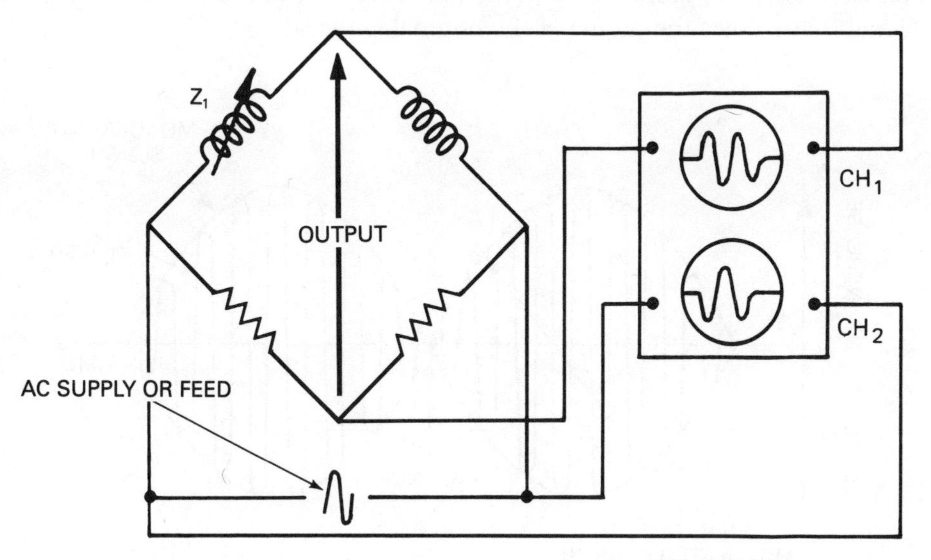

Figure 3.15
Relation of Signal and Feed in an AC Bridge

oscilloscope is a horizontal straight line, but as the value of Z_1 is increased, there appears on the screen an ac waveform with the same frequency as the excitation voltage. The variation in amplitude of this waveform is related to the variations in Z_1 and it demonstrates a fixed phase relationship with the excitation frequency. If now the bridge is returned to balance, the trace again returns to a straight line, and lowering the value of Z_1 further produces on the oscilloscope a waveform which at first sight appears identical to the one produced by increasing Z_1. Comparison of this trace with the ac supply signal, however, demonstrates that its phase has changed by 180°.

The output of the bridge, therefore, is an ac voltage whose amplitude follows the variation of the impedance of the bridge arm and which changes phase by 180° with respect to the bridge supply when it passes through its balance point. Such an output is of little value to the instrumentation engineer who is concerned with signals that vary in the same form as the original parameter. Consequently, the ac output, or *carrier* as it is called, must be turned into a dc level which is positive or negative, depending on its phase relationship to the bridge excitation voltage. Such a procedure is called *phase discrimination*, and the conversion circuit called a *phase discriminator*.

Phase Discrimination

Suppose in the bridge shown in Figure 3.15, Z_1, is varied in a sinusoidal manner round its balanced point. The output is as shown below in Figure 3.16.

The envelope of this carrier signal represents a double image of the variation in Z to the sinusoidal excitation producing it. To reproduce the original waveform it is necessary to remove the *positive* and *negative* parts of the envelope alternately, as shown in Figure 3.17, and then filter the carrier frequency, leaving only a signal that follows the profile of the original sinusoid.

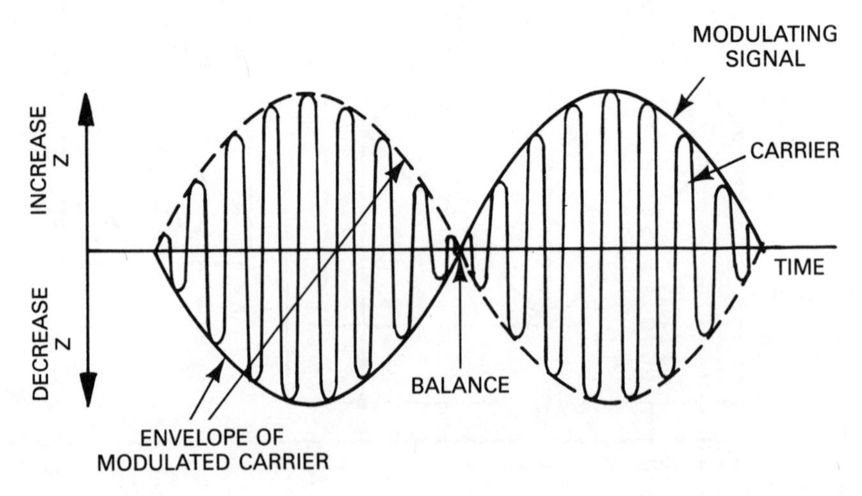

Figure 3.16
Carrier Output of AC Bridge

A simple phase discriminator to perform this function is shown in Figure 3.18 and it works in the following manner.

The reference voltage of the discriminator is obtained from the bridge supply and applied to transformer T_2 while the signal from the bridge is applied to transformer T_1.

If the original signal voltage is $E_1 \cos \omega t$ and the reference is $E_2 \cos (\omega t + \phi)$, the output at *AB* is the vector addition of these two voltages. Providing suitable values of *C* and *R* are chosen, the output at *xy* will be a direct current with a value $k\,(E_1{}^2 + E_2{}^2 + 2E_1E_2 \cos \phi)$ where *k* is a circuit constant. As the bridge passes through balance ϕ changes from 0 to π or vice versa and the output consequently follows the original signal profile.

CAPACITANCE IN CIRCUITS

If a voltage is applied to a circuit containing two metal plates in proximity, but insulated from each other, a current will flow that rapidly reduces to zero. If now the voltage is disconnected and the plates are connected externally, a discharge of current will take place in the reverse direction demonstrating that the plates have

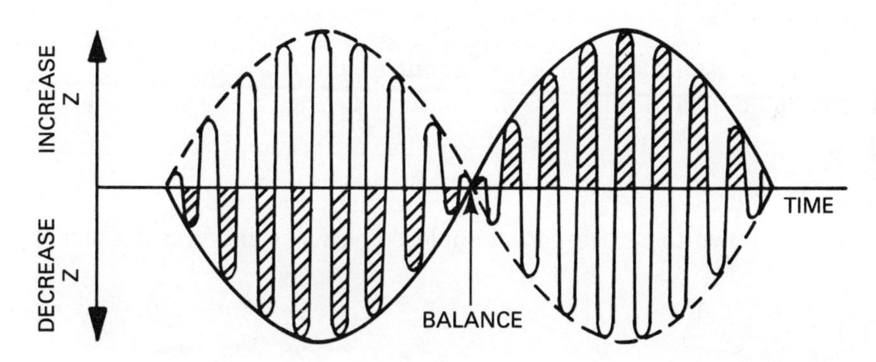

Figure 3.17
Phase Discriminated Output

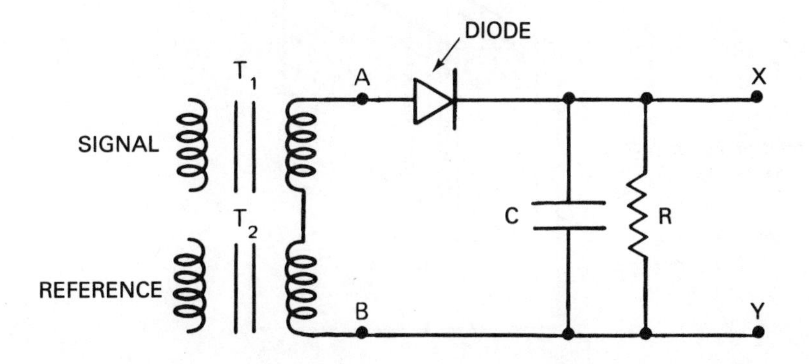

Figure 3.18
Simple Phase Discriminator

become electrically charged. Such a device is called a *capacitor*, illustrated in Figure 3.19, and the size of charge which it is capable of holding at a given voltage is a measure of its *capacitance*. The calculation of capacitance is based on a number of laws and theorems postulated by Gauss and Coulomb.

Gauss' Law

If an electric field is represented as lines of force acting perpendicularly to a surface, the total number of lines passing through the surface is:

$$N = EA \qquad (1)$$

where

N is the number of lines of force

E is the field intensity

A is the surface area

An electrical charge assumed to be concentrated in an infinitesimally small volume is called a point charge.

If a sphere with a radius r is considered with a point charge at its center in a vacuum, the field intensity at the surface will be uniform with a value:

$$E = \frac{q}{\epsilon_0 A} = \frac{q}{4\pi r^2 \epsilon_0}$$

where ϵ_o is a constant known as the *permitivity of free space.*

From (1) the total number of lines of force acting on the surface is:

$$N = \frac{q}{4\pi r^2 \epsilon_0} \cdot 4\pi r^2 = \frac{q}{\epsilon_0} \qquad (2)$$

This can be shown to be true for all enclosed surfaces and provides the general

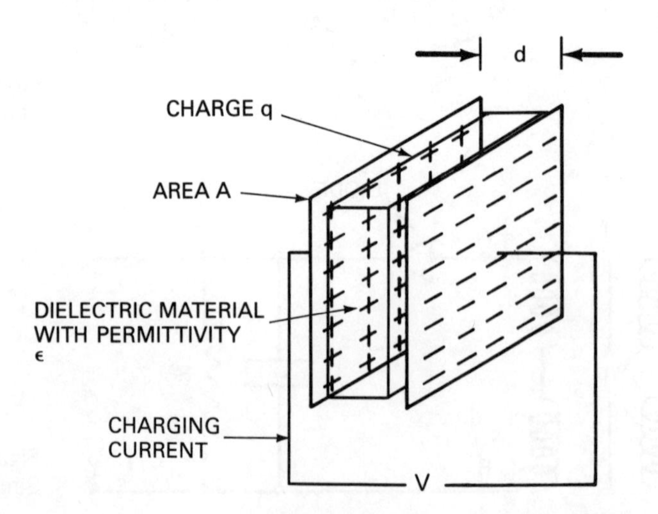

Figure 3.19
The Simple Capacitor

case for Gauss' law. This states that the total electric field acting through any closed surface is equal to the amount of charge enclosed.

Potential Difference

The electrical potential of a point in an electric field is defined as the potential energy of the field unit charge. If a unit charge is moved from a point in an electric field to another, work will be done on that field, and the work done will be a measurement of the difference in potential between the two points.

Suppose a unit charge is placed in a field of intensity E and it is moved by a small amount ds.

The work done will be Eds and in moving from a point S_1 to S_2 the total work will be:

$$\int_{S_1}^{S_2} Eds$$

The potential difference:

$$v = \int_{S_1}^{S_2} Eds \qquad (1)$$

Coulomb's Law

When two point charges of like or opposite polarity are separated by a distance in an insulating material, a repulsion or attraction force of:

$$F = \frac{q_1 q_2}{4\pi\epsilon r^2} \qquad (1)$$

is exerted between them, respectively, where

F is the force exerted on the charges in newtons

q_1 and q_2 are the magnitudes of the charges in coulombs

r is the separation between the charges in meters

ϵ is the permittivity of the material

Equation (1) may be written as:

$$F = E_1 q_2 \qquad (2)$$

where

$$E_1 = \frac{q_1}{4\pi\epsilon r^2}$$

E_1 is defined as the field intensity created by the point charge q_1 and is sometimes expressed in units of newtons/coulomb and sometimes in volts/meter.

Parallel Plate Capacitor

Probably the most common type of capacitor is one where two conducting plates face parallel to one another but are separated by an insulating medium. If a voltage v is connected across the capacitor, an electric field of intensity E will be created within it that will be uniform over the separation. (See Figure 3.19.)

Solving equation (1) in a previous section with E constant, we obtain

$$v = E(S_2 - S_1)$$

If the plate separation of the capacitor illustrated in Figure 3.19a is d, then $S_2 - S_1 = d$ and

$$v = Ed$$

Applying Gauss' Law, the electric field intensity E in this capacitor is given by

$$E = \frac{q}{\epsilon A}$$

where

q is the charge in coulomb

ϵ is the permittivity of the medium in coulombs per volt per meter or Farads per meter

A is the area of the plates in square meters

Therefore:

$$v = \frac{qd}{\epsilon A} \quad (1)$$

Defining capacitance C by the equation $q = Cv$ and substituting in (1) we obtain

$$C = \frac{\epsilon A}{d} \quad (2)$$

But by definition

$$\epsilon = \epsilon_o \epsilon_r$$

where

ϵ_o is the permittivity of free space and equals 8.854×10^{-12} Farads per meter
ϵ_r is the dielectric constant of the medium between the plates

Therefore:

$$C = \frac{\epsilon_0 \epsilon_r A}{d}$$

Concentric Cylinders

Consider two concentric cylinders, shown in Figure 3.20, having an inner radius r and outer radius R.

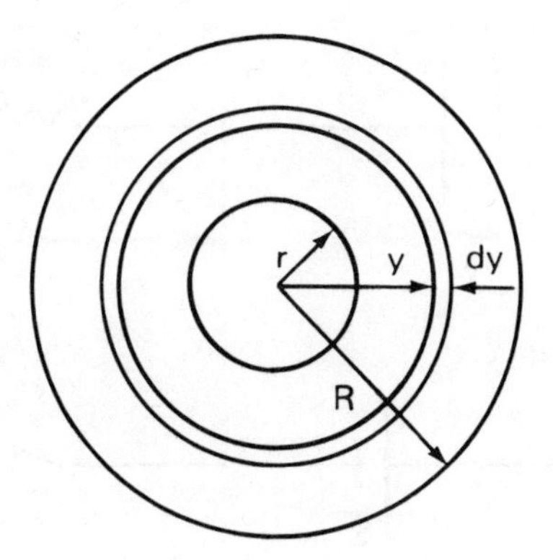

Figure 3.20
Elementary Concentric Cylinders

If the core is uniformly charged q Coulombs per unit length, then the total lines of force that will emanate from unit lengths of core will be q and will be radial.

If an elementary cylinder at radius y is considered, the surface area of unit length is $2\pi y$ and the field strength E, according to Gauss' Law, is:

$$E = \frac{q}{2\pi y \epsilon}$$

The potential difference V between two cylinders will be:

$$\int_r^R Edy = \int_r^R \frac{q\,dy}{2\pi y \epsilon}$$

$$= \frac{q}{2\pi\epsilon}\int_r^R \frac{dy}{y}$$

$$= \frac{q \ \log_e \frac{R}{r}}{2\pi\epsilon}$$

but

$$C = \frac{q}{V} = \frac{2\pi\epsilon}{\log_e \frac{R}{r}} \text{ per unit length.}$$

Charge Build-up of a Capacitor

Suppose a circuit with a capacitance C and a series resistance R as illustrated in Figure 3.21 is subjected to a step input of voltage. Suppose the voltage across the capacitor at any instant is v, the instantaneous current is i, and the instantaneous charge is q. The rate of change of the charge with respect to time is given by the relationship.

$$i = \frac{dq}{dt} = C\frac{dv}{dt} \tag{1}$$

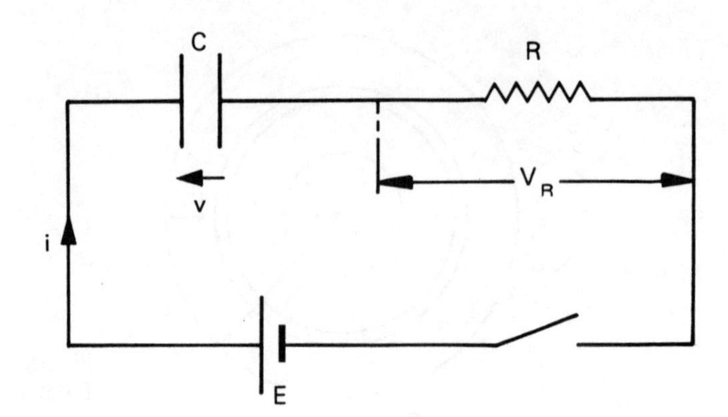

Figure 3.21
Charge of a Capacitor

But

$$i = \frac{V_R}{R} = \frac{E - v}{R} \qquad (2)$$

Combining (1) and (2):

$$\frac{E - v}{R} = C\frac{dv}{dt} \qquad (2)$$

$$v = E - CR\frac{dv}{dt}$$

The solution to this equation is:

$$v = E(1 - e^{-t/RC}) \qquad (3)$$

Decay of Voltage in a Capacitive Circuit

Consider a circuit in which a capacitor charged up to a voltage E across it with a resistor in series are short-circuited.

The capacitor discharges with a current

$$i = \frac{dq}{dt} = -C\frac{dv}{dt} \qquad (1)$$

with the minus sign indicating discharge and the current flowing in the circuit is

$$\frac{v}{R} = -C\frac{dv}{dt} \qquad (2)$$

or

$$v + RC\frac{dv}{dt} = 0 \qquad (3)$$

The solution to this equation is:

$$v = Ee^{-t/RC} \qquad (4)$$

Capacitance in an AC Circuit

Suppose an ac voltage is represented by:

$$v = E_m \sin \omega t \qquad (1)$$

If this voltage is applied to a capacitance C (Figure 3.22), and the charge on the capacitance at any instant is q, then

$$q = vC$$

If the instantaneous current is i, then

$$i = \frac{dq}{dt}$$

or

$$i = \frac{d\,(CE_m \sin \omega t)}{dt}$$

$$= C\omega\, E_m \cos \omega t$$

If maximum values are considered,

$$I_m = \omega C E_m$$

or

$$\frac{E_m}{I_m} = \frac{1}{\omega C}$$

The value ωC is known as the capacitive reactance of the circuit and is designated by the symbol X_c.

In comparing i and v it can be seen that the current and voltage have a phase difference of $\pi/2$ and to take this into account X_c is represented in its complex notation as $\frac{1}{j\omega C}$, where $j = \sqrt{-1}$.

Inductance in an AC Circuit

If an ac current

$$i = I_m \sin \omega t$$

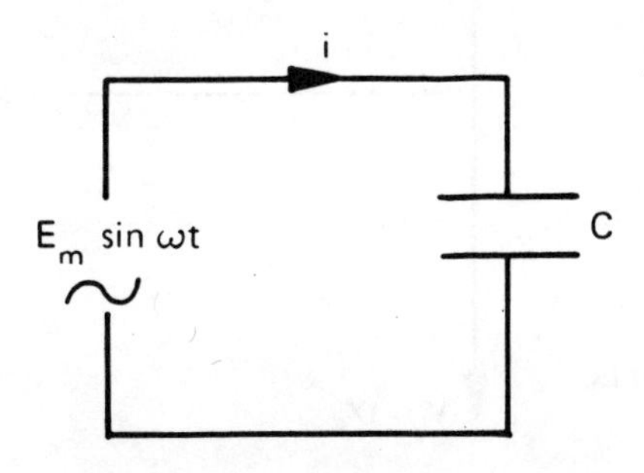

Figure 3.22
Capacitance in an AC Circuit

is passed through an inductance L, the instantaneous voltage v across the inductance is

$$v = L\frac{di}{dt}$$

or

$$v = \omega L\, I_m \cos \omega t$$

The amplitude V_m of this voltage is

$$V_m = \omega L I_m$$

The value ωL is the inductive reactance of the circuit and it is designated by the symbol X_L the volgate v leads the current i by $\pi/2$ or 90°, a relationship represented by the phasor equation

$$V = j\, \omega L\, I$$

where v and I are the voltage and current phasors of the inductance L.

INDUCTANCE AND CAPACITANCE IN AN AC CIRCUIT

In a purely inductive circuit the current lags the voltage by $\pi/2$, and in a purely capacitive circuit leads it by $\pi/2$. This may be represented by the vector diagram Figure 3.23 for a circuit containing an inductance L, and capacitance C. There is also a value of frequency where $X_L = X_c$; that is, when

$$2\pi f L = \frac{1}{2\pi f C} \qquad (1)$$

or

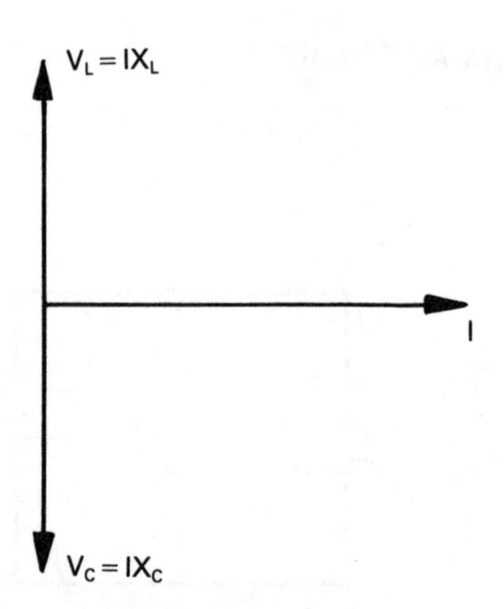

Figure 3.23
Vector Representation of Components in an LC Circuit

$$f = \frac{1}{2\pi\sqrt{LC}} \qquad (2)$$

This frequency is known as the resonant frequency of the circuit.

Capacitance and Inductance in Series

Consider an ac circuit, as shown in Figure 3.24, where a capacitor and inductance are placed in series. If the circuit parameters are expressed in the complex notation, at any instant the voltage v across the circuit and the current i will be given by:

$$v = \left(j\omega Li + \frac{i}{j\omega C}\right)$$

Multiplying the right-hand term by j/j:

$$v = i\left(j\omega L - \frac{j}{\omega C}\right) = ji\left(\omega L - \frac{1}{\omega C}\right)$$

but

$$\omega L = X_L$$

and

$$\frac{1}{\omega C} = X_e$$

Therefore:

$$i = -j\frac{v}{(X_L - X_C)}$$

At resonance X_1, equals X_c and the current becomes infinite. In practice there is always some resistance in the circuit and the current is limited to $I_m = V_m/R$.

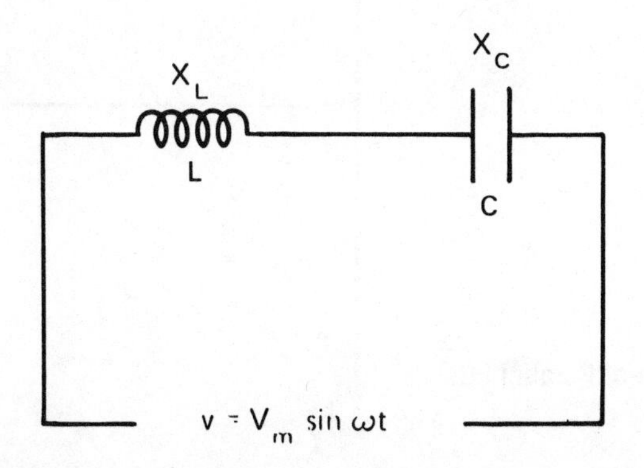

Figure 3.24
Series LC Circuit

Capacitance and Inductance in Parallel

If a circuit is constructed where a purely inductive circuit is connected in parallel with a capacitor (Figure 3.25), the currents I_c and I_L flowing in the capacitor and inductance are in phase opposition (Figure 3.26). At the frequency where I_c equals I_L the current flowing in the external circuit will be zero and the effective reactance (X_e) of the circuit will be infinity.

In this circuit:

$$\frac{V}{X_L} - \frac{V}{X_C} = \frac{V}{X_E} \quad (1)$$

or

$$X_E = \frac{X_L X_C}{X_L - X_C} \quad (2)$$

and when

$$X_L = X_C \qquad X_E = \infty$$

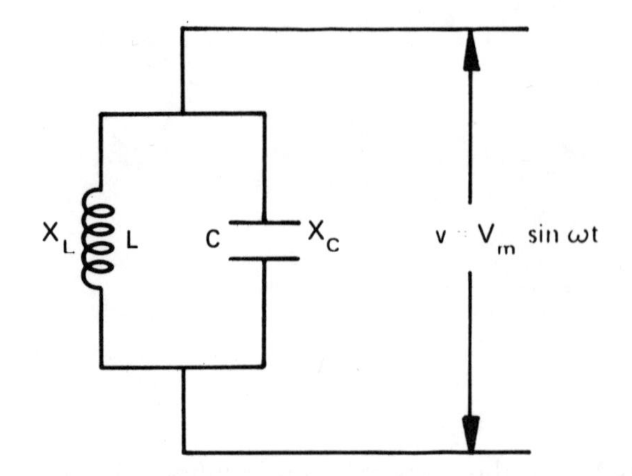

Figure 3.25
Parallel LC Circuit

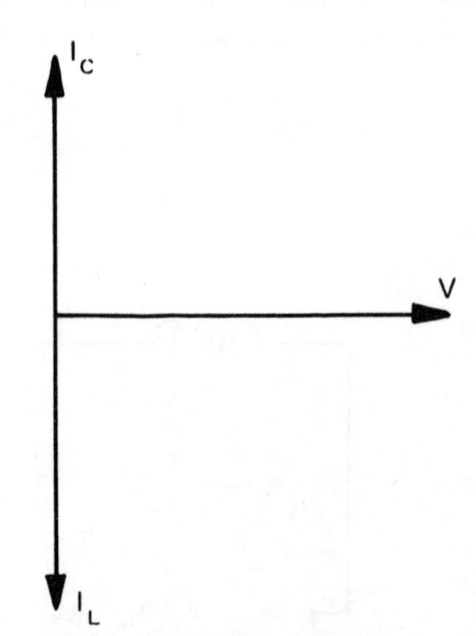

Figure 3.26
Vector Representation of Parallel LC Circuit

This again occurs when:

$$f = \frac{1}{2\pi\sqrt{LC}} \tag{3}$$

CAPACITORS AS TRANSDUCERS

If the expression for the capacitance of a parallel plate capacitor with a plate area a is examined:

$$C = \frac{\epsilon_0 \epsilon_r a}{d}$$

it can be seen that there are two possible ways of varying the capacitance.

It is possible to vary the area by Δa by adjusting the relative position of the plates to one another as illustrated in Figure 3.27. In theory the change in capacitance will be proportional to the change in area. In practice the linearity will be distorted by fringe effects. Where changes in area are used to vary the capacitance it is often preferable to use a differential arrangement and use the capacitor as two arms of a bridge. This is shown in Figure 3.28.

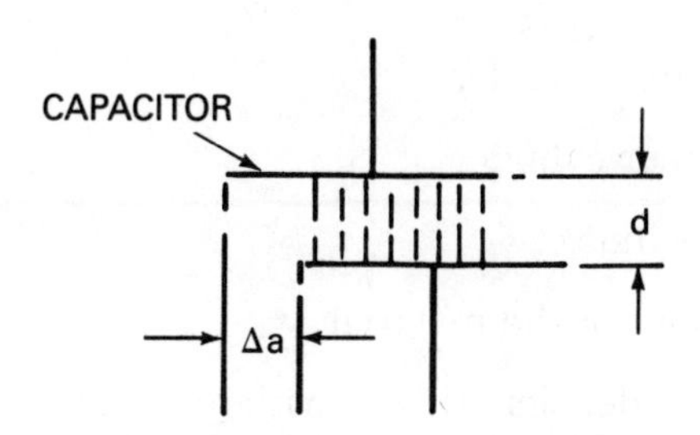

Figure 3.27
Variable Area Capacitance Transducer

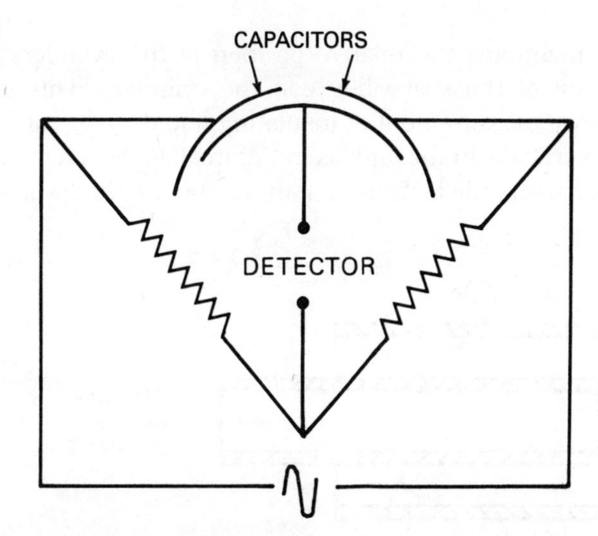

Figure 3.28
Differential Capacitance Transducer

It is also possible to vary the capacitance by varying the distance between the plates.

$$C = \frac{\epsilon_0 \epsilon_r a}{l} \qquad (1)$$

$$\frac{dC}{dl} = -\frac{\epsilon_0 \epsilon_r a}{l^2}$$

$$dC = -\frac{\epsilon_0 \epsilon_r a dl}{l^2} \qquad (2)$$

The output of this arrangement is a nonlinear function of l and the variation between the plates is usually severely restricted to maintain a simple calibration factor.

Concentric Cylinder Transducers

The capacitance C of concentric cylinders is given by the expression:

$$C = \frac{2\pi \epsilon_0 \epsilon_r l}{\log_e \frac{R}{r}}$$

where

R is the radius of the outer cylinder

r is the radius of the inner cylinder

l is the length of the cylinders

ϵ_r is the dielectric constant of the material.

It is possible to vary the cylinders lengthwise with respect to one another. (See Figure 3.29.) In this case the output is linear and restricted only by the geometry of the transducer.

An alternative method maintains the relative position of the cylinders and varies the dielectric constant of the space between the cylinders. This is an important concept in the level measurement of insulating liquids (e.g., oil). The two cylinders are mounted vertically in the tank as in Figure 3.30. When the tank is empty the dielectric is air with a dielectric constant of one. As the level of oil

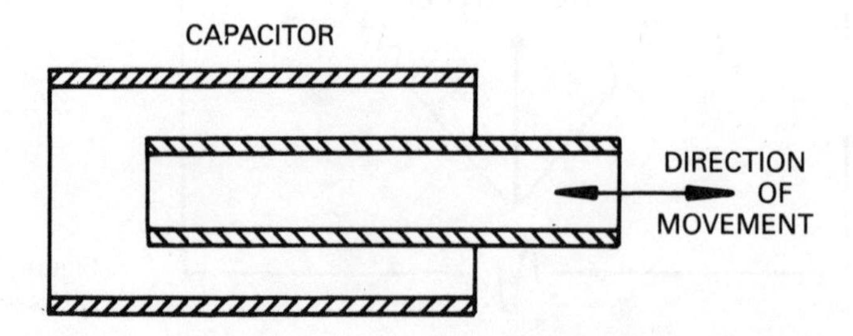

Figure 3.29
Variable Area Transducer

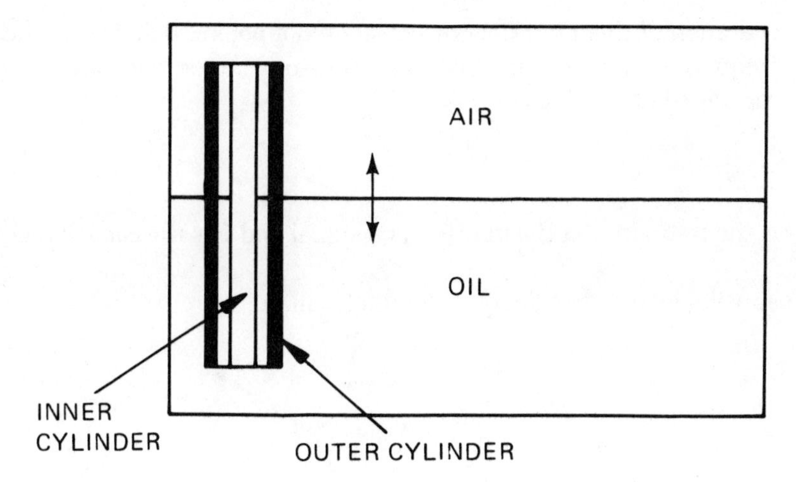

Figure 3.30
Variable Dielectric Transducer

rises, so the air is displaced by oil, which has a dielectric constant of about 2.1. The capacitance therefore increases linearly with the oil level.

As with the inductive transducers the change in capacitance is measured by an ac bridge. Changes, however, are smaller and it is usual to use a high excitation frequency. The measurement is also complicated by the presence of cable capacitance which is large compared with the sensor capacitance.

RC FILTERS

A problem encountered in many instrumentation measurements is that of electrical interference, or *noise*, as it is more commonly called. Noise may be induced into cables connecting the sensor to the measuring equipment and may mask the measurement signal. It may be reduced or *attenuated* by the introduction of *filters* in the cables between the sensor and the equipment. These filters may be complex circuits, and can now be purchased as complete units for insertion into the lines. In many instances, however, a much simpler circuit may be used which is described as an *RC filter*, as illustrated in Figure 3.31.

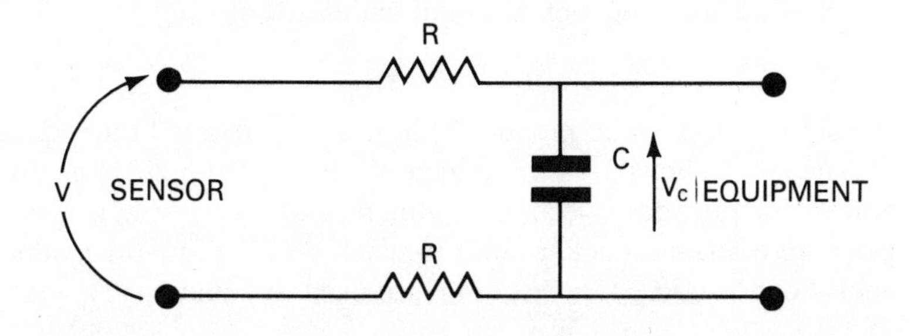

Figure 3.31
RC Filter

The operation of this filter may be considered a potential divider to the interference frequency, having little effect on a slow moving or *quasistatic* signal.

The reactance of the capacitor is:

$$X_c = \frac{1}{2\pi fC} \tag{1}$$

where f is the frequency of the interference signal and C is the capacitance of the filter.

The ac voltage appearing across the filter will be:

$$V_C = \frac{VX_c}{Z} \tag{2}$$

where

V_c is the ac voltage across the capacitor
V is the original unfiltered interference voltage
Z is the impedance of the circuit

$$V_c = \frac{VX_c}{\sqrt{4R^2 + X_c^2}} \tag{3}$$

If C is large X_c^2 will be small compared to $2R$ and

$$V_c \simeq \frac{VX_c}{2R}$$

Providing the filter has been designed with a value of R that is very large compared with X_c the ac interference voltage appearing across the capacitor will be very small.

Filter Response to a Step Voltage

The response of an RC circuit to a step voltage E applied at a moment $t = 0$ is given by the equation:

$$V = E(1 - e^{-t/Rc}) \tag{1}$$

where V is the voltage across the capacitor C after a time t, and R a series resistance.

It is convenient to express equation (1) in terms of a parameter described as a *time constant.*

The time constant τ of such a circuit is defined as:

$$\tau = RC$$

If the value equal to the product of R and C is inserted into equation (1) in place of t, it can be shown that the voltage V will rise to 63.2% of its final value in this period. This provides a second definition of time constant, i.e., the time that the voltage across the capacitor takes to reach 63.2% of its final value.

If values of $t = 2RC$, $t = 3RC$, etc. are used in equation (1), a table may be calculated that illustrates the voltage across the capacitor as a percentage of the final voltage at these intervals of time.

τ/RC	V_c/E
= 1	= 63.2%
2	86.46
3	95.02
4	98.17
5	99.325
6	99.752
7	99.909
8	99.967
9	99.988
10	99.996

It follows, therefore, that if a step function is applied to an RC filter a delay equal to 7 time constants must be allowed if a measurement is required to an accuracy of 0.1% of the true value. This effect also governs the separation that must be maintained between the signal frequency being measured and the interference frequency being attenuated.

COMMENT ON UNITS

It is convenient to express electrical and magnetic theory in SI units and these have been assumed through this chapter. A summary of SI units is listed in the appendix. In other aspects of instrumentation engineering, however, conventional units are still in general use, and in later chapters units appropriate to the subject and local usage may be applied.

CHAPTER 4

STRAIN GAGE THEORY

MECHANICAL STRAIN

Most materials exhibit elasticity when loaded; that is to say, the change in length of the material within certain specified limits is proportional to the load, and when removed the material returns to its original dimensions. The specimen shown in Figure 4.1 is assumed to be made of steel. It has a cross-sectional area of a and a gaged length of l.

If a load is added incrementally, a graph of load against change in length may be plotted which, for steel, takes the form shown in Figure 4.2.

Up to a certain point of loading the change in length is proportional to the load applied. This is sometimes described as the elastic range of the material. If loading is continued, an appreciable change of length of the specimen occurs without a corresponding proportional application of load. This point, marked on Figure 4.2 as point Y, is the *yield point* of the material, and beyond it the extension is no longer proportional to the applied load. Removal of the load demonstrates a permanent deformation of the material, while further application takes the material to an *ultimate load*, at point U, where it will continue to extend to its breaking point even though the load may be reduced to something less than ultimate.

Not all materials behave in the same way as steel. Some exhibit no marked yield point, whereas a few, such as aluminum, have no true elastic range.

If the elastic portion of the curve shown in Figure 4.2 is considered in detail,

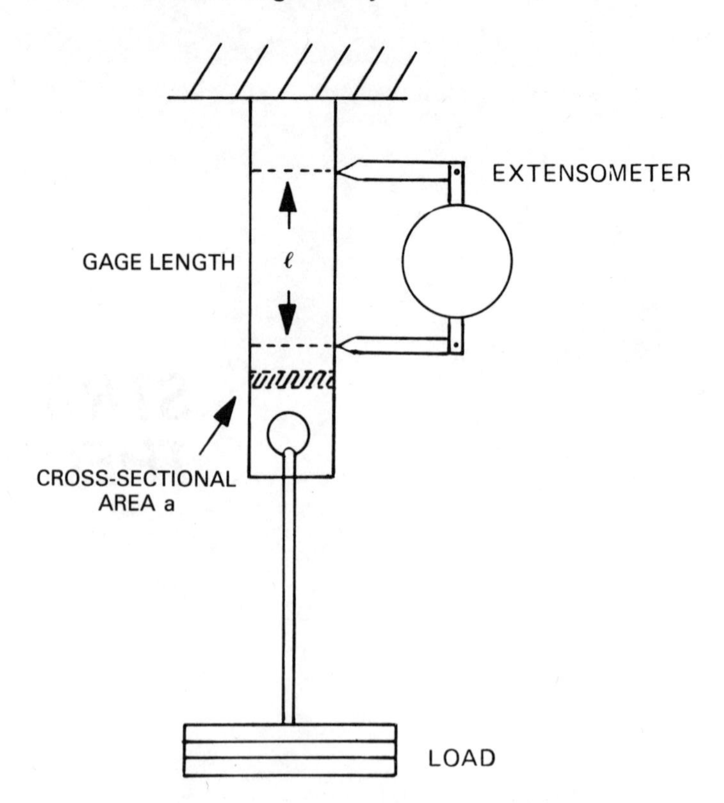

Figure 4.1
Mechanical Strain Testing

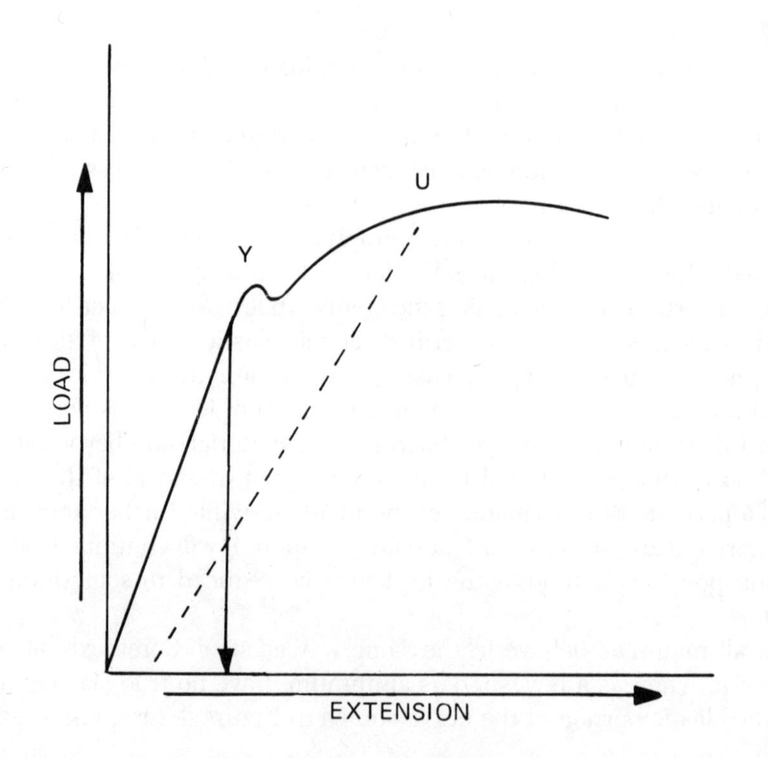

Figure 4.2
Stress/Strain Curve

the ratio of load to extension will be recognized as a constant for a given material and a specimen of given dimensions. If the constant is reduced to unit dimensions by dividing the load by the cross-sectional area of the specimen and the extension by its strained length, a constant for the material will result. This reduction is sometimes described as *normalizing* the expression.

- The ratio of load to cross-sectional area is known as the *stress* experienced by the specimen.
- The ratio of extension to original length is the *mechanical strain*. Strain is a dimensionless ratio and may be quoted as an absolute quantity or a percentage, although sometimes it is multiplied by a factor of 10^6 and described as microstrain.
- The ratio of stress to strain is known as *Young's Modulus* and is designated by the symbol $\mathscr{E}$. This ratio is a constant for any given elastic material.

ELECTRICAL STRAIN

If the electrical properties of the material under test are studied, there is a very close parallel between the electrical and mechanical characteristics. Suppose the length is changed by δl, there will be a corresponding change in resistance of δR and within the elastic limit δl is proportional to δR. A graph of δl versus δR is a straight line, and if these two values are normalized:

$$\frac{\delta l}{l} \alpha \frac{\delta R}{R}$$

$\delta l/l$ is the mechanical strain experienced by the material and $\delta R/R$ is given the corresponding name of *electrical strain*.

GAGE FACTOR

The ratio:

$$\frac{\text{Electrical strain}}{\text{Mechanical strain}} = \text{Constant}$$

and this constant, defined as *gage factor*, is independent of material to a first-order approximation.

Consider a wire of resistance R, length l, and cross-sectional area a:

$$R = \frac{\rho l}{\text{a}} \tag{1}$$

The effect of changes in l and a on the resistance of the wire can be expressed by differentiating the equation with respect to l, a, and p.

Therefore:

$$dR = \rho \frac{dl}{a} - \rho \frac{lda}{a^2} + \frac{l}{a} d\rho \tag{2}$$

Dividing (2) by (1):

$$\frac{dR}{R} = \frac{dl}{l} - \frac{da}{a} + \frac{d\rho}{\rho} \tag{3}$$

When a specimen is strained, not only does its length increase, but there is a corresponding reduction in dimensions at right angles to the line of strain. If these two changes are normalized, it will be found that the ratio of lateral strain to longitudinal strain is a constant for a material and lies between ⅓ and ¼ for most materials. The ratio is known as *Poisson's ratio*, usually designated by the symbol μ.

Consider a specimen with a square cross section of dimension t. If the specimen is strained, there will be a reduction in t of δt. In terms of cross-sectional area the change will be:

$$(t - \delta t)^2 - t^2 = \delta a$$

and if this expression is normalized the change will be:

$$\frac{t^2 - 2t\delta t + \delta t^2 - t^2}{t^2} = \frac{\delta a}{a}$$

Therefore:

$$-\frac{2t\delta t}{t^2} + \frac{\delta t^2}{t^2} = \frac{\delta a}{a}$$

The term $\delta t^2/t^2$ is small and can be ignored; thus the expression becomes:

$$\frac{\delta a}{a} = -2\frac{\delta t}{t} \tag{4}$$

But $\delta t/t$ is the lateral strain, and from Poisson's ratio:

$$\frac{\delta t}{t} = \frac{\mu \delta l}{l}$$

$\delta a/a$ may therefore be written as:

$$-2\mu \frac{\delta l}{l}$$

or in the limit

$$da/a = -2\mu \frac{dl}{l}$$

If this value is substituted in equation (3) for $\delta a/a$:

$$\frac{dR}{R} = (1 + 2\mu)\frac{dl}{l} + \frac{d\rho}{\rho} \quad (5)$$

Dividing (5) by dl/l:

$$\frac{dR/R}{dl/l} = 1 + 2\mu + \frac{d\rho/\rho}{dl/l}$$

If it is assumed that ρ is a constant, then the expression becomes $1 + 2\mu$. Since μ is about 0.3 for all metals, it would be expected that the ratio of electrical strain to mechanical strain would be a constant with a value of about 1.6. In practice, it is found to lie between about 1.9 and 2.1, and to explain the discrepancy it has been suggested that a change of resistivity must occur which is associated with the volume change of the wire. This change is assumed to be related by the expression:

$$\frac{d\rho/\rho}{dl/l} = m(1 - 2\mu)$$

where m is a constant for the material.

In practice, the relationship between mechanical and electrical strain, although not constant, is sufficiently consistent for a batch of wire to allow determination by sampling techniques. This allows the design of a practical strain gage.

Suppose a length of resistance wire is bonded to a specimen but electrically insulated from it. When the specimen is strained the wire will be strained by the same amount, and its resistance will change proportionally. From the derived constant, relating electrical and mechanical strain, it is possible to calculate the mechanical strain being experienced by the specimen.

Because of the significance of the electrical–mechanical strain relationship, the factor relating the two has been given the name of gage factor. Because of the unreliability of the factor's absolute value, it must be determined experimentally for each batch of gages produced.

THE PRACTICAL STRAIN GAGE

If the changes in resistance of a wire when subjected to a mechanical strain are to be turned into practical use, several conditions must be fulfilled.

1. The wire must be firmly bonded to a specimen so that when the specimen is strained the wire will experience the same strain as the material.
2. The bonding material must be capable of surviving the same strains as the gage and the specimen and must not change its characteristics with time, repeated stressing, or temperature variation.

3. If the specimen is a conductor, the wire element must be insulated from the specimen.
4. The change in resistance should be as high as can reasonably be attainable, which implies that the total length of wire in the gage should be as long as possible, and the resistivity of the wire be high.
5. The change in resistance of the gage due to temperature changes should be as low as possible.
6. The adhesion of the gage should not appreciably change the stiffness of the specimen.
7. The gage should be physically strong enough to survive handling.
8. The gage material should not be worked above its elastic limit.
9. The output of a series of gages of the same batch should produce repeatable results when subjected to identical strains on identical specimens.

These requirements often conflict, and strain gages are either designed as a compromise for general use or designed specifically for a specialized application. For some applications the *wire gage* shown in Figure 4.3a is still favored, but it is closely rivaled by the *etched foil gage* (Figure 4.3b). The wire-wound gage is produced by winding a number of turns of fine-resistance wire onto a form and then flattening the coil to make a grid. This grid is firmly bonded to an insulated

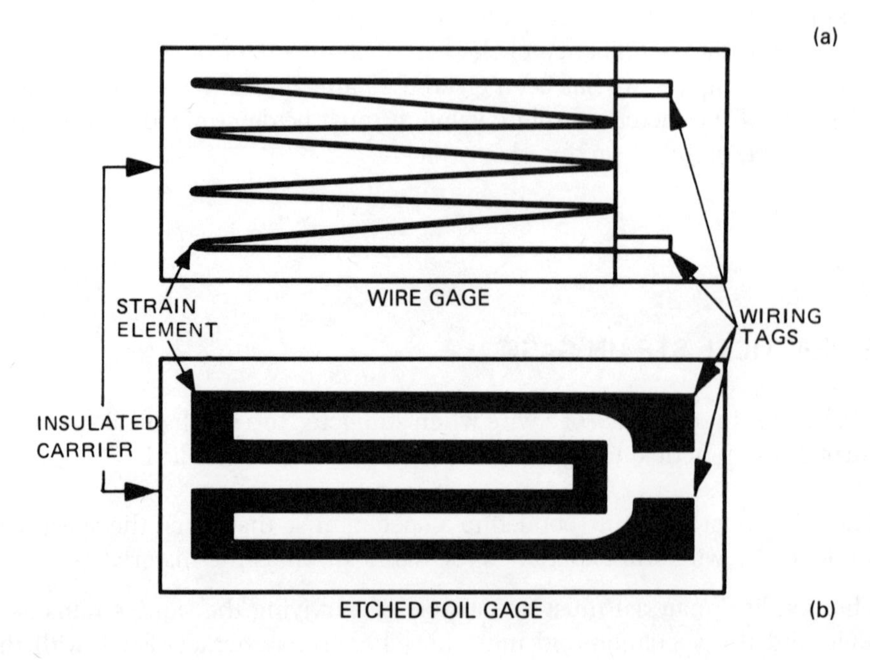

Figure 4.3
Wire Wound and Foil Strain Gages

backing; soldering tags are added to the ends of the wires, and the front is covered by a protective strip of insulation.

Foil gages are made by etching the gage from a metal-clad insulated backing in the same way that printed circuits are made. Foil gages were somewhat slow to be accepted. In the first place, they were considered difficult to install, but as installation techniques and skills improved, this ceased to be regarded as a problem. Also the early gages were made from low-resistance materials with correspondingly low bridge outputs. Since then improved manufacturing techniques have overcome the resistance problem, and amplifying techniques have also improved, which makes low output of less concern.

The traditional foil and wire gages are now challenged by the *semiconductor* gage, which provides much larger output than either. In these gages the wire element is replaced by a thread of semiconductor material. Resistance changes are due to changes in the crystal structure of semiconductors and is much higher than the equivalent changes in metal. Semiconductor gages with gage factors in excess of 100 are common.

Another recent development is the *deposited* gage, which is deposited in the same way as components in integrated circuits. Consequently, these gages are much more accurately positioned than the conventional gage, and the bonding problems no longer exist. Because of the special techniques required for their deposition, however, they are mainly used as tranducing elements in sensors. Although temperature effects still apply, transient temperature compensation is much more effective, because of the minute thermal mass and the integral bonding of the gage to the strained member.

STRAIN GAGE BRIDGES

Because the change in resistance of a strain gage is small and the base resistance is high, it is normal practice to use bridges as a means of measuring the changes accurately. These bridges may be excited by ac or dc supplies.

AC excitation was once considered advantageous because the amplifier is only required to amplify the carrier frequency. This allows the use of a simple amplifier which can have a high gain at that frequency. Providing this gain is stable, the amplification of modulated low-frequency signals will be accurate. Unfortunately the use of alternating current involves phase discrimination of the ac signal which loses some of the initial advantages.

Modern dc amplifiers sufficiently stable to use as instrumentation amplifiers have improved to the point where a high gain can be obtained that is constant over a wide frequency range. Consequently, ac systems are used for convenience rather than necessity. For example, in an instrumentation system involving inductive as well as strain gage bridges, it may be more convenient to excite all the transducers from a common power source and use one multichannel carrier system.

Bridge Balance

It is unlikely that the four arms of a bridge will balance exactly and because the changes in resistance to be measured are small, tolerance between the arms for balance is minute. To overcome this problem balancing circuits are used to "trim out" residual unbalance. The choice of balancing circuit is considerable but may be one of those shown in Figure 4.4.

It is possible to match the resistance of gages so that the bridge unbalance is less than 1% and circuits are required to cope with this amount. If the network in Figure 4.4a is considered, R_1 may be made 100 times the value of R_g, and the value of R to equal R_g. This allows about 2% change in balance which is adequate for most purposes without introducing unduly large errors caused by shunting effects of the lower arms of the bridge.

Figure 4.4b is a variation of 4a. In this case the potentiometer is connected across the supplies, and a limiting resistor R is placed in series with the wiper. The value of R_1 is often made comparatively low, possibly equal to the value of R_g, while R may have a value of 100 times the value of R_g to provide a balance range of 2%. This balance circuit is very useful. Apart from the economy in components, it is possible to change the balance range by changing only one resistor.

The third method involves opening the bridge and inserting a low-resistance potentiometer. This eliminates the shunting effect of parallel balance circuits where all arms are active gages.

Because of lead resistance problems, a balance circuit of this type is not satisfactory when the gages are remote from the balance unit. A similar principle is employed by some sensor manufacturers who add fixed resistors in series with gages to provide a high degree of balancing, to eliminate the necessity for external balance circuits.

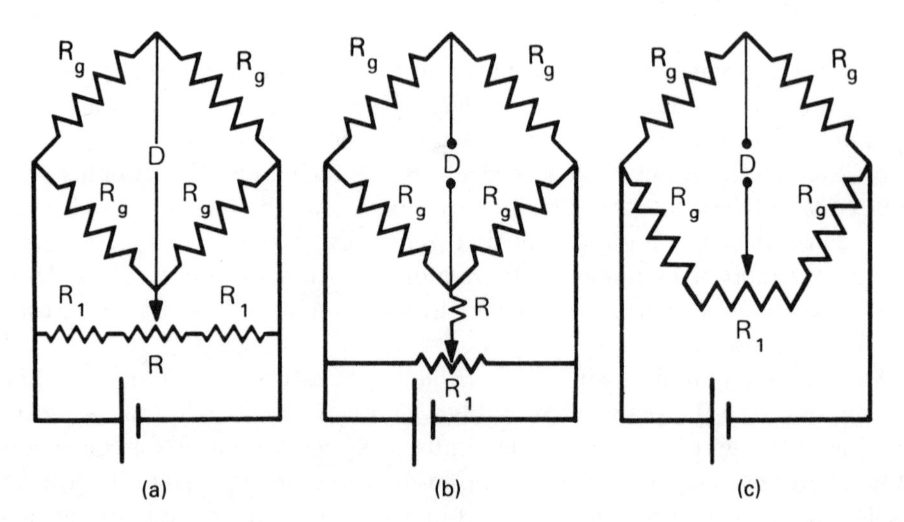

Figure 4.4
Strain-Gage Bridge Balancing Configurations

CIRCUITS FOR STRAIN-GAGE MEASUREMENTS

Single Active Gage

If a strain gage is bonded to a specimen as shown in Figure 4.5a and a second gage is bonded to an unstressed piece of identical material, called a *slip*, the two gages may be wired to form half of a bridge circuit shown in Figure 4.5b. The gage bonded to the specimen is called the *active gage*, and the one on the slip is called the dummy. The *sensitivity factor* of a bridge with more than one gage is the ratio between the output of that bridge as a result of a given change in the measured parameter, and the output of an equivalent bridge with only one gage. The sensitivity factor for the bridge in Fig. 4.5b therefore is that of a single gage (i.e., 1).

The gage on the specimen will respond to either bending or tensile strains. Providing the specimen and the dummy slip are in the same temperature environment, the resistive changes caused by variations in temperature will be equal, and the balance will be unchanged. This circuit is, within limits, *self-compensating* for temperature.

Bending Measurement

Two gages in Figure 4.6a are bonded to a cantilever to which a bending load L is applied. Provided the gages are symmetrically placed round the neutral axis of the specimen, gage A_1 will experience a tensile strain, and gage A_2 will experience a compressive strain of equal magnitude. The output of the bridge and its sensitivity factor, therefore, will be twice that of a single gage. A strain caused by a load E,

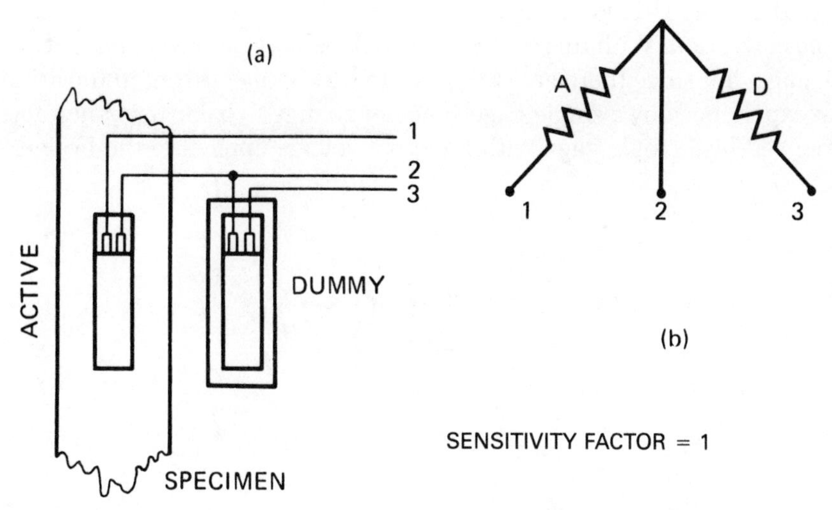

Figure 4.5
Half Bridge with Single Active Gage

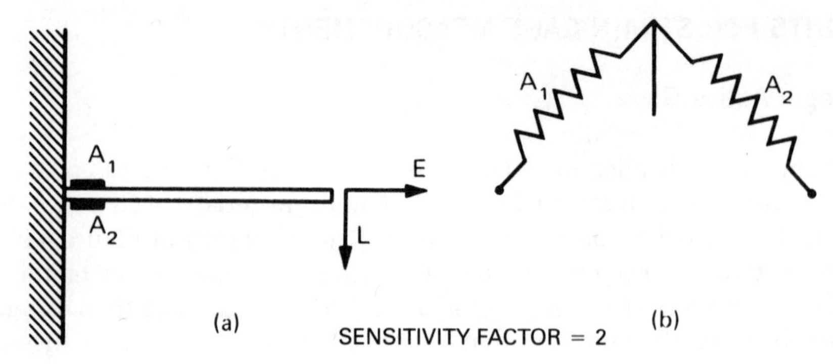

Figure 4.6
Half Bridge to Measure Bending Strains

however, will induce a tensile strain in each gage with the overall balance of the two arms unchanged. The arrangement is known as a bridge wired for *bending measurement* with *end-load correction.*

End-Load Measurement

Consider again the specimen in Figure 4.6a but this time with the gages wired as in Figure 4.7, where the active gages are wired in series and two slip dummies have been added.

The application of a bending strain due to load L still causes A_1 to increase in resistance and A_2 to decrease by an equal and opposite amount. The net effect, therefore, is to maintain the balance of the bridge. The application of load E will cause both gages to increase in resistance and produce a corresponding off balance of the bridge. This provides a *tensile* or *end-load measuring bridge* with *bending correction.* With this arrangement, the change in resistance is twice that of one gage, but since there are two gages in series, the current in the arm is half of that experienced by a single gage. The net change in balance is, therefore, that experienced by a single gage with an equal voltage applied to the bridge.

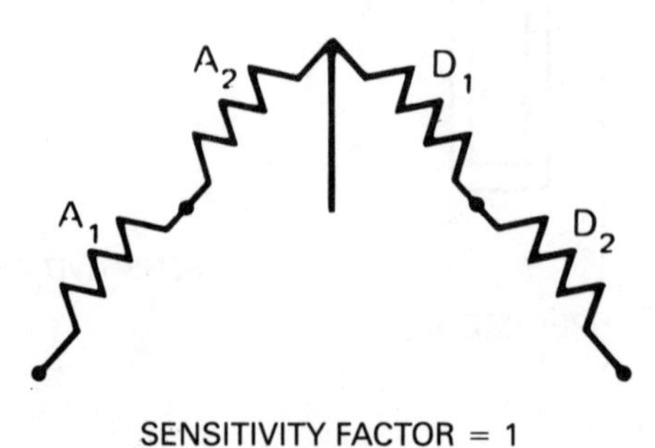

Figure 4.7
Half Bridge Measuring Tensile and Compressive Strain

Shear Strain Measurement

The measurement of shear strain is an important one because it is also used to measure torsional strains in shafts. It relies on the principle that when a shear strain is applied to a specimen, two complementary strains—one tensile, the other compressive—are produced at an angle 45° to the shear.

Consider a square block of metal in Figure 4.8 which is subjected to a shear stress q. The metal will deform such that a tensile strain is induced across the diagonal AB.

The shear strain produced will be:

$$\frac{B'B}{BD}$$

Since $B'B$ is small this ratio is numerically equal to ϕ, the angle $B'DB$ measured in radians. Shear stress divided by shear strain is defined as G, the modulus of rigidity. The relationship between the modulus of rigidity and the tensile modulus is given by the expression:

$$G = \frac{\mathscr{E}}{2(1 + 1/\mu)}$$

where: $\mathscr{E}$ is Young's modulus and μ is Poisson's ratio.

If a perpendicular is drawn from B to P, the tensile strain in AB is $B'B/AB$.

But $B'P = B'B/\sqrt{2}$ and $AB = \sqrt{2}\,BD$ (BB^1P and ABD are similar triangles); therefore:

$$\frac{B'P}{AB} = \frac{B'B \times 1/\sqrt{2}}{BD \times \sqrt{2}} = \frac{1}{2}\frac{B'B}{BD} \qquad (1)$$

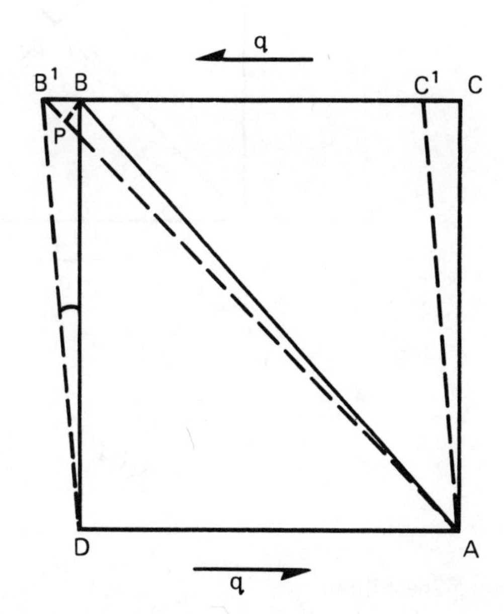

Figure 4.8
Shear Strain

The tensile strain induced across the diagonal is, therefore, half the shear strain, and it can be similarly shown that the corresponding strain across the other diagonal is compressive and numerically equal.

The existence of these tensile and compressive strains provides a means of applying strain gages to the measurement of shear strain. If the two gages are bonded to a specimen as shown in Figure 4.9a and wired as in Figure 4.9b, A_2 will measure the tensile strain developed by a shear strain and A_1 will measure the compressive strain. The output of the bridge in Figure 4.9b will be twice that of a single active gage, and since the tensile strain is only half the shear strain, the bridge output will be numerically equivalent to shear strain.

FOUR-ARM BRIDGES

The circuits described so far are effectively *half* or *two-arm bridges* that are connected to measuring equipment with three-conductor cables. The measuring equipment usually contains the two remaining arms of the bridge and the balancing circuits as illustrated in Figure 4.10.

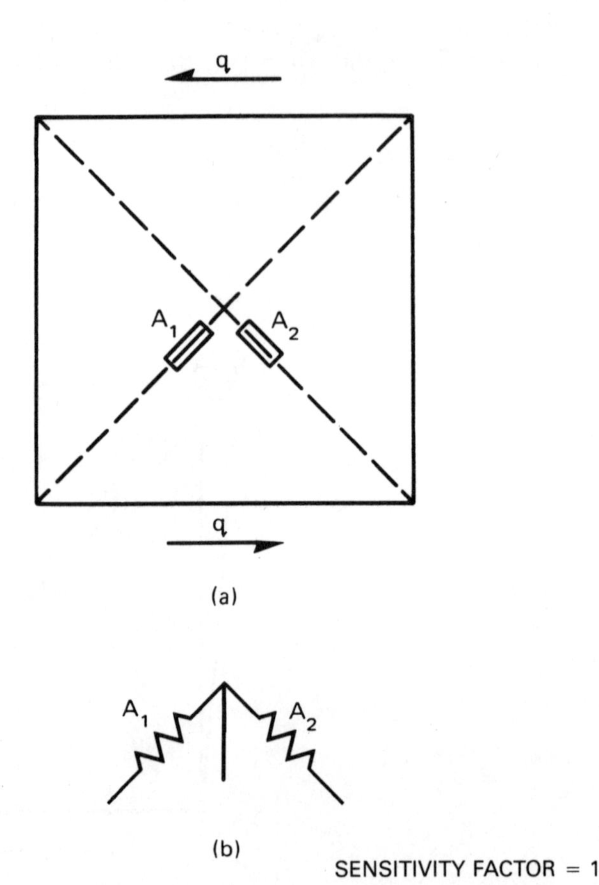

Figure 4.9
Half Bridge Measuring Shear Strain

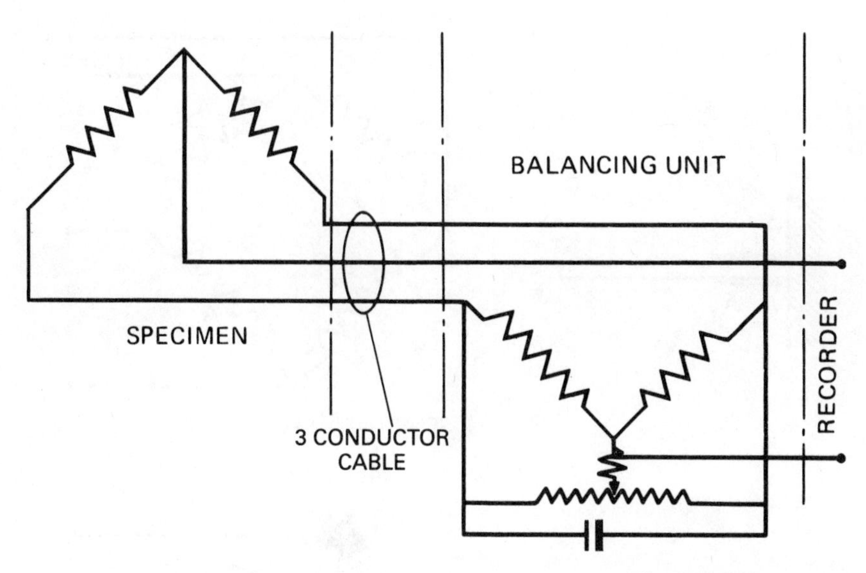

Figure 4.10
Half Bridge with a Remote Balancing Circuit

It is often possible to use an alternative configuration of gages connected as a complete or *four-arm bridge*. This provides maximum bridge sensitivity. The use of four conductor cables provides balanced conductors which improves the rejection of electrical interference and reduces errors due to temperature change. These bridges may contain four active gages or combinations of active and dummy, depending on the measurement. See Figure 4.11.

Balancing Four-Arm Bridges

The balance unit for the four-arm bridge does not require the resistors to complete the lower half of the bridge, but in practice it is usual to make provision in a balance unit for two- or four-arm operation. Selection of either mode is achieved either by switching or by using alternative input plug connections to cater for either situation.

ZERO METHODS OF STRAIN MEASUREMENT

Except as a piece of equipment for the demonstration of principles, the simple *potentiometric bridge* illustrated in Figure 4.12 has been replaced by the *data logger*. When using the elementary form of the bridge, connecting leads from the gages are brought back separately to a balance potentiometer P_1. Each separate gage circuit is brought back from measurement points and provided with its own

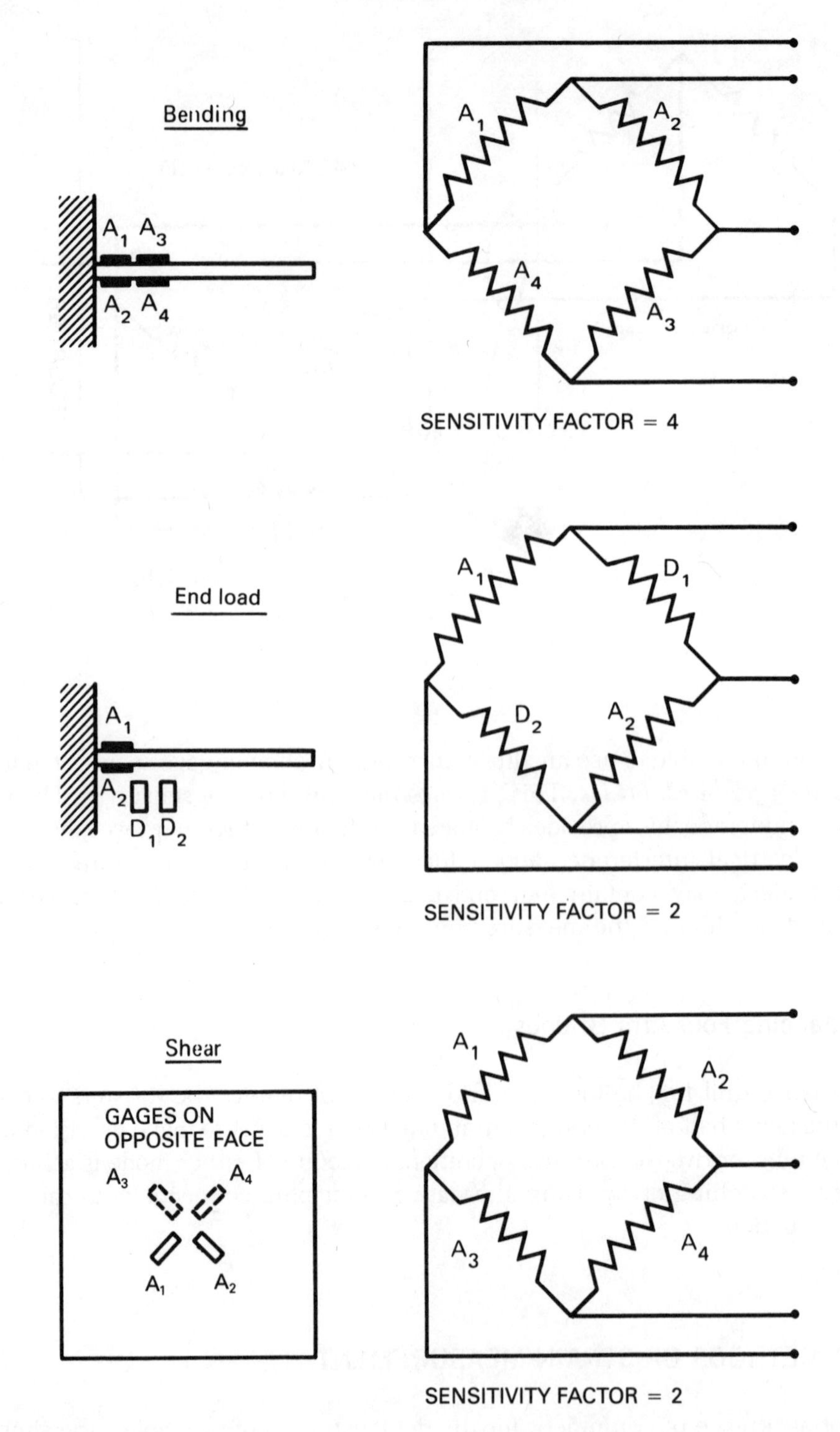

Figure 4.11
Four-Arm Bridge Configuration

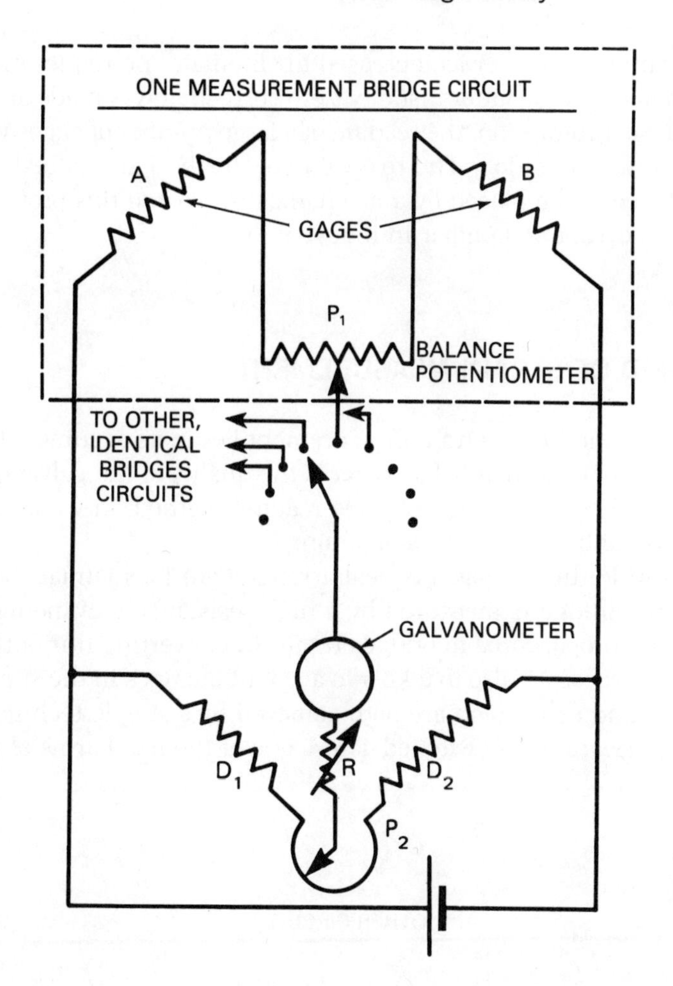

Figure 4.12
Simple Potentiometric Bridge

zero balance. Circuits are connected to the galvanometer in turn by the switch S_1 and the measurement made by adjusting P_2 until the galvanometer reaches a zero or a *null*. A final balance is obtained with the resistance R adjusted to its minimum value, to provide the bridge with maximum sensitivity.

Resistance of the potentiometer P_2 is usually made 1% of the values of D_1 and D_2 so that the scale calibration can be made directly in terms of $\%\delta R/R$. Conversion to mechanical strain is made by the application of the formula:

$$\%\ \text{mechanical strain} = \frac{\%\delta R}{RG_F N}$$

where

G_F = the gage factor,
N = the sensitivity factor of the bridge.

Development of the manual bridge has been limited. Adjustment to the resistance of the potentiometer P_2 provided adjustment for changes in gage factor. Sometimes the value of the slide wire was changed to give a calibration of $\delta l/l$;

sometimes the slide-wire resistance was increased to eliminate the requirement for balance potentiometers. In the latter instance zero correction was made arithmetically. Whatever the configuration, the reading of a large number of gages was a slow and tedious manual operation. The real value of this bridge was realized when the manual balance was replaced by a mechanical servo and this provided the basis of the chart recorder so familiar in industry.

DEFLECTION METHOD OF STRAIN MEASUREMENT

The advantage of the null method of strain measurement lies in its accuracy. The bridge supply voltage is not critical at balance nor is the quality of the galvanometer scale. The alternative to this is a direct scale reading method, in which the deflection of a meter is calibrated in terms of strain.

The bridge shown in Figure 4.13 is a typical arrangement for manual strain recording. Output of the bridge is measured by a high-resistance galvanometer which displays a reading proportional to bridge output. In converting this output into terms of strain it is necessary also to take into account changes in the supply voltage. The correction and conversion are both achieved by a simple technique in which a *calibration resistance* is switched across one of the fixed arms of the

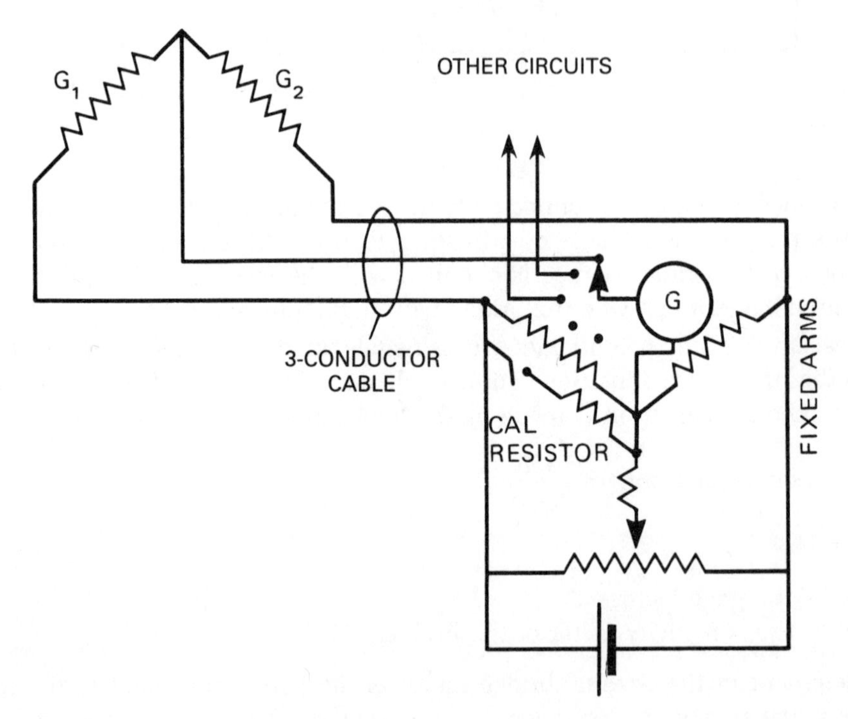

Figure 4.13
Deflection Strain Measurement Bridge

bridge causing a change in bridge output. It is common to make this change 0.1% of the resistance of the fixed arm to provide a simple calibration factor. This calibration is performed prior to making the measurement and the value of strain from the subsequent measurement is calculated from the formula:

$$\% \text{ mechanical strain} = \frac{0.1\, m_2}{m_1\, G_F\, N}$$

where

m_1 = the galvanometer deflection due to the introduction of the calibration resistor,

m_2 = the deflection of the galvanometer due to the strain applied to the bridge,

G_F = the gage factor,

N = sensitivity factor of the bridge.

To apply this method successfully, the calibration resistance must produce an accurately known change. This is possible in the case of two-arm bridges, because the balancing arms are fixed resistance. If four-arm bridges are used, the gage resistance is nominal, and the value of the calibration resistance to be meaningful must be selected to match the gage. This is not practical if a balancing network is to be used with a multiplicity of bridges, and it becomes necessary to calculate a calibration factor for each bridge. This will be discussed further in the chapter on calibration.

BRIDGE SENSITIVITY FACTOR

The values of N for two- and four-arm bridges are summarized in Table 4.1.

Table 4.1

	Four-Arm Bridge	*Two-Arm Bridge*
Tensile (end load)	2	1
Bending	4	2
Shear	2	1

EFFECTS OF TEMPERATURE ON STRAIN GAGES

So far, only the effect of mechanical strain on a strain gage has been considered in detail, although reference has been made to the self-compensating features of bridges subjected to temperature changes. Temperature produces two effects on a strain gage. The resistance of the element changes, because of the temperature coefficient of the wire, according to the relationship $R = R_0\,(1 + \alpha t)$. It also changes because of the difference between the coefficients of expansion of the gage materials and the specimen material.

Suppose a gage has a coefficient of thermal expansion λ_1 and the specimen a coefficient of λ_2.

$$\frac{\delta R}{R} = G_F \frac{\delta l}{l} \tag{1}$$

where G_F is the gage factor.

The effect due to differential expansion will produce a strain such that:

$$\frac{\delta l}{l} = (\lambda_2 - \lambda_1)\delta t \tag{2}$$

where δt is the change in temperature.

Therefore:

$$\frac{\delta R}{R} = G_F(\lambda_2 - \lambda_1)\delta t \tag{3}$$

Also, the effect due to the temperature coefficient of resistance is:

$$\frac{\delta R}{R} = \alpha_g \, \delta t$$

where α_g is the temperature coefficient of resistance of the gage element material.

The total effect of these two will be:

$$G_F(\lambda_2 - \lambda_1)\delta t + \alpha_g \, \delta t = \frac{\delta R}{R} \tag{4}$$

or

$$G_F(\lambda_2 - \lambda_1) + \alpha_g = \frac{\delta R}{R\delta t} \tag{5}$$

By definition, the temperature coefficient of a strain gage α_t is the change in $\delta R/R$ of a strain gage per degree centigrade, equaling:

$$G_F \, (\lambda_2 - \lambda_1) + \alpha_g = \alpha_t$$

Apparent Strain

The change in resistance of a gage caused by temperature changes, as shown in (5), is known as the *apparent strain*, and if a gage is subjected to mechanical strain as well as temperature changes, the combined signal is known as *indicated strain*. It follows that mechanical strain = indicated strain − apparent strain.

Because apparent strain is a function of temperature, it is necessary to apply correction factors derived from curves of apparent strain against temperature. This procedure implies that each time a measurement is made a corresponding measurement of gage temperature must be made. Because the gage is insulated from the specimen by a layer of glue and also by its backing material during transient conditions, appreciable differences in temperature are likely to occur which will introduce errors in the measurement. Because of the problems involved in this type of temperature correction, manufacturers have made considerable efforts to develop gages that are self-compensating for temperature.

Self-Compensation

It is possible, by choosing the correct materials for the strain-gage element, to produce a gage which gives no apparent strain over its working temperature range. This gage element material will have different characteristics, for gages adhered to test specimens with different coefficients of expansion. A gage will be steel compensated, or aluminum compensated, for example, and may only be used on the specified material.

Creep in Strain Gages

If a test specimen is subjected to a steady strain for a long period, strain gages measuring that change will demonstrate a gradual reduction in output. This is due to the gage backing and the bonding cements, which suffer plastic deformation when subjected to long-term loading. The process is accelerated at elevated temperatures. Much has been done in recent years to improve cements and backing materials, but this area represents one for continued development.

SEMICONDUCTOR STRAIN GAGES

In certain semiconductor materials distortion of the crystal lattice due to mechanical strain produces large changes in the resistance of the material. This effect is known as *piezoresistance* and provides a basis for designing gages with large gage factors. Silicon is one of the materials which not only produces large piezoresistive changes but also has the requisite mechanical properties.

Strain-gage elements are made from a single silicon crystal, which, although brittle in the original crystal form, becomes flexible when cut into very thin strips. These strips are very strong and capable of withstanding strains in excess of 0.006 (0.6%, or 6000 microstrain).

The addition of minute quantities of impurities to the original crystal can cause considerable changes to the electrical characteristics of the strain gage. This process is known as *doping*, and the effects of temperature on the element resistance and its gage factor are both adjusted by variation of the impurities. For example, gage elements doped with phosphorous exhibit a negative gage factor; that is, their resistance decreases with the application of tensile strain. Similarly, elements doped with boron display a positive gage factor; their resistance increases with the application of tensile strain. The two types of gages are known as *N* and *P* types, respectively.

Temperature Effects on Semiconductor Gages

The basic silicon element is capable of withstanding temperatures in excess of 300°C. In practice, the working temperature range is restricted by the properties of the backing and cements. Practically all creep and hysteresis exhibited by semiconductor gages are due to these materials. Some improvement may be made by the use of gages that do not have a backing, but unsupported elements are easily damaged during the bonding process.

Semiconductor gages suffer the same temperature effects as wire gages. The resistance of both types varies with temperature, and there is also a change due to differential expansion between the gage and the specimen.

The effect on the gage factor, however, can be much more dramatic. Whereas the wire gage demonstrates only second-order changes due to temperature, changes in most semiconductor gage factors are considerable and must be taken into account. If a *P*-type material is considered, it is possible to produce gages with very high gage factors by introducing a low level of doping. Unfortunately, these gage factors also are extremely temperature sensitive. As the doping level is increased, the gage factor and the temperature sensitivity both fall, until, beyond a certain concentration of impurity, the further reduction in gage factor no longer justifies the improvement in temperature sensitivity. Even then the compensation is not particularly good and works over a limited temperature range. In addition, the thermal expansion coefficient of the material is very low and this produces correspondingly large strains due to differential expansion between the gage and specimen materials. Consequently, *P*-type material is not the best to use when designing a self-compensating gage.

N-type material displays an interesting characteristic that while the temperature coefficient is positive, the strain sensitivity is negative. It is therefore possible to combine the features of *P*- and *N*-type gages to provide temperature compensation. This capability may be best applied where more than one gage is to be bonded to a specimen and wired into adjacent arms of a bridge. Figure 4.14

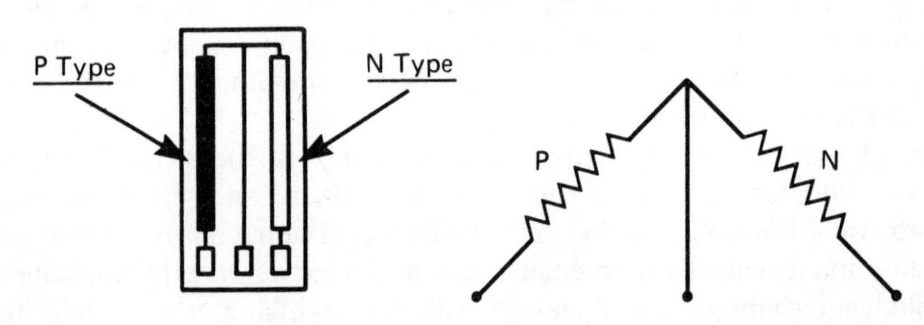

Figure 4.14
The P and N Temperature-Compensated Gage

illustrates *P/N* gage using this technique of bridge compensation using two carefully selected elements, one of *P*-type material, the other of *N*. The two elements are bonded to the same backing, lie parallel to each other, and are connected as a half bridge.

When subjected to a tensile strain the resistance of the *P*-type material increases while the *N*-type decreases. Wired as a bridge, these changes are additive. The temperature coefficients are both positive, and the two arms are subsequently self-compensating.

Gage-Factor Compensation

The gage factor of both *N* and *P* gages decreases with an increase in temperature. It is therefore necessary to compensate for this change. One method is to increase the bridge voltage as the gage factor decreases. A resistor with a negative temperature coefficient and connected in series with the bridge supply is often used for this purpose. The resistor must be located in the same temperature environment as the bridge. Consequently, the method is more adaptable to sensors containing a strain gage bridge than it is to strain gages bonded to an open specimen.

Constant-Current Compensation

If the current through a bridge is maintained at a constant level it automatically follows that the voltage across the bridge must change as the resistance of the gage changes. Since it is usual for semiconductor gages to increase their resistance with temperature, a constant current maintained in the gage by an increase in bridge voltage tends to compensate for the decreasing sensitivity. This is known as *constant-current compensation*. By suitable choice of materials a gage type may be obtained in which the gage factor is almost corrected over the working range of the gage when a constant current is maintained.

It is possible to obtain power packs with constant-current capabilities, but it is equally possible to use a constant-voltage power supply and introduce a high series resistance into the bridge circuit. The value of this resistance should be at least 20 times the bridge resistance. When a large series resistance is used, the voltage of the power source may require increasing to maintain a reasonable bridge output.

Another method of compensation incorporates the high series resistance into the arms of the bridge by making the dummy arms large compared with the gage resistance.

The constant-current technique has a second function when applied to semiconductor strain-gage bridges, which makes it an attractive method of excitation. As shown in Chapter 3, if one arm of a Wheatstone bridge is changed by δR the output of the bridge is:

$$V_g = \frac{V\delta R}{2(2R + \delta R)}$$

In wire-strain-gage circuits δR is very small compared with R, and the bridge output is for practical purposes:

$$\frac{V\delta R}{4R}$$

In a semiconductor-strain-gage bridge, however, δR may be of the order of 10%, and the value of $(2R + \delta R)$ can no longer be reduced to $2R$. The output of this bridge has been calculated assuming a constant voltage supply. Suppose now the current is to remain constant, the voltage across the bridge must change to compensate. This new voltage V' will be $i \times Rc$ where Rc is the new value of

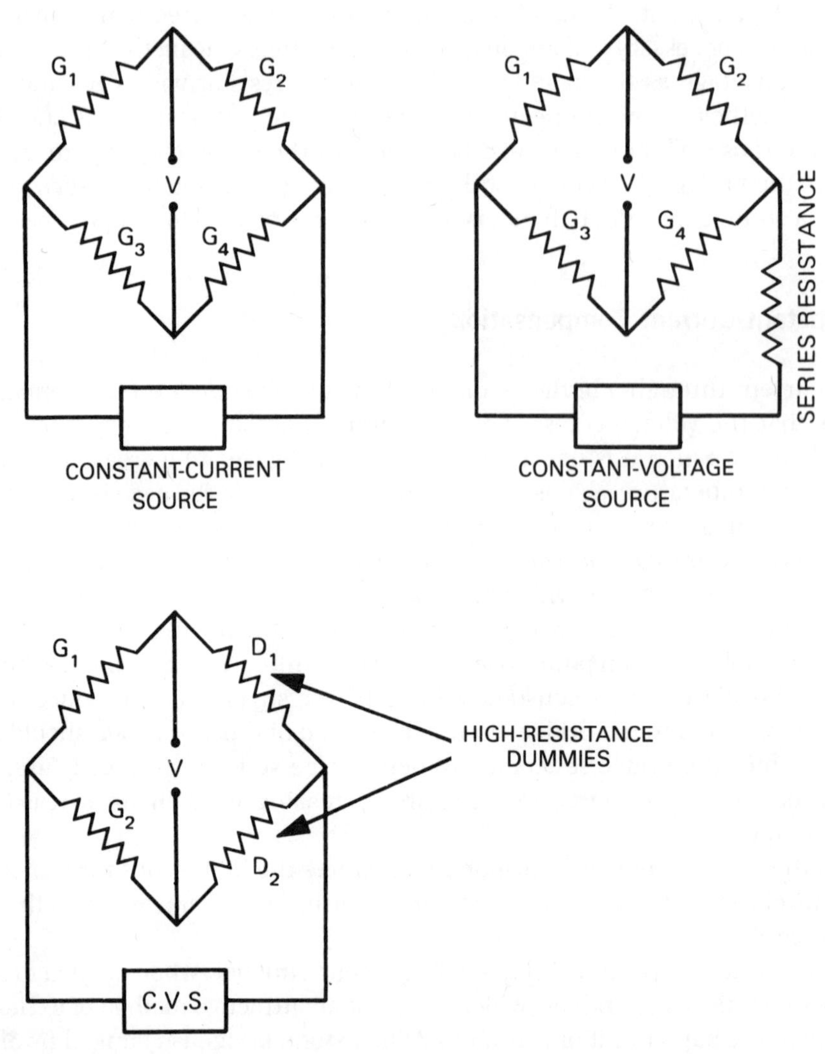

Figure 4.15
Constant Current Temperature Compensation of Semiconductor Strain Gage Bridges

bridge resistance and is equal in value to a parallel circuit where the resistance of one branch is $2R + \delta R$ and the other $2R$. Therefore:

$$Rc = \frac{2R(2R + \delta R)}{(2R + \delta R) + 2R}$$

and

$$V' = \frac{2iR(2R + \delta R)}{4R + \delta R} \tag{1}$$

The output of the bridge V_g will now be:

$$V_g = \frac{V'\delta R}{2(2R + \delta R)} \tag{2}$$

Substituting (1) in (2):

$$V_g = \frac{2iR(2R + \delta R)}{(4R + \delta R)} \times \frac{\delta R}{2(2R + \delta R)}$$

$$= \frac{iR\delta R}{4R + \delta R} \tag{3}$$

but $iR = V$. Therefore:

$$V_g = \frac{V\delta R}{4R + \delta R}$$

If this is compared with the constant-voltage case, in which

$$V_g = \frac{V \times \delta R}{4R + 2\delta R}$$

it becomes obvious that the δR term in the denominator has been halved. Since this is the term responsible for the nonlinearity in the bridge, it follows that bridge linearity is considerably improved.

CHAPTER 5

PRESSURE MEASUREMENT

RELATIONSHIP BETWEEN FORCE AND PRESSURE

The total force exerted by a pressure is the product of pressure and the area to which it is applied. This principle is often used to provide mechanical advantage in systems.

Consider two interconnected cylinders in Figure 5.1. If a force F is applied to a piston of area a the pressure created is:

$$\frac{F}{a} \tag{1}$$

Providing friction losses are ignored, the pressure will be reflected on the ram, which will move unless it is opposed by a load L.

To balance the two the pressures must be equal:

$$\frac{L}{A} = \frac{F}{a} \tag{2}$$

or

$$F = \frac{La}{A}$$

If a is small and A is large, the force to move the load is correspondingly small.

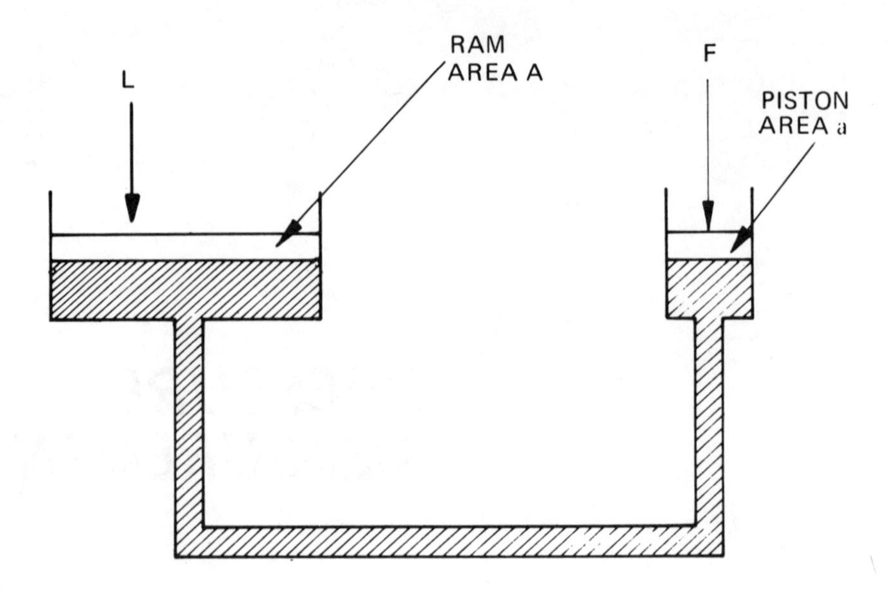

Figure 5.1
Relationship between Force and Pressure

ATMOSPHERIC PRESSURE

If a tube sealed at one end is filled with mercury and the open end inverted into a pool of mercury, a column of liquid supported by atmospheric air pressure will remain in the tube. The height of this column is approximately 30 inches (760 mm), but it will vary from day to day and with its height above sea level. This device is called a *barometer*. In spite of the variations it is possible to use the barometer as a pressure standard by defining the parameters that constitute standard atmospheric pressure. Consequently, the *standard atmosphere* is defined as that pressure which will support a column of mercury 760 mm high at sea level and at a temperature of 0°C.

From this basic definition it is possible to derive the other units in common use. The density of mercury is 13.60 g/cm². Therefore the pressure exerted by a 760 mm column is 13.60×76 g/cm² = 1033 g/cm² (approximately 15 psi).

ABSOLUTE, GAGE, AND DIFFERENTIAL PRESSURES

The measurement of a pressure is qualified by the reference point to which the measurement is made. The atmospheric pressure acting against the mercury in the barometer is being referenced to the vacuum above the column and is, therefore, an *absolute measurement* of pressure. If a barometer is modified to measure

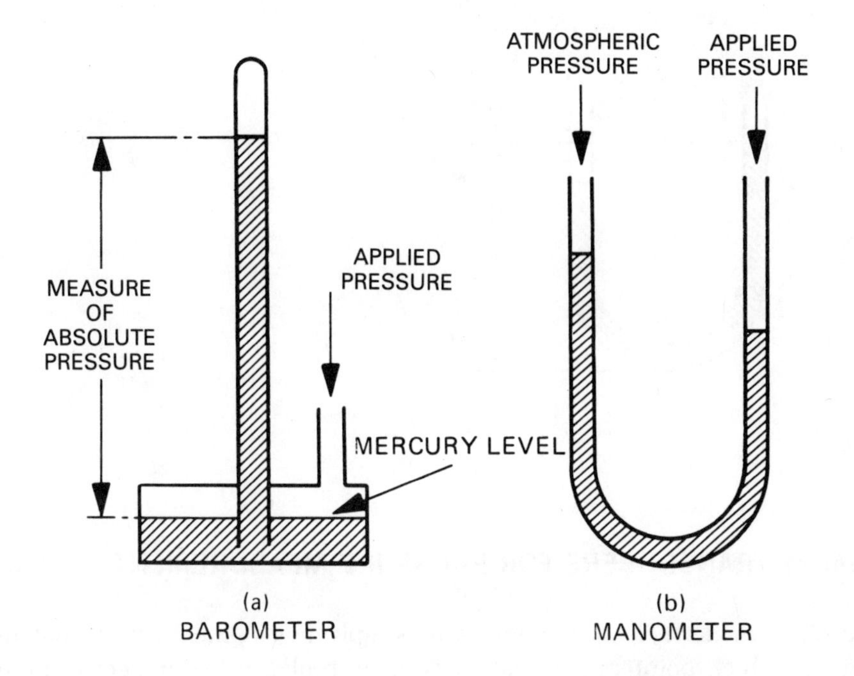

Figure 5.2
Pressure Measurement with Barometers and Manometers

absolute pressure, it takes the form shown in Figure 5.2a. *Gage pressure* is the measurement of pressure above atmospheric pressure. If one end of a U-tube shown in Figure 5.2b is left open to the atmosphere, the difference in levels of the column is a measurement of the pressure above atmospheric pressure. Since atmospheric pressure can vary from day to day and with the height above the sea level, it is therefore an unstable reference which must introduce an error into the measurement. If high pressures are involved, this is negligible, but when an accurate measurement is required around the atmospheric level, errors can be appreciable. It is possible under these circumstances to seal the reference port at a standard atmospheric pressure and use that as reference. This type of sensor is known as a *sealed gage*. Although sealed gage sensors eliminate errors due to changes in atmospheric pressure, sealing of the port will in itself introduce substantial errors, unless the sealed volume is large compared with the displaced volume of the column.

Differential pressure measurement is the measurement of one pressure with reference to another. In effect, absolute and gage pressure measurements are special cases of differential pressure measurement. Absolute uses zero pressure as reference, and gage uses atmospheric pressure.

The manometer in Figure 5.3 has a pressure P_1 applied to one port and P_2 to the other. The differential pressure $P_2 - P_1$ is proportional to the difference in height $h_2 - h_1$.

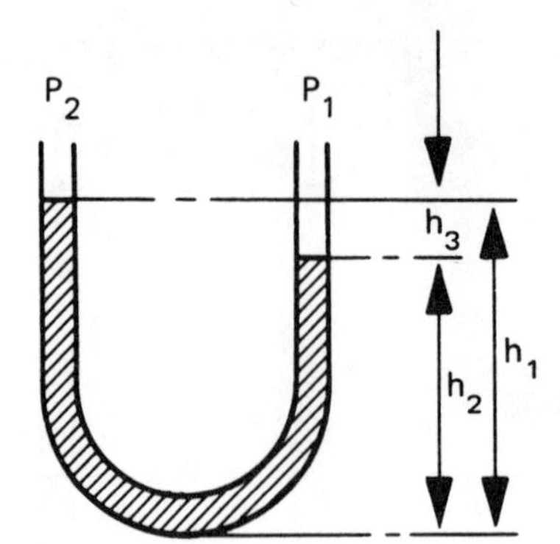

Figure 5.3
The Manometer

PRIMARY TRANSDUCERS FOR PRESSURE MEASUREMENT

Many of the modern electrical sensors are simply developments of original instruments, in which pointers and dials have been replaced by an electrical output. These primary transducers are usually rugged and reliable and are capable of moving the actuator of a secondary transducer through an appreciable displacement. In their conventional dial form, their potential accuracy was often restricted by the limitations of the gearing and mechanical linkages. With an electrical output the resulting sensor is often more accurate than the original.

The Bourdon Tube

The C-type *Bourdon tube* (Figure 5.4) consists of a tube with an elliptical cross section that has been bent into a shape resembling a letter C. One end of the tube is connected to a pressure port; the other end is sealed. As pressure is applied to the port, the cross section tends to become more circular and, in the process, attempts to straighten out the C. This causes a movement at the free end which is proportional to the pressure applied. If the pressure is below atmospheric, the tube attempts to flatten, and the radius of the C is reduced.

The materials and dimensions of the tubes are determined by the fluids to be measured and the pressure range. Phosphor bronze and beryllium copper are used for pressures up to about 5000 psi (350 kg/cm^2) and solid drawn-steel tubes above this. If corrosion is a problem, monel metal or stainless steel may be used.

Bourdon tubes are basically accurate devices with repeatabilities of the order of 0.1% FS, but they are sensitive to temperature, which causes zero and span errors. Zero errors are principally due to changes in dimensions caused by the coefficient of expansion of the tube material; span errors are due to changes in Young's modulus of the material.

In situations in which high accuracy is essential, it is possible to use nickel alloy tubes having a value of Young's modulus independent of temperature.

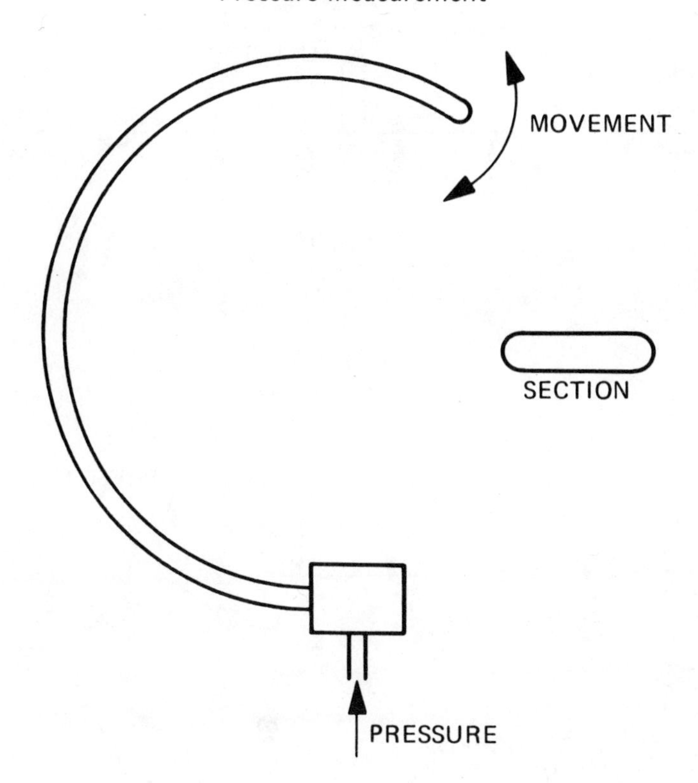

Figure 5.4
The Bourdon Tube

Extended Bourdon Tubes

If large movements are required to drive the secondary transducer, the length of the Bourdon tube can be extended by turning it into either a spiral or a helix. These tubes are ideal for pointer drives, as they eliminate the need for gearing, but for the most part it is possible to obtain adequate electrical output with the smaller movements of a *C* type at less cost. Spiral Bourdon tubes are used effectively in calibration equipment; these are discussed in detail in a later section.

The Stiff Diaphragm

The most common early application for the *stiff diaphragm* was probably in the aneroid barometer, but it is now used in a wide variety of applications. In its simplest form it is a single, concentrically corrugated disk to which pressure is applied. Deflection of the diaphragm causes movement of a push rod (Figure 5.5). The push rod, in turn, actuates a secondary transducer, which may be used to provide an electrical output. The basic diaphragm can be turned into a capsule which is effectively two diaphragms welded or soldered round their edge. These capsules can again be joined together, or stacked as it is called, to increase sensitivity (Figure 5.6).

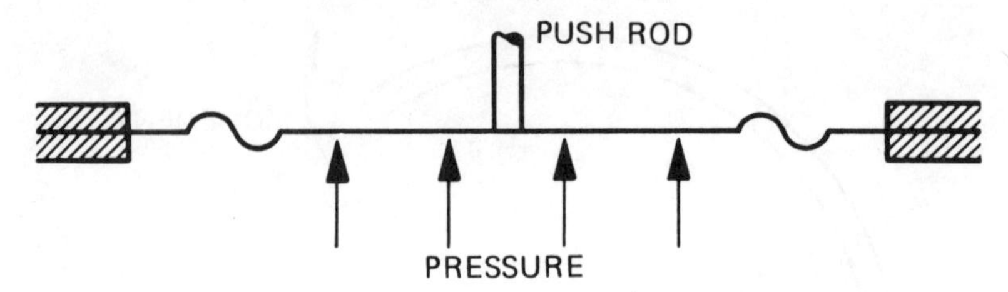

Figure 5.5
The Stiff Diaphragm Pressure Transducer

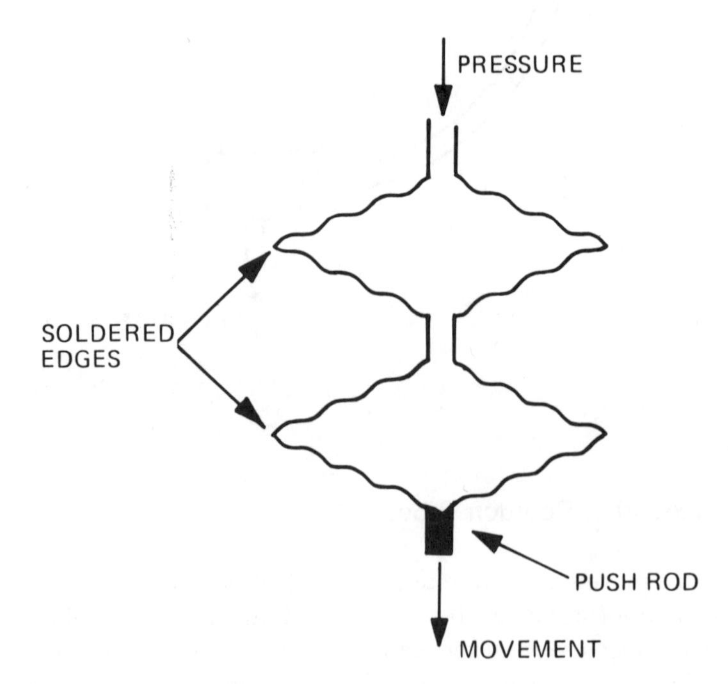

Figure 5.6
Stacked Pressure Capsules

Bellows

A variation of the stacked capsule is the *bellows element* (Figure 5.7), which is often made from tubing and formed hydraulically.

Pressure inside the bellows causes an increase in length of the unit, which is used to actuate a secondary transducer. Pressure applied to the outside opposes the extension, and the unit may therefore be used as a differential device. The addition of stops as shown in Figure 5.8 will prevent the destruction of the element in the event of a failure of pressure on one side. Absolute measurements of pressure are also possible with bellows by using them in matched pairs, one bellows being evacuated and sealed, the other fitted with a port to receive the measurement pressure. With this arrangement, changes in atmospheric pressure act on both the evacuated and the pressure-sensitive elements, thus cancelling changes.

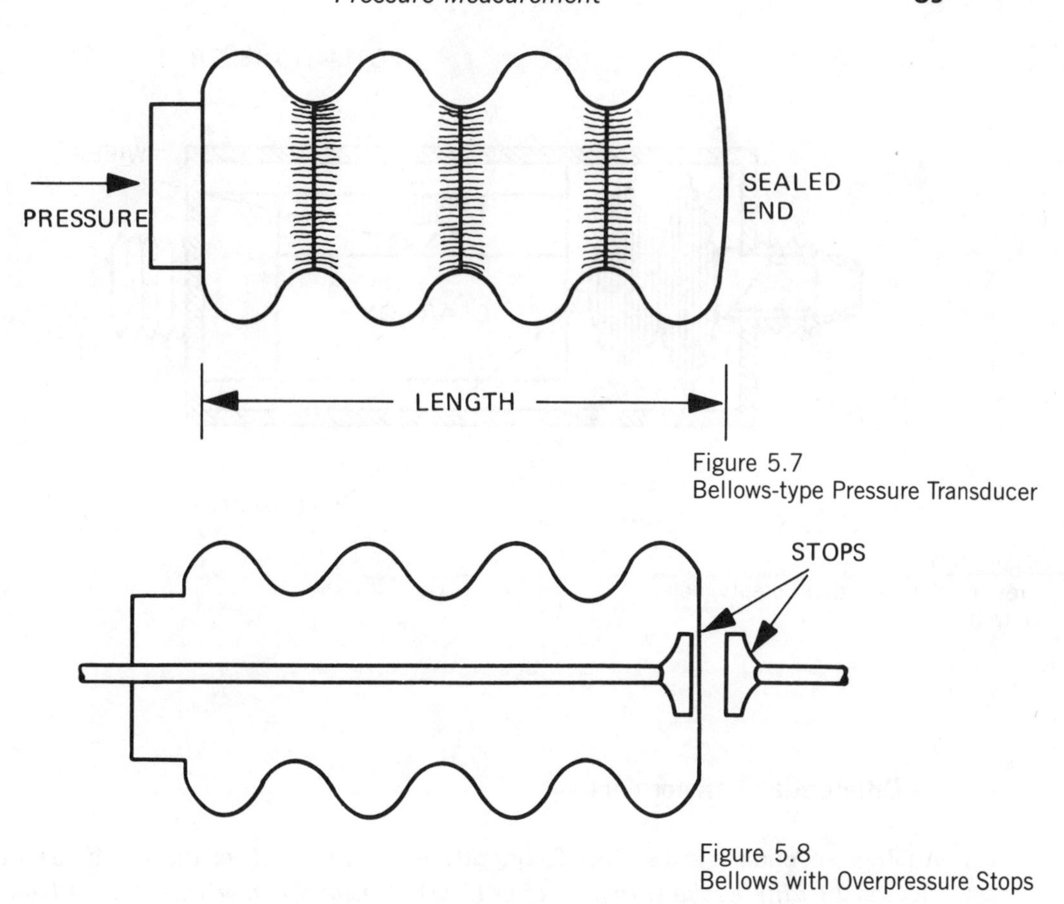

Figure 5.7
Bellows-type Pressure Transducer

Figure 5.8
Bellows with Overpressure Stops

SECONDARY TRANSDUCERS

The primary transducing elements described are representative of those in use for many years in industry and which usually used a pointer drive for output. It follows that the pointer drive may be easily replaced by any one of the number of electrical transducers.

Potentiometers

Potentiometers may be used easily with stacked stiff diaphragm capsules or bellows. The unit shown in Figure 5.9 is fairly typical of this type of device.

The speed of response of such a device is obviously limited, and pulsating pressures can very quickly damage the windings of the potentiometer. On the other hand, large output signals are obtainable from the potentiometer which can be extremely useful to drive meters without amplification. In areas of high electrical noise the high output is also invaluable.

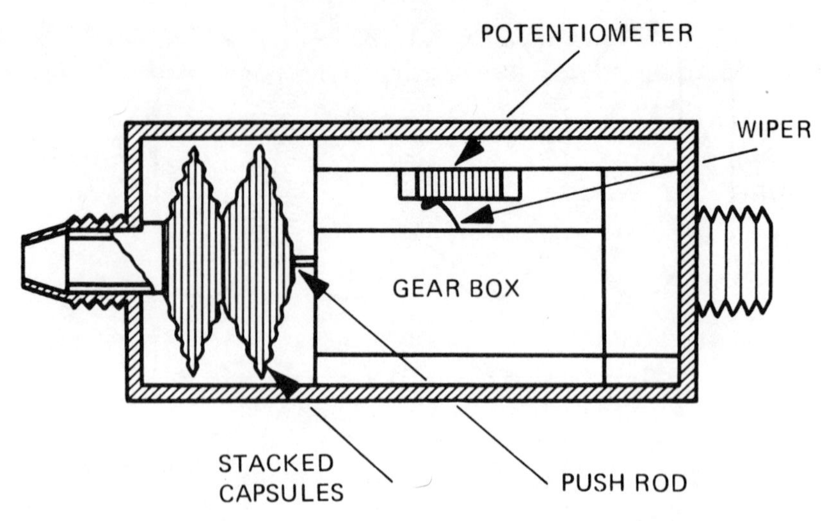

Figure 5.9
Pressure Sensor with a Potentiometer Output

Differential Transformers

Another secondary transducer frequently used with each of these primary devices is the *differential transformer* or LVDT (*Linear Variable Differential Transformer*) as it is often known (Figure 5.10). This consists of a primary coil P and two secondary coils S_1 and S_2 which are connected in opposition; that is to say, voltages induced in the secondary windings oppose one another. The ferrite core is moved by the operation of the primary transducer and in its zero state is adjusted so that the voltages across S_1 and S_2 are equal.

As the core is moved by a change of pressure the magnetic flux linking one coil is increased and the flux linking the other is decreased. Consequently, there is a change in the output of the two coils, one now inducing a higher voltage than the other. This output is an alternating current and the displacement of the core to one end or the other determines the phase of the output signal. The output signal is therefore of the form that must be phase discriminated.

The LVDT may be connected as a bridge, illustrated in Figure 5.11. It is an extremely rugged device, and since there is no necessity for the core to touch the coil form, there are no losses in the movement due to friction. Discriminated output is substantially linear over the working range, flattening off at the ends, as shown in Figure 5.12.

An interesting design of pressure sensor using an LVDT in conjunction with a *stiff diaphragm capsule stack* is shown in Figure 5.13. The capsules are built into a substantial nonferrous alloy body and are coupled to the core of the transformer without any supporting bearings or flexures. A signal from a platinum resistance element embedded in the reference chamber is returned to the electronic unit to provide temperature compensation. Good accuracy is therefore obtained from the sensor, and its construction makes it particularly suitable for maritime applications.

Because of the importance of the LVDT it will be discussed in detail in Chapter 7.

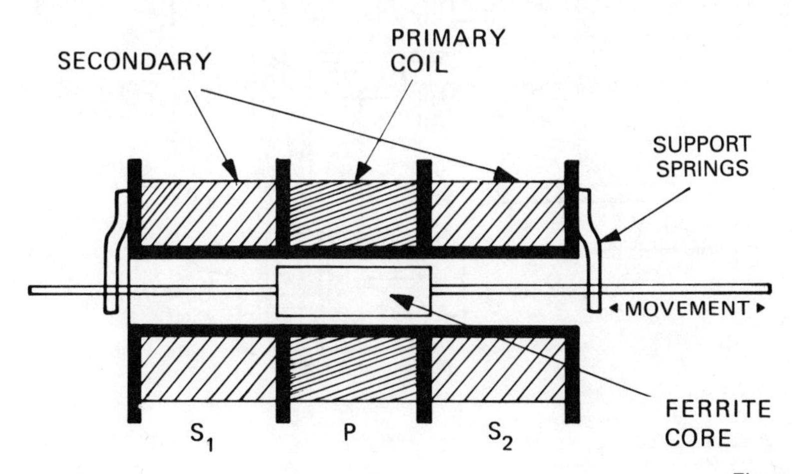

Figure 5.10
Differential Transformer

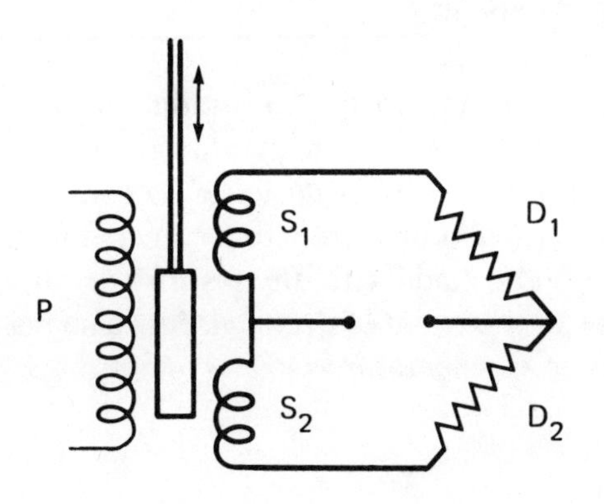

Figure 5.11
Differential Transformer Bridge

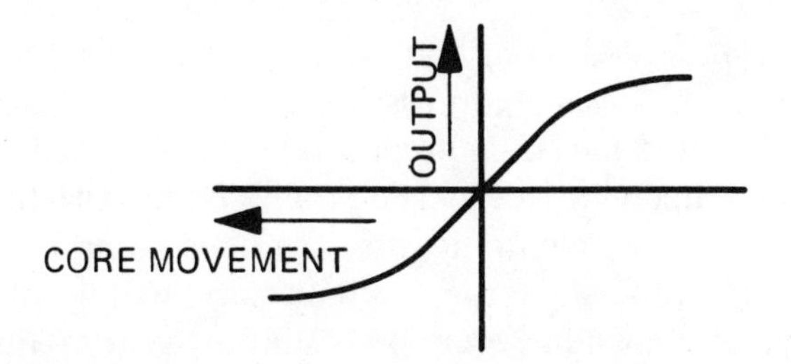

Figure 5.12
Output of Differential Transformer

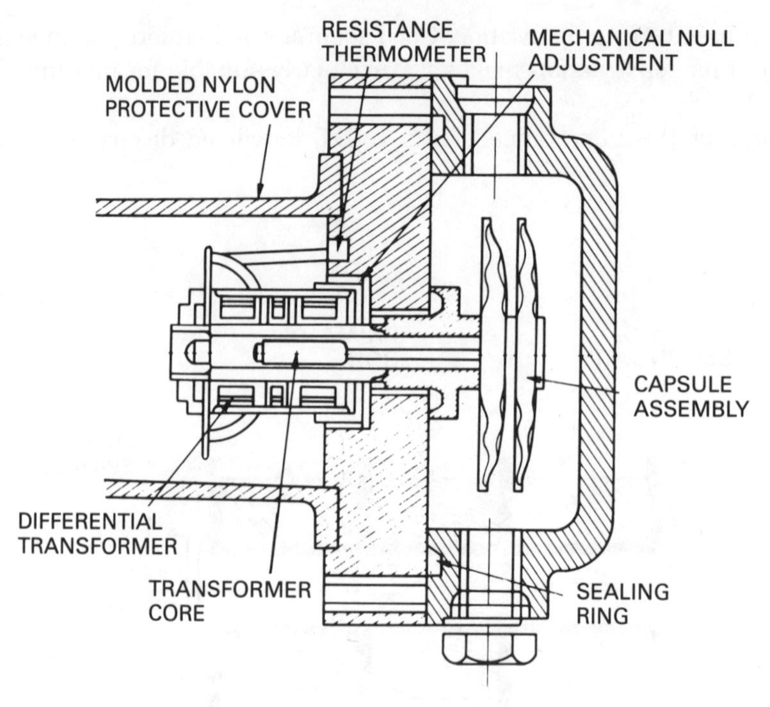

Figure 5.13
Pressure Sensor
(KDG Instruments Ltd.)

MODERN PRESSURE SENSORS

Sensors so far discussed are modifications of traditional units which originally drove a pointer and which were modified to provide an electrical output. There is, however, a selection of pressure sensors designed as complete units. These sensors take advantage of electrical transducers that can operate from small displacements. As a result they can be made with much stiffer diaphragms which reduce the likelihood of rupture in the event of overload. It is also possible to reduce the overall size of the sensor considerably, a factor of prime importance in some applications.

Bonded Strain-Gage Sensors

If a pressure is applied to a tube sealed at one end, strains are set up in the walls of the tube that may be used as the basis of a pressure sensor. These strains have two components: One is a longitudinal strain produced by pressure acting on the end of the tube; the other component is a tensile strain acting at right angles, known as the *hoop strain*. It is possible to measure the hoop strain by bonding strain gages to the walls of the tube along the strain axis and wiring them to an end-load measuring bridge; as shown in Figure 4.11. Dummy gages required for this measurement are usually bonded to strips of the tube material, which in turn are *soft bonded* to the tube. This means that they are bonded to the tube in a way which does not transmit strain.

To complete the sensor the tube is fitted with a pressure union to allow easy connection to the measuring point, an outer case to protect the strain gages, and an electrical cable socket to facilitate wiring. Figure 5.14 illustrates a typical bonded sensor of this design. As it stands, the sensor is a gage pressure-measuring device, but it is possible to make the case gas-tight and fit a reference port to measure differential pressure. Evacuation of the space now turns the sensor into an absolute pressure-measuring device.

When the sensor is used for differential pressure measurement it is possible to waterproof the gages and use liquids in the reference chamber, but it is more usual to confine the reference pressure to noncorrosive gases.

The disadvantage of this type of transducer is the low output, although this can be offset considerably by the use of semiconductor gages. Its advantage lies in its simplicity, and in an emergency they can be manufactured, in plant, at short notice.

Suppose a tube (Figure 5.15) has an internal radius r and a wall thickness of t. The hoop stress in the walls of a thin-walled tube due to a pressure p will be

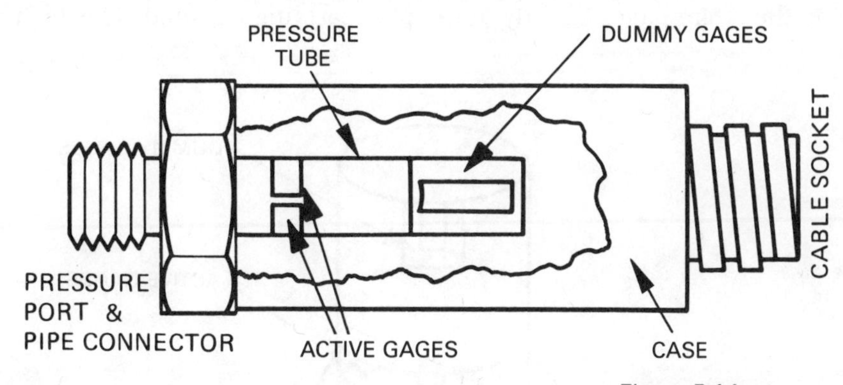

Figure 5.14
Bonded Strain-Gage Pressure Sensor

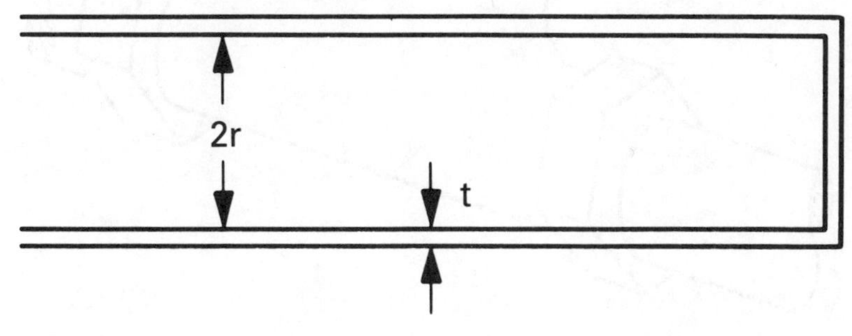

Figure 5.15
Design of Pressure Tube for a Bonded Strain-Gage Sensor

$f = p \times r/t$. The strain resulting from this stress will be $f/\mathscr{E}$ where $\mathscr{E}$ is Young's modulus of the tube material. It is usual to work gages at strains about 1000 microstrain ($\delta l/l = 0.001$), and from these figures a suitable size of tube can be selected.

The strain gages should be bonded at least one diameter from each end to avoid errors due to end effects. This makes the length of the tube at least three diameters, and it is preferable to increase this figure.

Mounting on smaller pipes may be possible by fitting the end of the tube with clamping into a standard *T*-fitting as illustrated in Figure 5.16. When mounting larger pipes it is necessary to fit the tube with standard pipe fittings and then the unit begins to approximate the commercial sensor.

Dummy gages are most easily mounted on a piece of similar tube which has been split along its length to make strips a little wider than the gages. After mounting, the whole sensor should be insulated to allow the most effective temperature compensation.

Strain-Gaged Pipes

A simplification of the system described in the previous section is possible by applying the strain gages directly to the pipe carrying the fluid. This technique

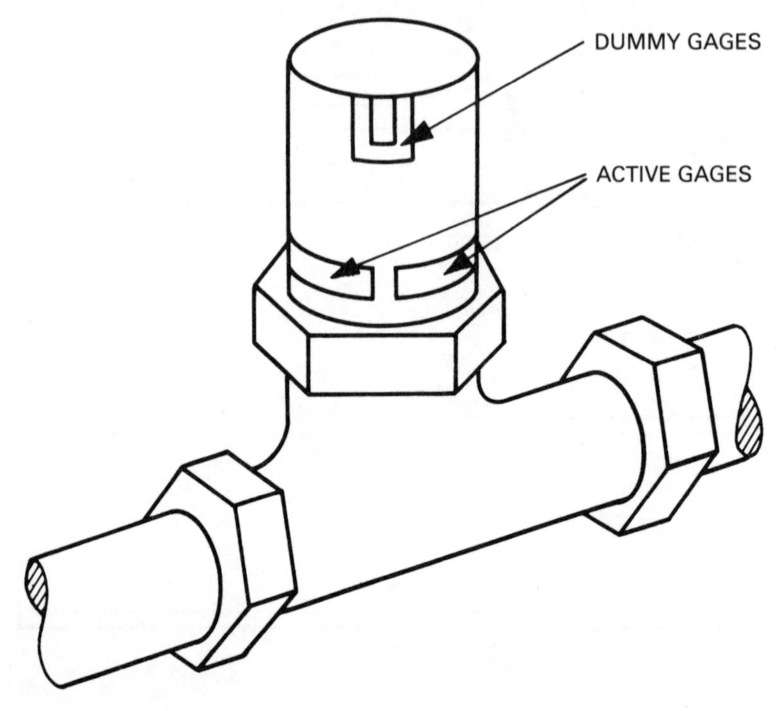

Figure 5.16
Simple Bonded Strain-Gage Pressure Sensor

has inherent problems. Calibration of the gages may not be possible, and then calculation of pressure from the pipe dimensions is the best that may be achieved.

Unbonded Strain-Gage Sensors

Bonded strain-gage sensors of the type described have several limitations, apart from their low output. Their thermal mass, for example, is relatively high, which makes the response of their temperature compensation slow; their size, too, in some applications, may be regarded as a problem. A third limitation is the frequency response of the sensor. When it is required to measure surge pressures, the sensor must respond to the highest frequency component present in the surge. The highest frequency to which the sensor responds depends on the volume of its chamber. The larger the volume, the lower the frequency response. The chamber volume of the sensor described is comparatively large, and its frequency response is therefore correspondingly restricted. These limitations are overcome in the unbonded strain-gage sensor, which in many ways is superior to the bonded type.

The transducer shown in Figure 5.17 may be used in a sensor with a *flush diaphragm*. That is a diaphragm which is exposed directly to fluid pressure so that the chamber volume is virtually zero. The diaphragm in turn operates a push rod which moves an armature relative to a yoke. The movement of the armature is restrained by four resistance windings *A*, *B*, *C*, and *D* which join the yoke to the armature. These windings are insulated from the yoke and armature and are wound having a carefully controlled prestrain.

When pressure is applied to the diaphragm movement of the armature relative to the yoke is produced, causing a reduction of strain in windings *A* and *B* and an increase in *C* and *D*. These are the conditions required for the operation of a bridge with four active arms.

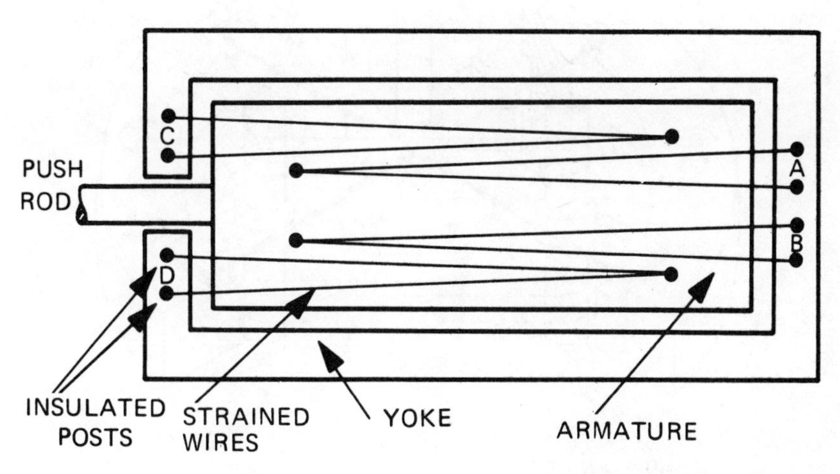

Figure 5.17
Unbonded Strain-Gage Sensor Element
(Statham Instruments)

This sensor therefore fulfills the requirements of low chamber volume and relatively high sensitivity, and because it can be made into a very small unit, its thermal mass is also small. An alternative design is shown in Figure 5.18 which fulfills the same requirements. The push rod in this case operates a star spring which when deflected tensions the windings on the lower posts and reduces the tension on the upper windings. This again produces the conditions required for four-arm bridge working.

Strain-Gaged Diaphragm Sensors

One method of minimizing chamber volume with a bonded strain-gage pressure sensor is to strain-gage the diaphragm. Unfortunately conventional bonded gages introduce added stiffness to parts of the diaphragm causing nonlinearity of output. It is in fact possible to vacuum deposit thin-film strain-gage circuits directly onto a diaphragm and such sensors are used. An alternative method is often used where the diaphragm is connected to a cantilever beam on which the gages are

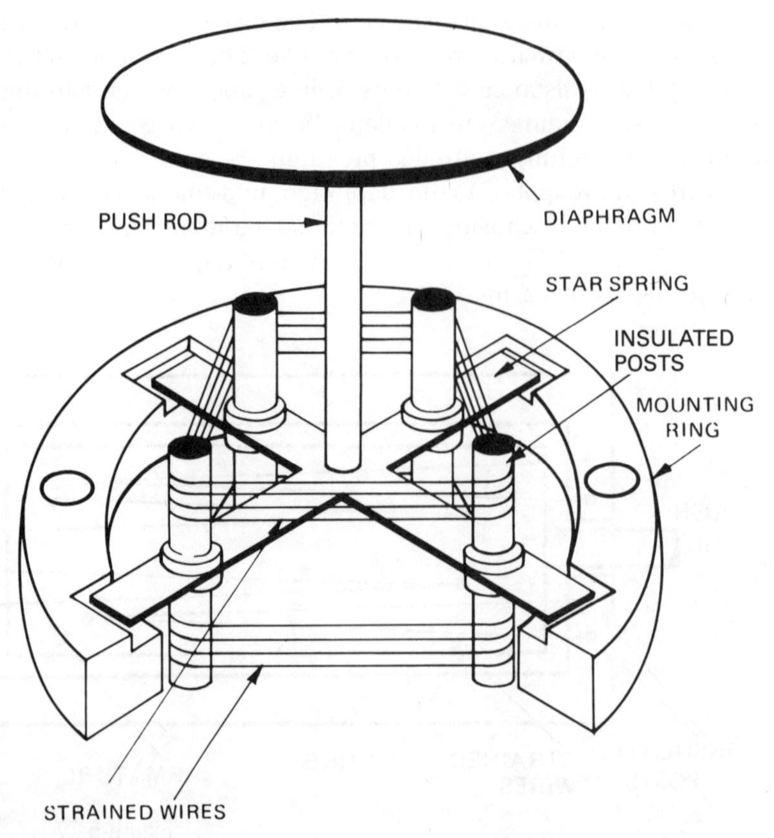

Figure 5.18
Pressure Sensor (Bell and Howell)

deposited. This procedure isolates them from sudden temperature changes that may be experienced by the diaphragm and allows the use of efficient overload stops.

Thin-film strain gages are claimed to have exceptional linearity, hysteresis, and creep characteristics and are particularly valuable in nuclear radiation environments. Zero drift and sensitivity change over a temperature range of 100°C are claimed to be less than 1%. Their output signal is that of a four-arm strain-gage bridge.

Strain-Gage Pressure Sensors with Amplified Output

The low output of strain-gaged pressure sensors has detracted from their use in many industrial control applications, in spite of many obvious advantages. In an effort to make them compatible with traditional sensors, manufacturers of both thin-film and unbonded sensors are offering units with built-in amplifiers. These amplifiers usually supply a standard, full-range output of 4–20 mA, although most manufacturers will produce specials if requested.

Silicon Diaphragm Pressure Sensors

A fairly recent development in the field of strain-gaged sensors is the use of a silicon diaphragm on which are diffused *N*- and *P*-type semiconductor gages. These sensors are very useful for specialized applications. They can be designed with diaphragms 0.1 inch (2.5 mm) in diameter which still produce an output of several hundred millivolts. Repeatability, hysteresis, and creep characteristics of these sensors are also exceptionally good.

Inductive Pressure Sensors

The sensor shown in Figure 5.19 demonstrates the robustness of inductive pressure sensors.

In this type of sensor two coils wired in opposition form two arms of an ac bridge. The diaphragm placed between the coils is made of a magnetic material and changes the relative inductance of the coils as it moves toward one or the other.

The air gaps between the cores and the diaphragm are kept very small; consequently, the chamber volume is also small. Frequency response is often limited not by the geometry of the sensor but by the frequency of the ac excitation current. The small air gap also allows the diaphragm to touch the core face in the event of gross overload, thus preventing its destruction.

It is possible to produce such sensors for absolute pressure measurement by evacuating one side of the diaphragm and sealing off the reference port.

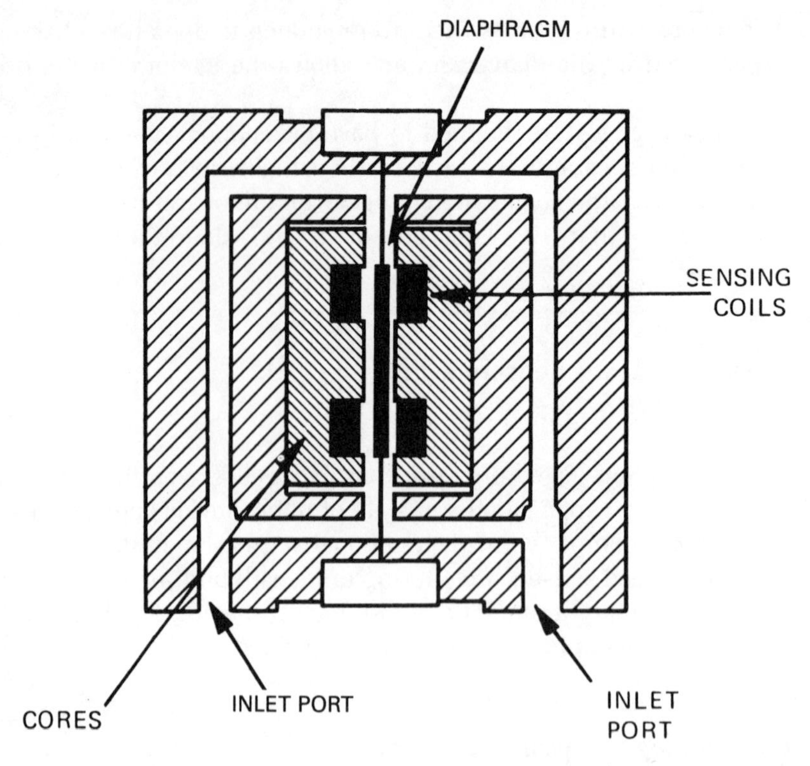

Figure 5.19
Inductive Pressure Sensor
(SF Laboratories)

The construction of these sensors makes them extremely rugged, and they are very stable in their performance. Their output is of the order of 100 mV full scale, and their overload protection is excellent. Their disadvantages are that their frequency response is limited to approximately 0.25 of the carrier frequency, and they suffer some nonlinearity of output.

Capacitance Pressure Sensors

A capacitance sensor can be made by varying the distance between two plates of a capacitor. The pressure sensor is an obvious application of this principle. One plate of the capacitor is used as the diaphragm which varies its distance from a fixed plate with the application of pressure.

Capacitive pressure sensors may be made very small, but they suffer from inherent problems which make their application somewhat difficult. Small capacitance changes require high-frequency excitation to turn the changes into workable signals. Handling high frequencies over long cable lengths can cause problems as cable capacitance is often many times that of the change in the sensor, and capacitance imbalance can also be experienced between conductors of the cables which is much larger than the desired change. This does not mean that capacitance sensors may not be used; they are indeed used widely, but before using them the reader should discuss their application carefully with the manufacturer.

Vibrating Diaphragm Pressure Sensors

The stiffness of a diaphragm increases with the application of pressure. As the stiffness increases, so does the resonant frequency. This has made possible the design of a vibrating diaphragm sensor which is excited mechanically by an ac magnetic field, whose frequency is adjusted by a servo oscillator until the resonant frequency of the diaphragm is found. At resonance the amplitude of vibration of the diaphragm increases. This is reflected by increases in strain in the diaphragm material. These strains are usually sensed by a strain gage and are used to lock the oscillator to the resonant frequency. Frequency can be measured very accurately, and this sensor can be made to provide pressure readings of a high standard.

Piezoelectric Pressure Sensors

If certain crystals are sandwiched between two metal plates and are charged at several hundred volts, the crystal becomes polarized and maintains this condition after the voltage is removed. This is called the *piezoelectric* effect. Barium titanate is an example of a piezoelectric material. After polarization the transducer behaves like a charged capacitor, producing a voltage between the plates proportional to the physical pressure exerted on them. The charge that appears across the plates is not capable of delivering any current into a load and must therefore be conditioned by special high-impedance amplifiers. Cable capacitance shunts the charge and has a nonlinearizing effect as well as attenuating the signal. Piezoelectric sensors have much more important applications in vibration measurement and are discussed further in that section.

PRESSURE SURGE DAMPING

Where a diaphragm is exposed directly to the fluid in a pipe, shock waves caused by valve operation or obstructions in the pipe can overpressurize the diaphragm. When the measurement of these transients is of interest the sensor must have the necessary range and fequency response to cope with the situation. In many instances, however, only the mean pressure is of interest and then the sensor need not be mounted directly in the pipe wall, but at the end of a tube connected to the pipe. This mode of mounting may be desirable when the fluids in the pipe are hot, or where space around the measurement point is restricted. It also provides a means of removing the sensor to a safe area if the environment around the pipe is hazardous due to the presence of explosive gases.

Under some circumstances the use of a tube to mount a sensor can also introduce problems. Shock waves in the pipe and local turbulence around the pipe tube junction are made up of many frequencies, one of which may coincide with a frequency equal to $v/4l$, where v is the velocity of sound in the pressure medium and l is the length of the pipe. A disturbance with this frequency content will set up a standing wave in the pipe as illustrated in Figure 5.20 with a ¼ wavelength equal to the length of pipe. The pressure experienced by the sensor when this occurs may be several times higher than expected, and even if damage is not caused to the diaphragm, an incorrect measurement may be made. This effect is sometimes described as *organ piping* and is eliminated by the introduction of a restriction in the tube called a *pressure snubber* (Figure 5.21)

PRESSURE MULTIPLEXING

When it is necessary to measure a large number of pressures, the cost of individual sensors can become very high. In the lower pressure ranges it has been traditional to use banks of manometers, either mercury or water. If these pres-

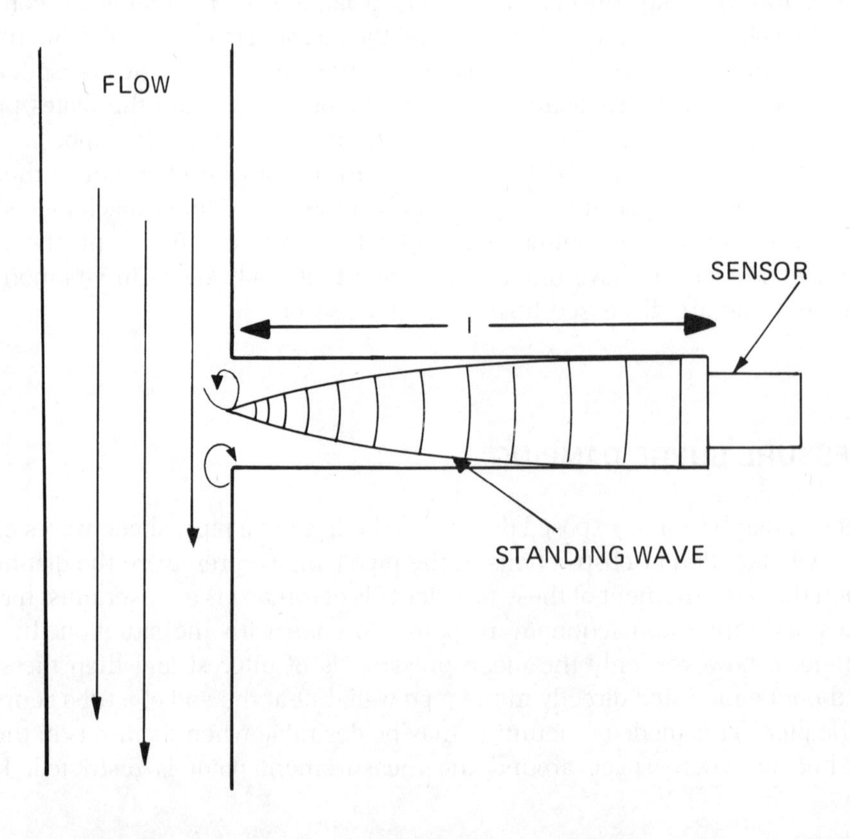

Figure 5.20
Organ Piping

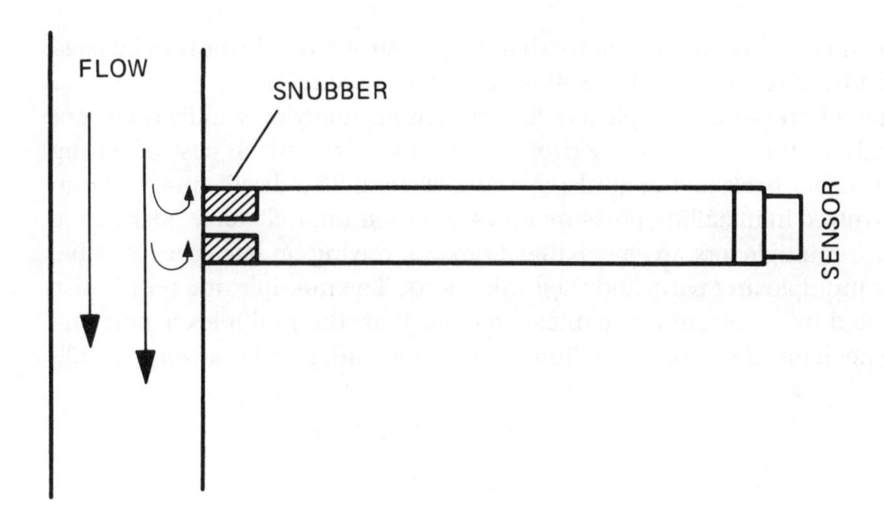

Figure 5.21
Pressure Snubbing

sures must be read frequently, the reading costs and the subsequent manipulation of data become very expensive. The method is also extremely prone to human error.

As an alternative to the high cost of using individual sensors, on the one hand, and high labor costs, on the other, techniques of sampling a large number of pressure sources by a single sensor have been developed. This process is known as pressure multiplexing.

The pressure multiplexer (Figure 5.22) consists of a motor-driven pressure port which connects each of the pressure sources in turn to a sensor. It is possible in this way to measure 50 or so pressures from one sensor at a scan rate of up to

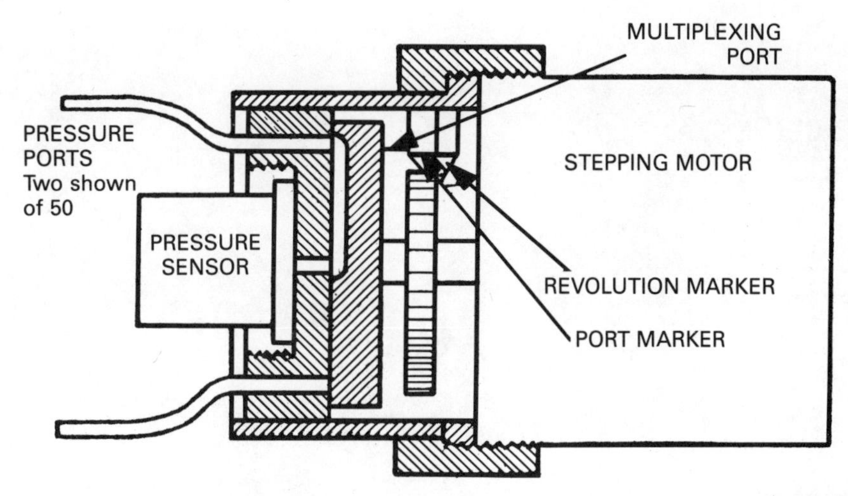

Figure 5.22
Pressure Multiplexer

about five points per second. It follows that if a pressure multiplexer is to be used to the best advantage, automatic reading techniques are required.

The value of pressure multiplexers lies in their economy over individual sensors, not only in the cost of the sensors themselves, but in the cost of wiring individual sensors back to a central recording system. The disadvantages lie in the work involved in installing pressure pipes and their unpredictable operational life. Pressure multiplexers are mechanical devices relying on gastight faces between their multiplexing ports and their inlet ports. The multiplexing port is also usually rotated by an electromechanical stepper. Both the multiplexer port and the electromechanical stepper have limited lives and suffer under adverse conditions.

CHAPTER 6

TEMPERATURE MEASUREMENT

TEMPERATURE SCALES

In any discussion of temperature, three scales predominate, centigrade (or *Celsius*), *Fahrenheit*, and absolute (or *Kelvin*). There is a general movement away from the Fahrenheit scale to the Celsius, but at present both are in widespread use. The formula to convert Celsius reading to Fahrenheit is:

$$t_F = \left(\frac{9t_C}{5} + 32\right) \tag{1}$$

and conversely:

$$t_C = \frac{5}{9}(t_F - 32) \tag{2}$$

where t_C is the temperature in degrees Celsius and t_F is the temperature in degrees Fahrenheit. Both these scales are based on the boiling and freezing points of water. In the Celsius scale the freezing point is that of distilled water; the Fahrenheit uses salt water in proportions that produce the lowest freezing point. Lord Kelvin moved away from this concept when he defined his absolute thermodynamic (or Kelvin) scale in terms of mechanical work. To understand the Kelvin scale it is necessary to consider the properties of gases which are governed by Boyles' and Charles' laws.

Boyles' law states that if a certain mass of gas occupies a volume V_0 at a pressure P_0 and a volume V_1 at a pressure P_1, provided the temperature is constant:

$$P_0V_0 = P_1V_1$$

or

$$V_1 = \frac{V_0P_0}{P_1} \tag{3}$$

Charles' law states that if a certain mass of gas occupies a volume V_0 at an absolute temperature T_0, then its volume V_1 at an absolute temperature T_1 is given by the expression:

$$\frac{V_0}{V_1} = \frac{T_0}{T_1}$$

or (4)

$$V_1 = \frac{V_0T_1}{T_0}$$

By combining Charles' and Boyles' laws, the general gas law is obtained:

$$\frac{P_1V_1}{T_1} = \frac{P_0V_0}{T_0} = \text{a constant} \tag{5}$$

The constant known as the gas constant is represented by $\mathcal{R}$, and (5) is usually written:

$$PV = \mathcal{R}T \tag{6}$$

This law only holds for ideal gases, which do not exist. Nevertheless, the elemental gases such as nitrogen, hydrogen, oxygen, and helium obey the law very closely within limited excursions of pressure.

Kelvin used the relationship $PV = \mathcal{R}T$ as the basis for a thermometer and developed his scale accordingly. The Kelvin degree is defined as 100th of the interval between the freezing and boiling points of water, so that 1° Kelvin represents the same temperature interval as 1° Celsius. The thermometer itself employs an elemental gas in a chamber of constant volume. Variations of temperature produce a corresponding variation of pressure in the chamber which may be measured and used to calculate the temperature. Such themometers are called *constant-volume gas thermometers.* By choosing suitable materials they may be used over a very wide range of temperatures, from near absolute to in excess of 1500°C.

An alternative form of this thermometer is one in which the pressure is kept constant, but thermometers based on this technique are much more difficult to use than the constant-volume devices and is therefore less favored.

The gas thermometer is not an instrument particularly suited to industrial use. Although it is used, other methods have been developed which also provide high degrees of accuracy.

EXPANSION THERMOMETERS

The most commonly used expansion thermometer is the *liquid in glass*, whose value to the instrumentation engineer lies mainly in it's value as a calibration standard. In many instances temperature measurement is required only to within one or two degrees, and a good laboratory quality mercury thermometer is usually an adequate standard in the range −20°C to about +200°C. Beyond this range, special high-temperature thermometers are available in which inert gases are introduced above the mercury column at sufficient pressure to prevent the mercury boiling.

The obvious disadvantage of the liquid-in-glass thermometer in industry is its fragility and the difficulties experienced in reading its scale. To the instrumentation engineer a further difficulty is added, namely, that of applying a secondary transducer to obtain an electrical output.

MERCURY-IN-STEEL THERMOMETER

The mercury-in-steel thermometer (Figure 6.1) replaces the glass bulb and stem with steel counterparts, and because the filling is no longer visible, a Bourdon tube is used to produce a movement which traditionally drove a pointer. The process of converting these instruments into electrical sensors is identical to the processes described for converting Bourdon-tube pressure gages.

Mercury-in-steel thermometers are often used with long capillary tubes to remove the Bourdon tube from the hot zone. In such installations, temperature variations along the length of the capillary will cause reading errors because of differential expansion between the mercury and the steel. There are a number of ways of achieving compensation, one of the most convenient of which is the temperature compensation chamber (Figure 6.2).

The capillary, from the thermometer to the reading head, is joined through an expansion chamber. Inside the chamber is an Invar core which has a negligible coefficient of expansion. The outer case is usually made of low-carbon steel with a coefficient of expansion of about ⅙ that of mercury. If the Invar core is designed to fill up ⅚ of the space in the chamber and the mercury ⅙, then the increase in volume of the air space due to expansion of the case will be exactly that of the increase in volume of the mercury. If temperature variations are experienced along the length of the capillary, it is necessary to insert a number of compensators at critical points along the length. As a filling liquid, mercury may only be used down to a temperature of −40°C, and thus for these low temperatures various fillings are used as alternatives. Alcohol is probably the most familiar, but a whole range of organic liquids is used. By using such liquids the lower temperature range may be extended down to about −200°C. In the higher ranges, mercury may be used up to about 650°C while the boiling points of organic substances restrict their ranges to something less than 300°C.

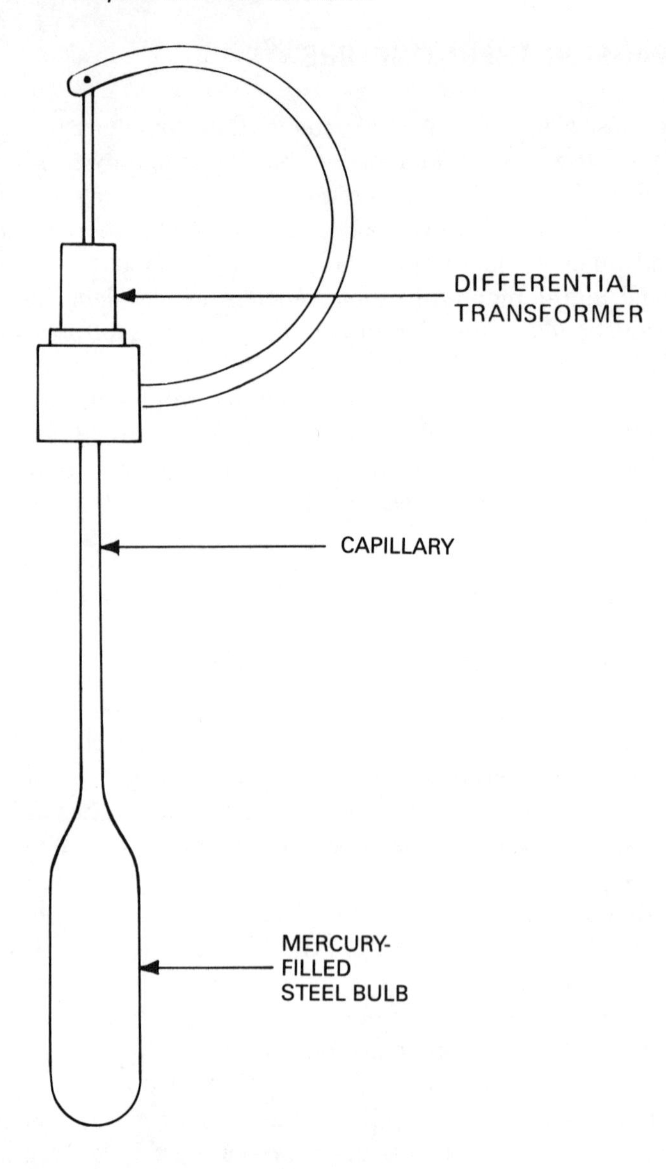

Figure 6.1
Mercury-in-Steel Sensor with
Differential Transformer Output

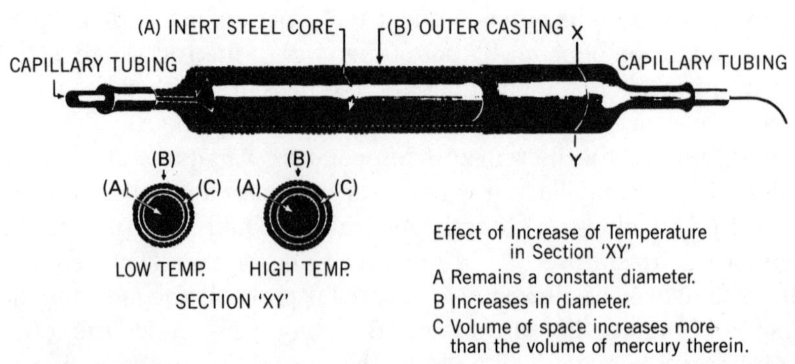

Figure 6.2
Temperature Compensation Chamber
(British Rototherm Co. Ltd.)

CHANGE-OF-STATE THERMOMETERS

The *change-of-state* or *vapor-pressure* thermometers (Figure 6.3) are partially filled with a liquid above which is a space containing only vapor of the filling liquid. The vapor in the space exerts a pressure that is a function of temperature and which may be used to operate a Bourdon tube.

In common with the liquid-in-steel thermometer, the vapor-pressure thermometer is made up of a steel container connected to the Bourdon tube by a capillary. Unlike the liquid-filled counterpart, the pressure exerted on the Bourdon tube is unaffected by changes in volume of either the filling or the container. Consequently, this type of thermometer does not require temperature-compensation devices, but it suffers from *cross-ambient effects* which can cause errors in

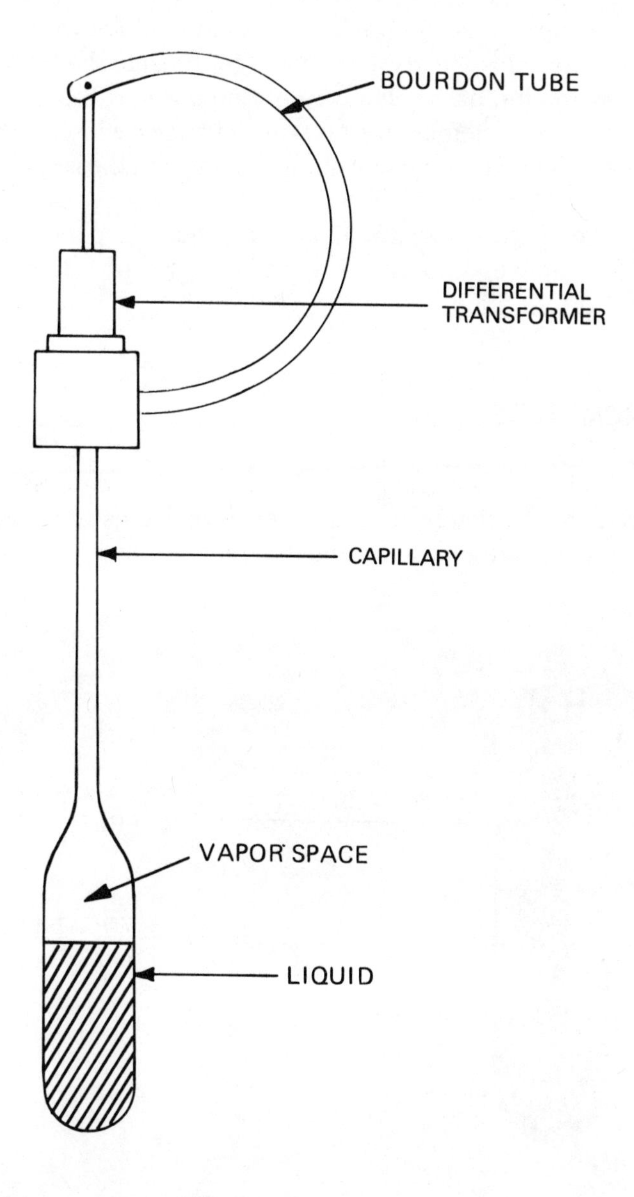

Figure 6.3
Change-of-State Sensor with Differential Transformer Output

the readings. This effect occurs because the Bourdon tube is at a lower temperature than the bulb, which causes vapor to condense in the cool end. An increase in ambient temperature will cause some of the liquid at the Bourdon-tube end to vaporize, causing an increase in pressure and hence the temperature reading of the thermometer. The increase in pressure is eventually offset by recondensation of the vapors at the bulb end, and the system returns to stability. The problem can be particularly troublesome when temperatures around ambient are being measured, and it becomes more serious as the length of the capillary increases. To overcome the problem, vapor-pressure thermometers have been developed into filled-capillary vapor-pressure thermometers shown in Figure 6.4.

In this device, the Bourdon tube and the capillary are filled with a nonvolatile liquid, and the capillary is extended to the bottom of the bulb. Floating on the pool of nonvolatile liquid is the vapor-pressure-producing liquid, and above that is a space containing only vapor. By this means the vapor is confined to the bulb, and the cross-ambient effects are eliminated. Temperature changes in the Bourdon tube and capillary can still cause temporary pressure changes which give erroneous readings, but these depend on the rate of the change and the response of the instrument.

A wide range of filling materials is available, and the choice depends on the application. Toluene, for example, is used for a range between −150°C and 250°C; argon is used down to −250°C.

RESISTANCE THERMOMETERS

The accuracy and repeatability of the change in resistance of certain metals to changes in temperature is good enough to make them suitable as temperature standards. Platinum and nickel are typical metals that may be used to construct

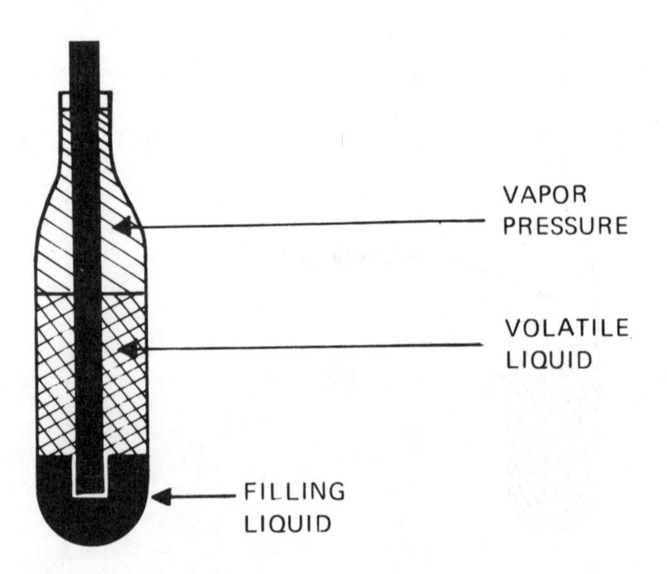

Figure 6.4
Filled-Capillary Vapor Pressure Thermometer

resistance thermometers, and it is interesting to note that they can be made for a very low cost. Although these low-cost thermometers are obviously not standard grade, they are still accurate devices.

The design of resistance thermometers depends very much on their application. Some of the simpler types are manufactured in an identical manner to strain gages, the resistance wire being bonded to a backing, which in turn is bonded to a specimen whose temperature is to be measured. Where the temperature of a fluid in a pipe is to be measured, the element is mounted in a probe (see Figure 6.5). And if the fluid is a gas, the sheath surrounding the element is often perforated. The mounting for the element inside its sheath varies from manufacturer to manufacturer. Each one has a preferred design, but all are attempting to solve the same problems. All designs seek to:

- Protect the resistance wire from contamination, mechanical damage, and moisture
- Minimize response time to a change in temperature
- Secure the mounting
- Facilitate easy replacement of the thermometer
- Maximize signal
- Minimize self-heating due to excitation current
- Maximize mechanical support for the element

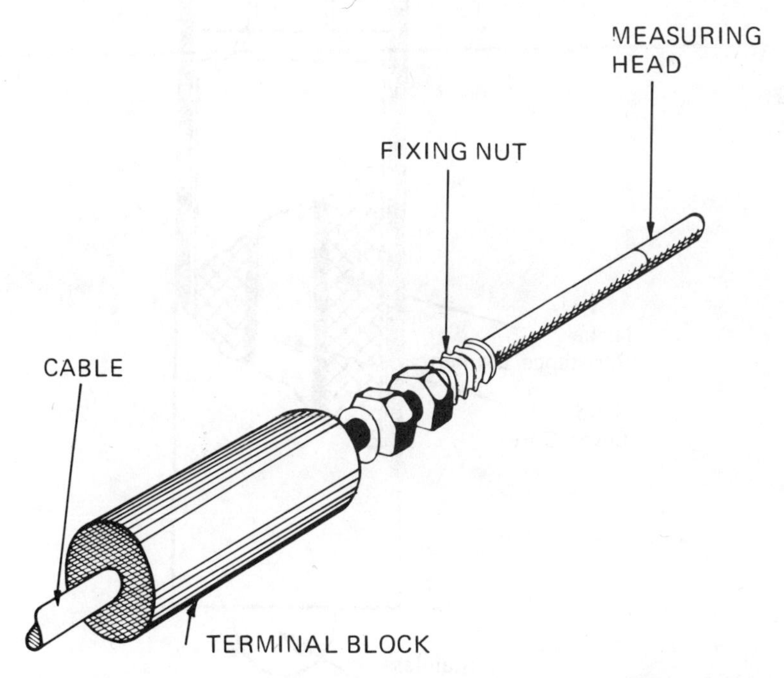

Figure 6.5
Typical Resistance Thermometer

- Minimize thermal EMFs due to junctions of dissimilar metals
- Make the cost competitive

It is not surprising that in attempting to reconcile these often opposing requirements, there are as many solutions as there are manufacturers, and each solution must inevitably be a compromise. A typical probe design is illustrated in Figure 6.6.

The most accurate method of using a resistance thermometer uses a nulling bridge to measure the resistance change and then the temperature is calculated from basic parameters. In practice, this is not a practical method of approach if a large number of sensors require frequent reading. The alternative approach uses a bridge to measure off-balance voltage calibrated in terms of temperature.

If a resistance temperature sensor is used at a location remote from the bridge, not only will the sensor change resistance with temperature, but so also will the leads connecting it to the bridge. To overcome these errors the simple bridge is often modified into a three-wire configuration as shown in Figure 6.7 so that the

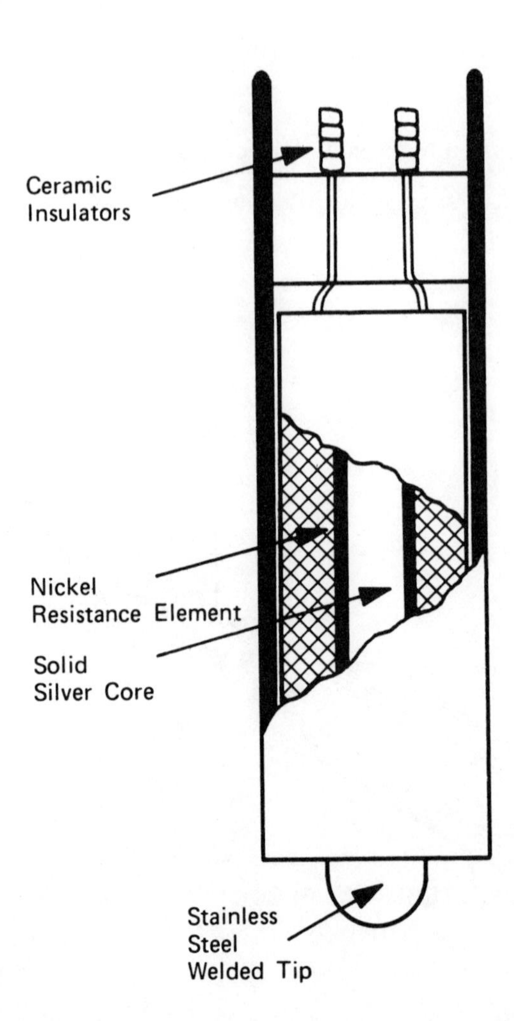

Figure 6.6
Resistance Thermometer
(Foxboro-Yoxall Ltd.)

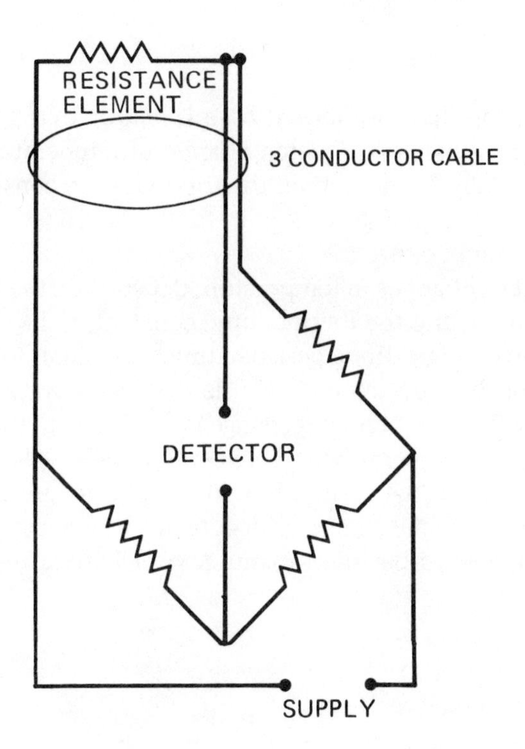

Figure 6.7
Three-wire Temperature Compensation of Cable to Remote Sensor

two wires are connected from one side of the sensor. This then introduces wires equally into the two adjacent arms of the bridge, with the detecting instrument connected remotely to the sensor through the third wire. Providing the wires are of equal resistance, no change will be experienced in the bridge because of temperature changes in the wires. For most purposes this type of cable compensation is adequate, especially for the less accurate types of sensor.

SELF-HEATING OF RESISTANCE SENSORS

Temperatures measured by resistance sensors may be in error because of effects introduced by the measurement technique as well as the construction of the sensor itself.

If a sensor is to be used in a simple bridge, the current flowing through the sensor will vary considerably over a temperature range when a constant voltage supply is used. This is due to the large changes in resistance of the sensor which could double its resistance over a temperature change of about 250°C. Such changes will cause nonlinear bridge output which may be partially compensated by the use of a constant current source.

Another serious effect introduced by variation in excitation current is its effect on the self-heating properties of the sensor. The power dissipated by the sensor resistance will be I^2R. This power dissipation causes self-heating of the element which increases its resistance and causes an error in the measurement.

THERMAL TIME CONSTANT

All temperature-measuring sensors have a *thermal time constant* that governs their applicability. If a sensor is subjected to a step change of temperature, its resistance will change exponentially. By definition, the thermal time constant is the time required for the resistance to change to $1/e$ of its final value (63.2%) after experiencing a step change in temperature.

The ability of a sensor to follow changes in temperature depends on the rate at which the temperature is changing and the thermal time constant of the sensor. Normally, the response to a step change after a period of time equivalent to seven time constants will be 99.9% of the true value. This rule does not always apply, however, and sensors that obey the exponential response very closely to the 63% point may break away from it at some point lying between 70% and 90% of the final value. Above this point the time constant may be considerably longer than at the lower end of the range. When response is a critical factor it is necessary to obtain test data of the step response either from manufacturers' literature or by calibrating the sensor.

THERMOCOUPLES

The principle of the thermocouple was demonstrated by Seebeck, who discovered that if two copper leads of a galvanometer were joined by an iron wire and one of the joints was heated, a current flowed in the circuit. He further demonstrated that as the temperature of the joint was increased, the current also increased to a certain point and then began to fall, eventually reaching zero, and even becoming reversed. The maximum point on the curve is called the *neutral temperature* and the point at which it crosses zero is called *the temperature of inversion.*

Closer examination of the curve shows that it is a parabola which may be described by the expression:

$$E = a(t_2 - t_1) + b(t_2 - t_1)^2$$

where a and b are constants and $t_2 - t_1$ is the temperature difference between the hot and cold junctions.

The range over which a thermocouple may be usefully employed depends on its neutral temperature, which varies with the materials used. The useful range for a number of combinations of materials is shown in Figure 6.8.

If a thermocouple is to provide a useful output it must have two junctions, one of which is used to sense the temperature and the other which must be controlled if it is not to introduce errors into the measurement. From a measurement point of view, it is convenient to keep this second junction at the temperature of melting ice, which allows the reading on an indicator to be calibrated directly in degrees Celsius (see Figure 6.9). In the industrial environment, where a large number of thermocouples are used, maintaining a reference of 0°C is difficult. Refrigeration devices are available which provide highly stable termination of groups of thermocouple junctions and which maintain ice-point temperatures.

For many applications they are ideal, especially in semilaboratory environments, but when thermocouples are scattered over wide areas, the number that can be terminated at one point becomes less as the scatter grows. The choice of termination often becomes a difficult compromise. To use a refrigerator for single couples is unreasonably expensive, while routing groups of wires a considerable distance to a central termination point can also be expensive in cable costs.

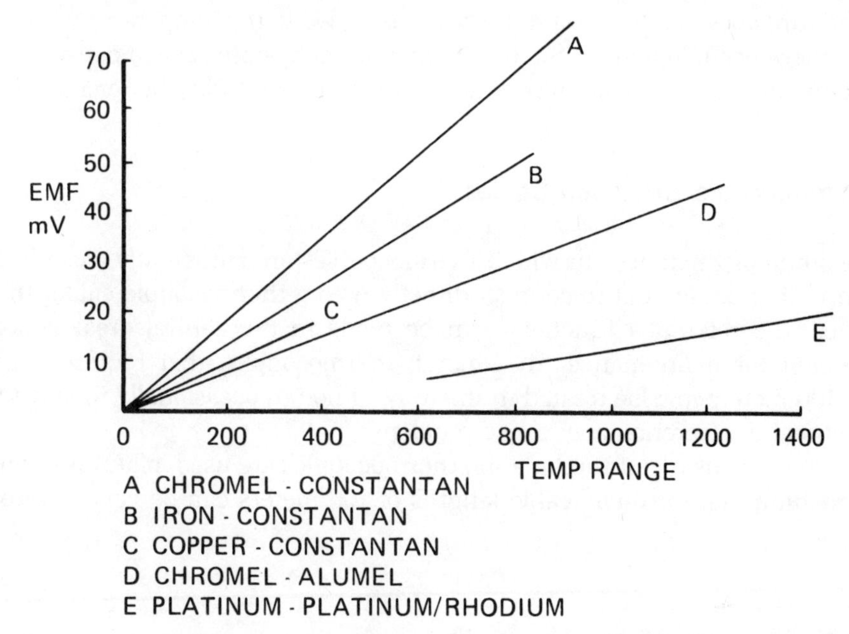

Figure 6.8
Temperature Operating Range of Typical Thermocouple Combinations

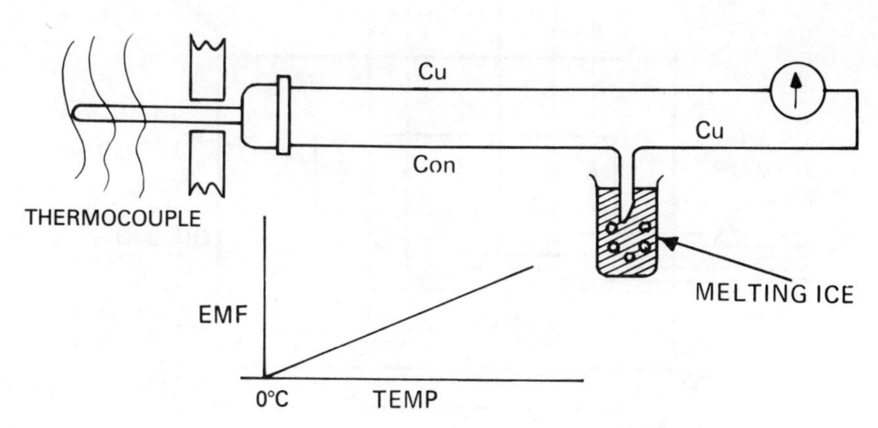

Figure 6.9
Thermocouple with Reference Junction at 0°C

Above-Ambient Reference Junctions

When long thermocouple cable runs are necessary, although the refrigeration unit provides an ideal termination, cost may preclude its use. An alternative is inexpensive *above-ambient* junctions. One type illustrated in Figure 6.10 shows the iron/copper thermocouples brought to a reference junction box where the iron junctions are terminated with copper leads. These terminations are made at terminals mounted on but insulated from a substantial metal block and immersed in an oil bath which is temperature controlled. The bath is set at some convenient temperature above ambient, and this value is added to the temperature indicated by the thermocouple output. Such reference junctions are easy to construct and maintain, and can be made to provide a temperature stability better than 1°C.

Thermocouple Extension Cables

There are many instances in which thermocouples are sufficiently close to central recording equipment to connect directly with a thermocouple cable. In this instance, the reference junctions can be made in the central area, which is convenient for maintenance. In general, thermocouples that require runs of about 100 meters may be treated in this way, although occasionally the presence of electrical interference may cause problems.

When precious metal combination thermocouples are used, platinum or platinum/rhodium, for example, cable lengths of 100 meters can be very expensive.

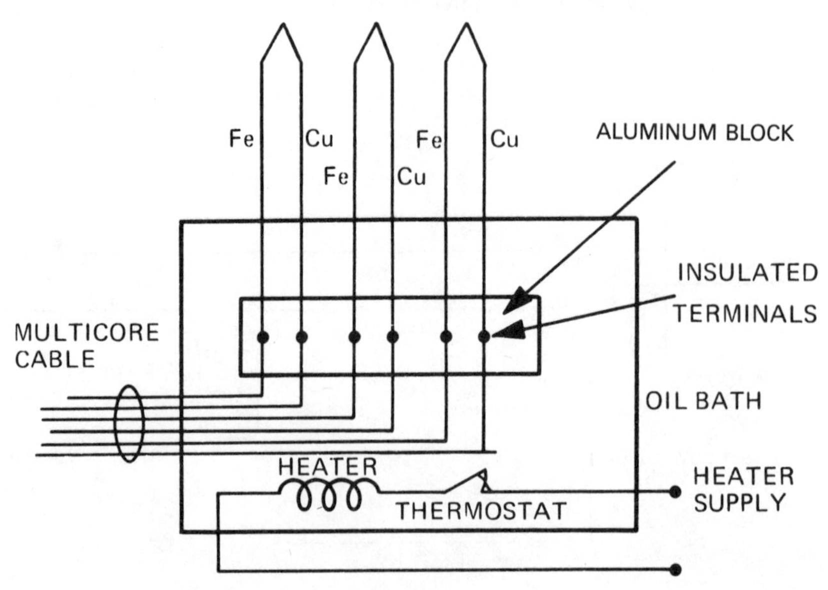

Figure 6.10
Above Ambient Thermocouple
Reference Junction

Consequently, it is normal to terminate precious metal couples as close as possible to the thermocouples and use an extension thermocouple cable with similar thermoelectric characteristics.

Uncompensated Reference Junctions

However simple the reference junction, its cost is inflated by the installation costs and subsequent maintenance. Elimination of the constant temperature reference junction as such can achieve savings in this direction. Where computer based logging systems are used the constant temperature junction may be replaced by one that is free to change with ambient conditions. The temperature of the junction is measured, and the computer calculates a correction factor for each reading.

The junction is usually made on a substantial metal block enclosed in a container to protect it from drafts. Temperature of the block is usually measured by a resistance element, and this value is used by the computer to correct each output.

COMMON AND SERIES MODE VOLTAGES

Suppose two wires connect an external circuit *CD* with a resistance *R* to the terminals *A* and *B* of a recording device, illustrated in Figure 6.11. Suppose futher that each of the wires has an electrical resistance *R*, that there is a common point in the external circuit which has a potential *V* with respect to ground and that the resistances of the circuit between the common point and *C* and *D* are R_1 and R_2 respectively. Providing R_1 and R_2 are equal, the voltage measured between *A* and *G* and *B* and *G* will be equal. Although the potential of both lines is above ground, a meter across *AB* would not indicate any voltage. Even if the voltage *V* is varying, the values at *A* and *B* to ground will remain equal, and no resultant voltage will appear across *AB*. Voltages which appear between *A* and ground and *B* and

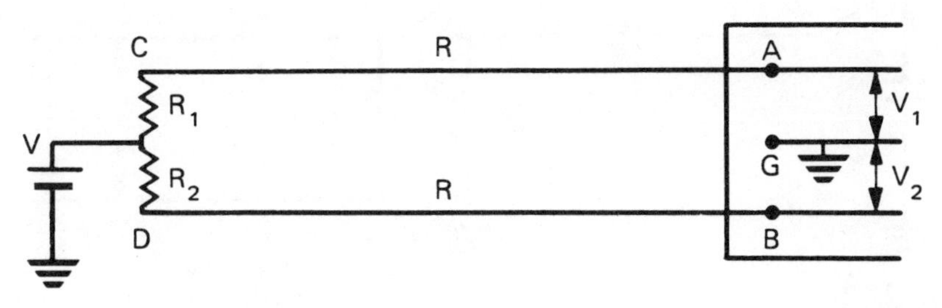

Figure 6.11
Common Mode Voltages

ground are known as *common-mode voltages*. These voltages may not be detected by the measuring instruments, but if their value is sufficiently high, they cause the instrument input to fail.

Suppose now terminal B is connected to ground. The voltage between B and G will be zero, while the voltage between A and G will still be at its original value. The signal appearing across A and B, therefore, will now equal V. Such a circuit is said to have no capability for common-mode rejection.

In practice, thermocouple circuits approximate Figure 6.12. The two wires from which the thermocouple is made will have different values R_1 and R_2, and in this example it is assumed that some leakage to both will occur along the cable length. This is represented by the two resistances R_3 and R_4.

It is common practice to weld the hot junctions of thermocouples to the metal whose temperature is being measured, and while this provides excellent thermal contact, it also provides electrical contact. The potential of this point may be a true ground potential, or at some value V above ground potential.

Under these circumstances the voltage at A is:

$$\frac{VR_4}{R_1 + R_4}$$

and at B the voltage is:

$$\frac{VR_3}{R_2 + R_3}$$

The resultant voltage between A and B will be:

$$V\left(\frac{R_4}{R_1 + R_4} - \frac{R_3}{R_2 + R_3}\right) \tag{1}$$

This resultant voltage is known as *series-mode* or *normal-mode* voltage, and it is indistinguishable from the signal. It can be a problem in all low-level analog circuits, but particularly in thermocouple installations in which unequal line resistance is inevitable.

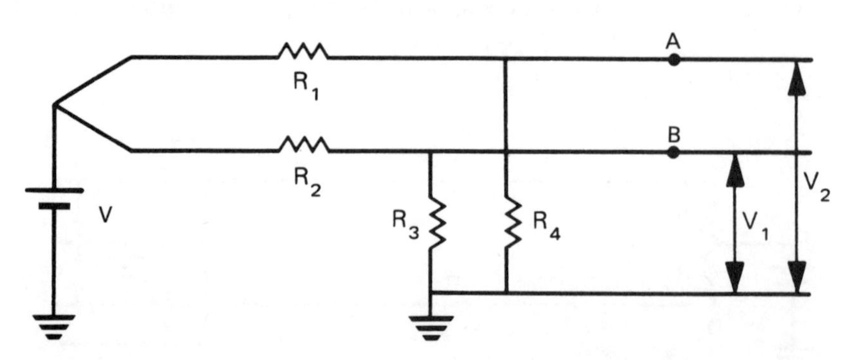

Figure 6.12
Common and Series Mode Voltages in Thermocouples

THERMOCOUPLE TRANSMITTERS

Although the temperature-controlled junction allows conversion of thermocouple leads into copper leads, it is a passive unit with no amplification facilities. The output of thermocouples is often only a few millivolts, and if this signal is to be transmitted long distances, electrical noise may be of far higher value than the signal. To overcome this problem the *thermocouple transmitter* has been developed which is located close to the thermocouple and which produces a large transmission signal from the sensor output.

The schematic drawing of the transmitter illustrated in Figure 6.13 is based on a circuit called a *modulator* which is excited by an oscillator. Depending on the type of modulator, either the frequency of the oscillator or its amplitude may be modified proportionally to the thermocouple signal. The output of the modulator is fed into a correction circuit where its span and its zero may be adjusted and where the output is corrected for changes in the reference junction temperature. This latter function is usually accomplished by using a temperature-sensitive resistance in the circuit which changes the zero datum of the modulator to compensate for changes of the junction. The corrected output is then demodulated and used to drive an amplifier stage, which supplies a current through transmission lines at low impedance. The output of the transmitter is often set at 4–20 mA, representing the full span of the measurement. The 4 mA represents the zero datum and is known as a *live zero.* This makes possible detection of a transmitter or signal-line failure, even when the indicator is operating around its zero point. The 4–20 mA range is an IEC standard widely accepted in the process

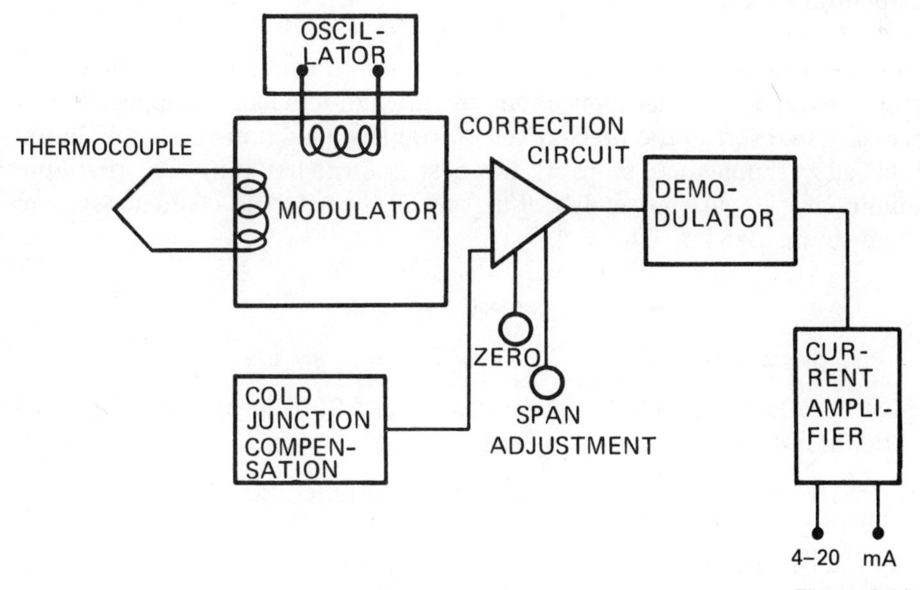

Figure 6.13
Thermocouple Transmitter

industries, although 10–50 mA is often used as an alternative. In addition to these values, instruments are to be found with 0–10 mA, and most manufacturers offer a voltage option on their current output instruments.

INDUSTRIAL THERMOCOUPLES

Industrial thermocouples fall into two classes, base metal and precious metal. Some of the base metal couples and their properties are listed in Table 6.1.

Table 6.1

Thermocouple	*Remarks*
Copper and constantan Cu and (40% Ni/ 60% Cu) alloy	Operates to 400°C and has high resistance to corrosion by condensation
Chromel and constantan, 90% Ni/10% Cr alloy and 40% Ni/60% Cu alloy	Operates up to 700°C as well as in an oxidizing atmosphere
Iron and constantan Iron and (40% Ni/ 60% Cu) alloy	Operates up to 850°C and is suitable for use in oxidizing or reducing atmospheres
Chromel and alumel 90% Ni/10% Cr alloy and 94% Ni/ 2% Al 1% Si and 3% Mn approximately alloy	Operates up to 1100°C in oxidizing atmospheres

Thermocouples of these combinations represent the majority of couples used in industry today. For special applications and for high-temperature applications it is necessary to resort to the precious metal couples. Platinum versus platinum/rhodium alloy couples are probably the best known, but tungsten, rhenium, palladium, silver, gold, and molybdenum are also used. The characteristics of some are summarized in Table 6.2.

Table 6.2

Thermocouple	*Remarks*
60% Au/40% Pd alloy and 90% Pt/10% R alloy	Operates up to 1000° in an oxidizing atmosphere
W and 75%/W/25% Re alloy	Operates from 900 to 2800°C
Pt and 87% Pt/13% Rh alloy	Operates up to 1500°C
95% Pt/5% R alloy and 80% Pt/20% Rh alloy	Operates up to 1000°C

Except where high thermal response is required, all types of couples are usually mounted in protective sheaths. A typical arrangement is illustrated in Figure 6.14.

The thermocouple is terminated at the end of the sheath in one of three ways. The junction may be exposed to the temperature which obtains the maximum thermal response but leaves it open to mechanical or chemical damage. Environmental protection may be obtained by closing the sheath, but this gives protection at the expense of thermal response. A compromise between these two is obtained by welding the couple to the sheath, but this improves the response with the sacrifice of electrical insulation. The choice of junction termination will therefore depend on the application.

The open ends of the couple are terminated in the head of the assembly in some type of terminal block. The type of block varies between manufacturers, but the types least likely to cause errors are those in which the compensating cables are clamped firmly against the thermocouple wire.

It is necessary to insulate the thermocouple wire along the length of the sheath. Many units use wires threaded into ceramic beads or tubes, but in those operating at high temperatures the wire is often protected by magnesium oxide or alumina packing.

THERMOPILES

If thermocouples are connected in series, the EMF produced by each couple is additive. The series arrangement is used to detect small changes in temperature, usually associated with radiant heat. Such clusters of thermocouples are called

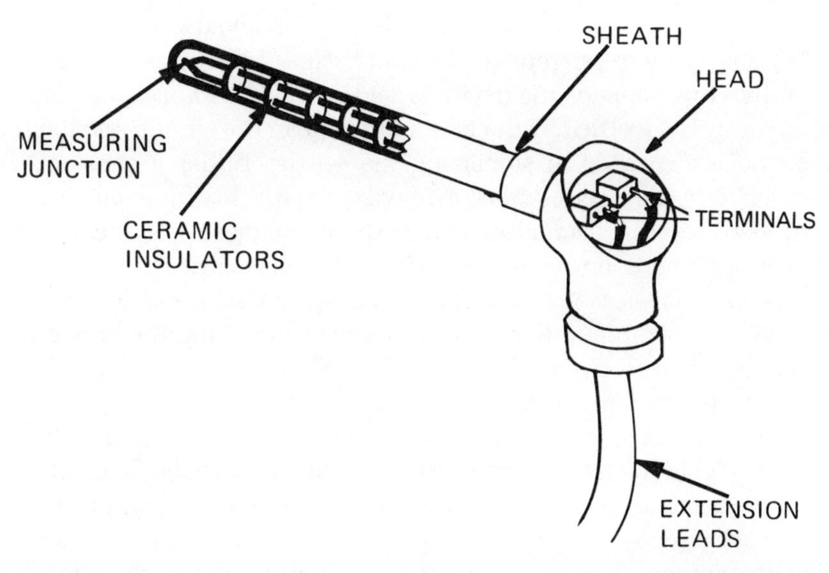

Figure 6.14
Typical Thermocouple Heads

thermopiles and are often used in the steel industry where the temperature of moving slabs may be measured by measuring the radiant heat they emit.

THERMISTORS

Thermistors or *negative-coefficient* resistors are resistors whose resistance value decreases as their temperature increases. In effect, they operate in the opposite way to a platinum resistance element. Thermistor materials are commonly semiconductors made by sintering mixtures of metallic oxides such as manganese and cobalt. For temperature-measurement applications they are usually made in the shape of small beads, sometimes covered in a glass envelope. Their resistance is a function of their absolute temperature and obeys the law:

$$\frac{R(T)}{R(T_0)} = e^{k\left[\frac{1}{T} - \frac{1}{T_0}\right]}$$

where

$R(T)$ = the resistance at absolute temperature T

$R(T_0)$ = the resistance at absolute temperature T_0

k = a thermistor constant.

Changes in resistance over a given temperature range are very high, which makes the thermistor a very sensitive device for the measurement of temperature. A typical thermistor will change its resistance from 10,000 ohms to about 5 ohms over a temperature range of 0–300°C.

To take advantage of the thermistor characteristics it is necessary to measure its resistance, and this requires a current being passed through it. This current causes self-heating of the thermistor and above a threshold current its resistance falls. This allows more current to flow, and causes the resistance to fall still further. When this happens the device is said to be in a *runaway condition*. The limit of runaway is governed by the power available in the circuit, and limiting the power provides a method of stabilizing the sensor. Either a constant-current power supply or a high series resistance will perform this function.

The use of a series resistance to limit the power suggests that the thermistor is ideal for bridge operation where the adjacent bridge arm is, in fact, a series resistor. In operation it is not desirable to allow the thermistor to run at a temperature much above the temperature being measured, and the bridge resistors should be chosen accordingly.

For many applications thermistors make extremely useful sensors. They are small, have a fast response, are inexpensive, and have high output. Their negative-resistance characteristics present some problems, and they are not as stable as might be desired. If care is taken in bridge design and the nonlinearity of the bridge is not regarded as a handicap, they make useful sensors in the range −50 to 300°C. In addition, their sensitivity makes them suitable for monitoring small changes in temperature.

THERMOSTATS

A very wide range of temperature switches are available which change their state at preset temperatures. These switches are very familiar devices in their bimetal form and are given the generic name of *thermostat*. A bimetal thermostat as represented in Figure 6.15 relies on the differential coefficient of expansion between two metals. If two metals *A* and *B* are firmly fastened together and metal *B* has a higher coefficient of expansion than *A*, in a heated condition, the strip will tend to curl upward, causing a break in the contacts.

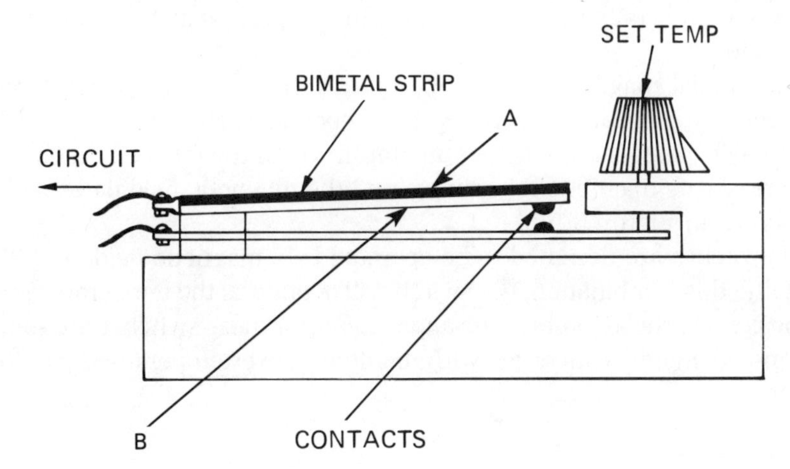

Figure 6.15
Bimetal Thermostat

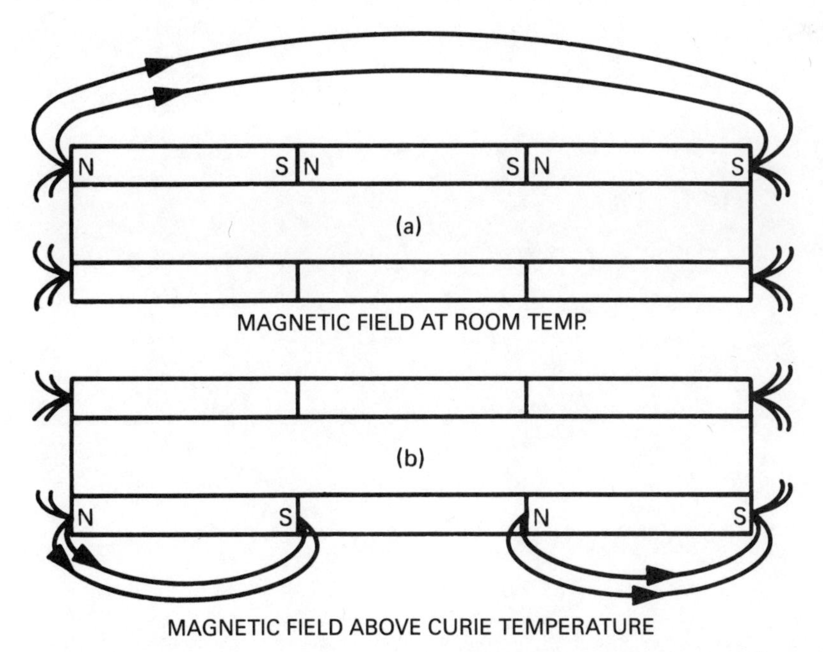

Figure 6.16
Curie Point Thermostat

The disadvantage with this basic design is the slow contact break experienced which can cause arcing when heavy currents are being switched. To overcome this weakness many types of snap-action devices are used, often based on double-curvature designs of the bimetal strip.

In addition to the bimetal switch there are several other types in use. The simplest is the mercury-in-glass thermometer with two platinum wires inset into the mercury column. Provided the current is low, this type of switch is very accurate and the switching temperatures very consistent.

A further design is based on the *Curie point* of a material. Certain materials exhibit sharp changes in magnetic reluctance when their temperature rises above a value known as the Curie point. This property may be made to break a magnetic circuit.

If two toroidal magnets are set up with a ferrite material between them at normal room temperature, the lines of force operate as shown in Figure 6.16a. If the assembly is heated, there comes a point in which the ferrite material dramatically increases its magnetic reluctance, and the magnetic field pattern changes to that shown in Figure 6.16b.

Reed switches are designed to be operated by a magnetic field, and if one is inserted into this combination, the switch will operate as the temperature reaches the Curie point of the assembly. The applications for these switches are somewhat limited and are mainly concerned with monitoring overtemperature conditions of machinery.

CHAPTER 7

DISPLACEMENT MEASUREMENT

LINEAR VARIABLE DIFFERENTIAL TRANSFORMER

Displacement sensors have already been discussed as the secondary transducing elements in pressure and temperature sensors, and these same techniques will also be found repeated in many other types of measurement. There are, however, many occasions when displacement itself is required as a measurement, and in these situations differential transformers, potentiometers, and capacitance sensors are all used as appropriate.

An important transducer extensively used for many applications is the linear variable differential transformer, usually referred to as LVDT, or sometimes differential transformer. These devices allow very flexible design and may be built with ranges as low as 0.00002 mm or upward of 800 mm. The cross section of a typical LVDT is illustrated in Figure 7.1.

The device consists of two secondary windings $S1$ and $S2$ wound on a hollow form on either side of a primary winding P. An *armature* made of magnetic material moves inside the form to produce the displacement signal. The primary winding P is excited by an ac carrier which in certain circumstances may be as low as 60 Hz but is more generally in the region of 1–10 kHz. The two secondary windings are carefully balanced and connected so that the voltages induced are in opposition. With the armature in its central position, these voltages are equal and opposite, and the output of the device is therefore zero.

Movement of the armature to the right from center will increase the flux linking S2 and decrease the linkage to S1. The result is an output voltage proportional to the displacement from the center. Similarly, when the armature is moved to the left, the voltage in S1 increases and decreases in S2. This voltage output is an ac signal with an amplitude proportional to movement of the core either side of the center point. A number of circuits are possible to allow the detection of this voltage; two are illustrated below.

The first method illustrated in Figure 7.2 shows the secondary coils connected to form an ac bridge. (The theory of ac bridges has already been summarized, and this is typical application.)

The method illustrated in Figure 7.3 connects the center tap of a transformer supplying the LVDT primary to one side of its secondary. This center tap connection may or may not be grounded.

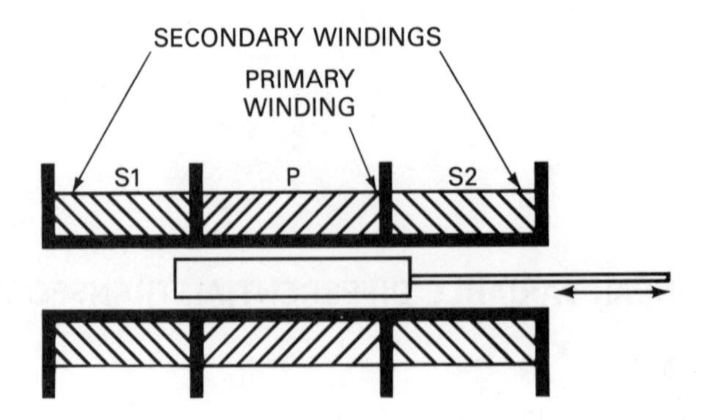

Figure 7.1
Cross Section of a Typical Linear Variable Differential Transformer (LVDT)

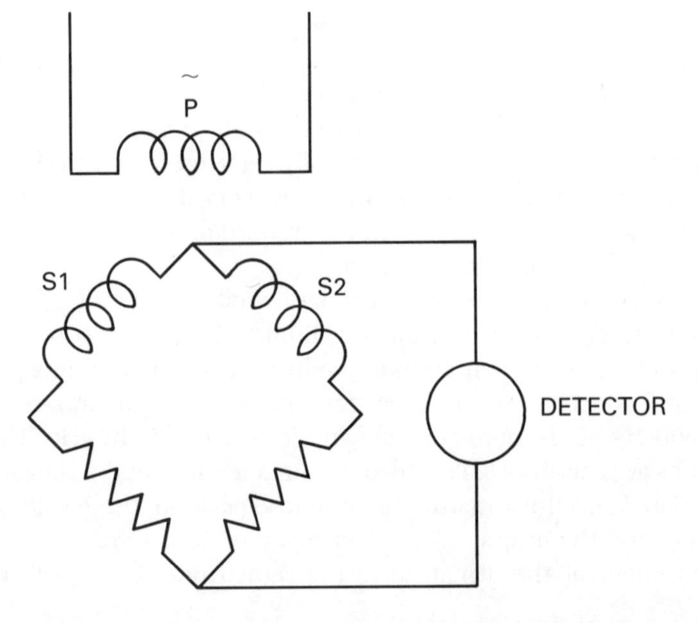

Figure 7.2
Bridge-Connected LVDT

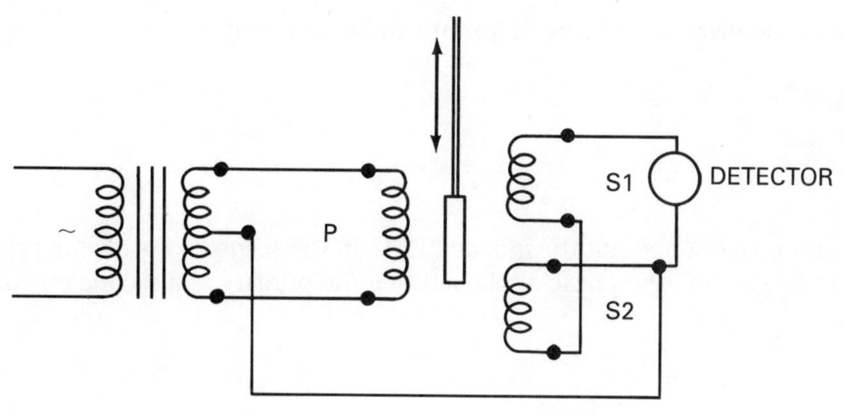

Figure 7.3
Alternative Connection of LVDT

LVDT Output Signal Form

The output of the LVDT is an ac signal with a frequency which is that of the excitation voltage usually described as the *carrier*. As the armature moves through its balance position the phase of the output signal changes abruptly through 180°. This provides a means of determining the position of the core relative to the center point. Depending on the inductance, resistance, and capacitance of the windings and the load on the secondary, the primary and secondary voltages will have an asymmetric relationship relative to zero. If the capacitance of the primary winding is considered small compared with its inductive reactance X_L and resistance R, the relationship between the resistive and inductive components of its voltage may be shown by the vector Figure 7.4 if I is the current flowing in this winding.

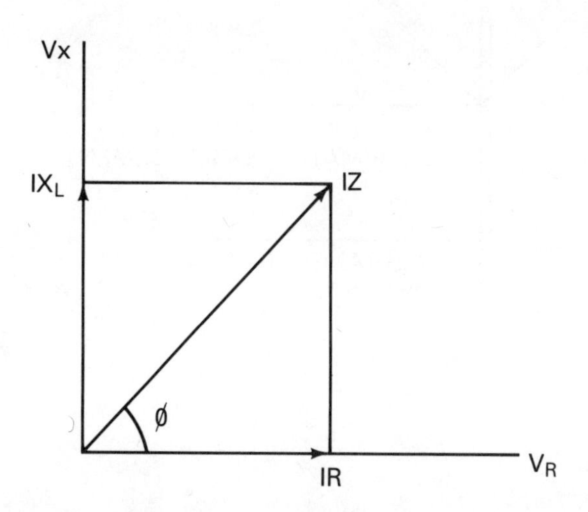

Figure 7.4
Relation Between Inductive and Resistive Components of the Voltage (IX_L and IR) in an AC Circuit

From the geometry of this Figure it can be seen that

$$\tan \phi = \frac{X_L}{R} = \frac{2\pi fL}{R}$$

$$\phi = \tan^{-1} \frac{2\pi fL}{R}$$

In a transformer circuit the induced EMF in the secondary leads the primary current by 90°, and the phase angle between the primary and secondary voltage becomes

$$\phi = 90 - \tan^{-1} \frac{2\pi fL}{R}$$

The changing of the output voltage amplitude and phase with the armature's displacement is illustrated graphically in Figure 7.5 for positive and negative displacements.

Residual Null Voltage

The output of the LVDT in a central condition is nominally zero. In many applications the residual output is so small as to be regarded as zero; but where very small displacements are to be measured may become significant. This residual

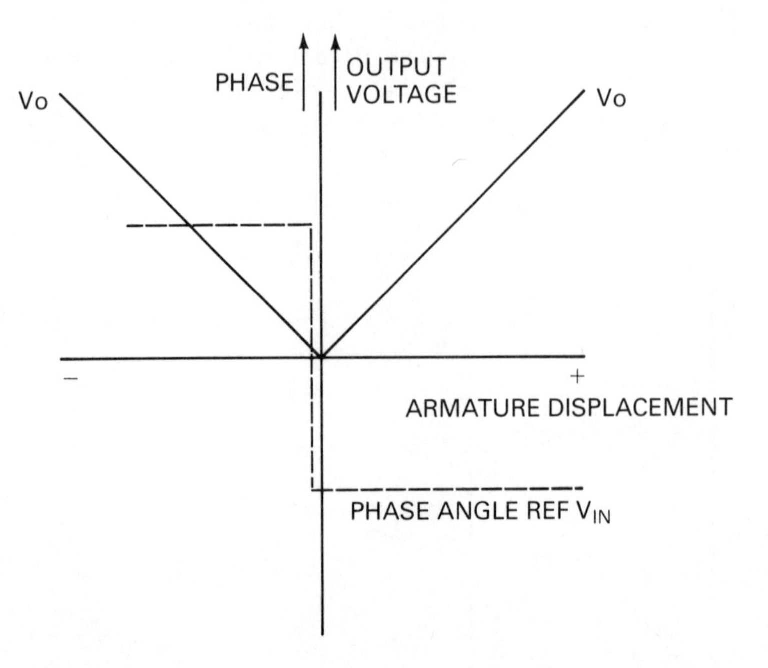

Figure 7.5
Output Voltage and Phase Relationships for LVDT

voltage may be due to a number of causes, and it may be due to combinations of more than one.

If the excitation voltage has a high harmonic content, the harmonics will manifest themselves at the balance point, resulting in a residual signal. If this signal is significant, it can only be reduced by filtering the excitation voltage. The accuracy with which the coils can be wound is also limited. This may result in secondary coils having small differences in inductance, resistance, and capacitance. Similarly, the capacitance between the primary and each secondary may also be different. All these differences may produce residual null voltages, and may be balanced to a minimum by a balance circuit. Two of these are illustrated in Figure 7.6.

LVDT Output Considerations

The linear range of an LVDT depends not only on its design, but also on its excitation frequency and the load into which it operates. Normally, a unit will be specified as having a nominal linear range with an armature designed to operate at a certain frequency, and the expected full-range accuracy will be quoted for these conditions. If reduced accuracy can be accepted, the range may be extended, and if improved accuracy is required, restriction of the linear range and operation on one side of zero often makes this possible.

It is normal to specify output linearity when the LVDT is operating into a high-resistance load, typically of the order of 100 kΩ. When a low-resistance load is used, the linear range may be significantly affected but this may be restored by introducing buffer amplifiers. The output of the LVDT is proportional to the primary excitation voltage. The accuracy of the output signal depends, therefore, on the stability of this supply.

The frequency at which the armature may be oscillated is determined by the carrier frequency. A minimum ratio of carrier frequency to armature frequency of 4 is required to allow satisfactory discrimination and filtering, but most manufacturers would recommend a higher ratio, even as high as a factor of 10.

Long-Stroke LVDTs

Using conventional winding configurations, the ratio of measurement range to device length is about 0.3 to 0.4. This implies that an LVDT with a range of 300 mm would have a winding length of about 1 m. One method of reducing this device length uses the tapered winding technique shown conceptually in Figure 7.7.

This design requires that the primary and two secondary windings S1 and S2 be wound concentrically along the full length of the former. The secondary windings are *taper wound* to allow a continuous change as the armature moves along the entire length of the windings. Such a winding allows a range-to-device-length

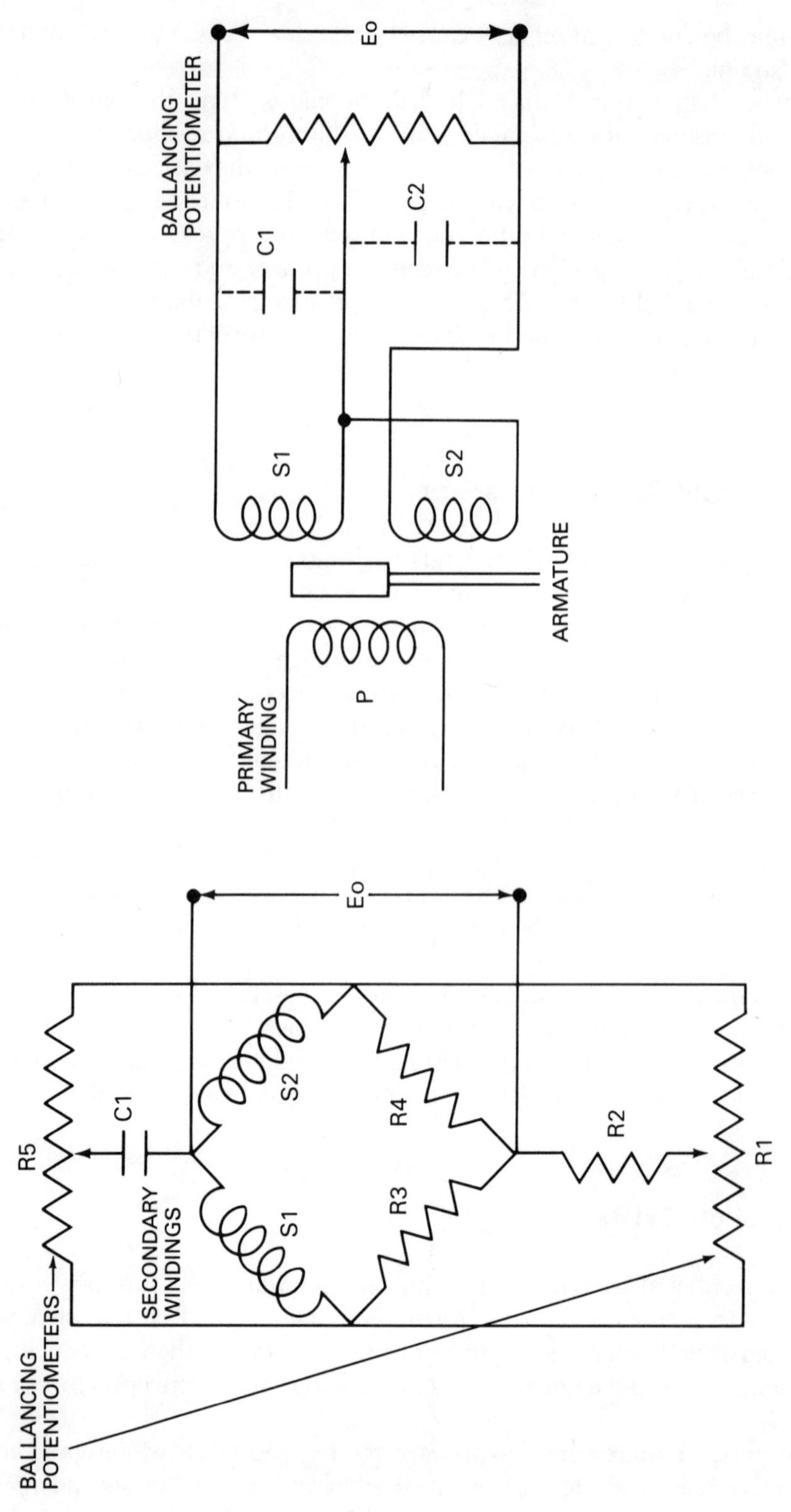

Figure 7.6
Null Balancing of LVDT

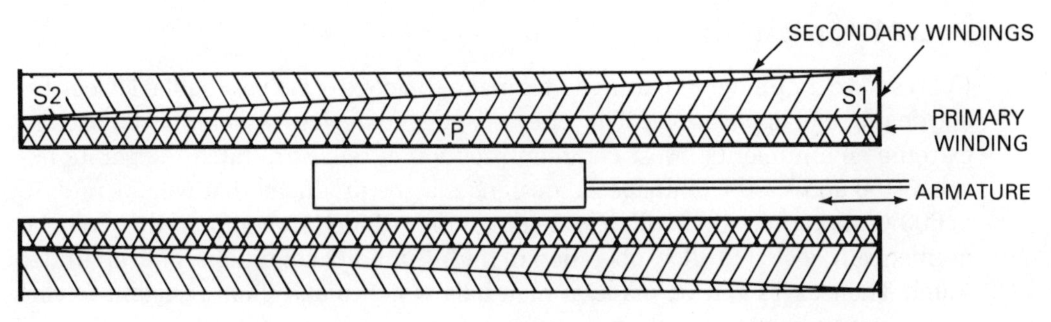

Figure 7.7
Taper Wound LVDT. (Schaevitz Engineering)

ratio of about 0.75 to 0.8, with a consequent halving of body length when compared with the traditional configuration.

The production of devices with continuous tapered windings presents winding problems. Therefore, it is normal to segment the windings, as illustrated in Figure 7.8.

DC-to-DC LVDTs

In some applications the use of ac carrier supplies are not readily available, and to meet these situations the dc-to-dc version has been developed. These devices provide the carrier oscillator, the discriminator, and filtering circuits built into the body. This allows dc supplies to be used and produces a signal which is a dc analog of the armature movement. Electronics in early LVDTs of this type were rudimentary and suffered severe limitations, but the introduction of microcircuits has represented a considerable improvement. DC versions now rival the ac versions in all but extreme environmental considerations.

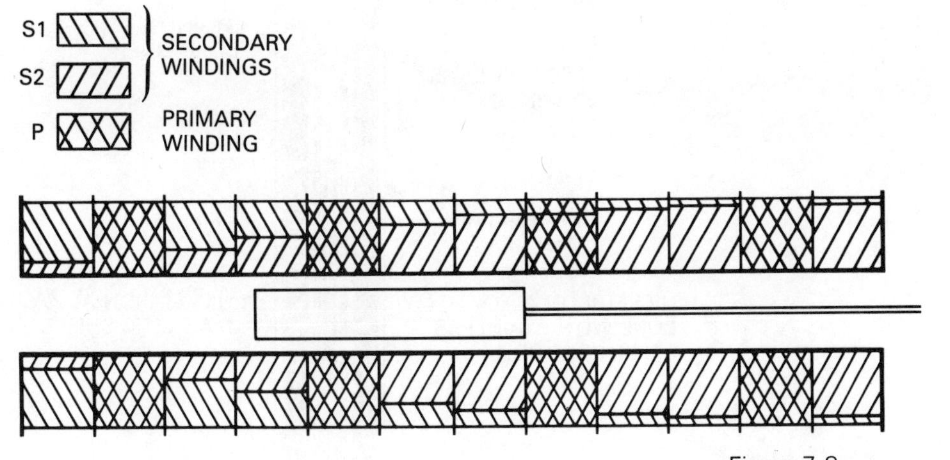

Figure 7.8
Segmented Taper Wound LVDT. (Schaevitz Engineering)

LVDT Applications

LVDTs have a number of features that make them extremely valuable devices for a wide range of applications. They are very rugged devices which may be used in extreme environments. Most common production types are rated to operate between 150 and −20°C, but special designs may be produced that will operate up to 600°C and others that will operate down to −265°C. In addition, they may be hermetically sealed to allow operation in high-humidity conditions, or even under water. Their cases can be made of materials which make them resistant to corrosive atmospheres, and they may be designed to survive in the presence of nuclear radiation. Figure 7.9 illustrates the construction of the LVDT designed to operate in radiation environments.

In normal operation the armature is not in contact with the body, and as a consequence, no friction is introduced into the measurement. Apart from the improvement in accuracy which may be expected where transducing forces are low, the absence of contact increases the mechanical life indefinitely. Another useful feature of the design of the LVDT is the complete isolation of the core from the windings. This isolation allows the use of simple seals where the armature must be operated in pressure vessels and the body exposed to low pressures. The isolation is also extremely valuable in explosive atmospheres.

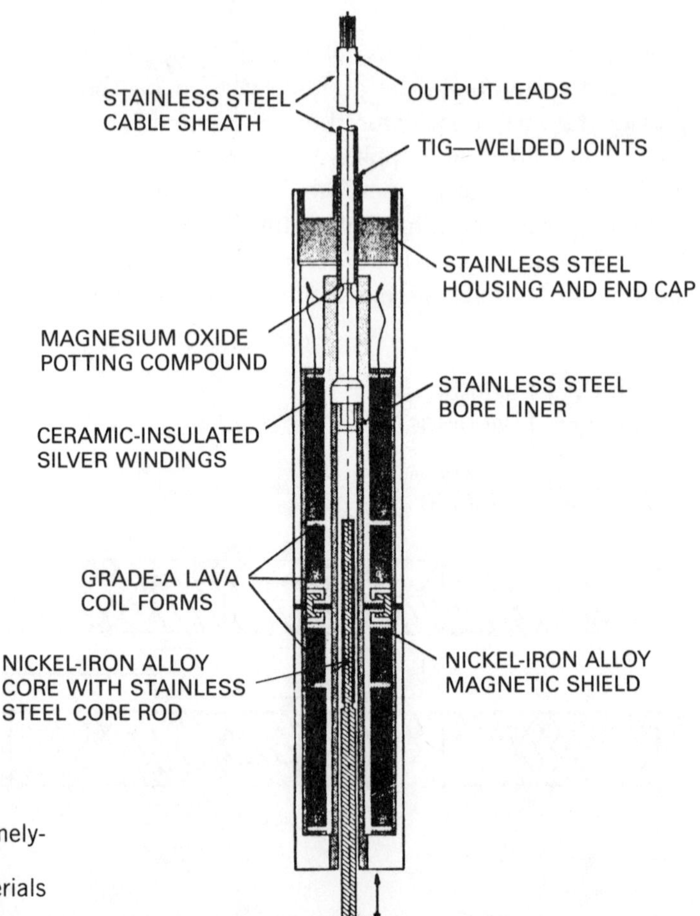

Figure 7.9
Cross-section of an LVDT for Extremely-High-Temperature, High-Radiation Hostile Environment Showing Materials and Techniques of Construction

POTENTIOMETERS

Potentiometers can be easily made to accommodate a very wide range of displacements from fractions of an inch to many feet. They may also be designed to measure angular displacement, and they can be designed with nonlinear characteristics. The main disadvantage of the potentiometer lies in its finite life, due to wearing of the wiper and its resistance winding. This wear is accelerated if repeated measurements are being made over a limited portion of the winding or if the sensor is subjected to severe vibration.

Many commercial potentiometers are designed for displacement measurement. These may be constructed with wire-wound resistors or, alternatively, conductive plastic. Conductive plastic is more rugged and can be expected to have longer life.

All commercial potentiometers are designed with the criteria of precision and minimum friction in mind. Bearings, wipers, and gearing, therefore, are all precision engineered and tend to raise the cost higher than might be expected when considering the basic construction. The choice of a commercial potentiometer, however, is not difficult. The specifications are easily understood, and the choice is often governed by the suitability of the unit to fit into the geometry of a situation.

In many cases, a displacement measurement is required where wiper friction is of no consequence and the constraint imposed is that of geometry into which no standard commercial unit will fit. In these cases the potentiometer is probably one of the easiest sensors to build "in house" and, provided the wiper and winding are protected from excess dirt and grease, should not present any particular problems in operation. Often, in these circumstances the resistance of the potentiometer is a single strand of wire rather than a winding. This implies a low-resistance potentiometer. Although low resistance can cause problems with resistance to wiper contact, the sensor can be used to advantage in a bridge configuration.

DIGITAL DISPLACEMENT SENSORS

In addition to analog displacement-measuring sensors, there are a number of types of sensors which provide a *digital output*. These outputs are usually *binary* or are variations of a *binary code* which make the sensors ideal for operation with digital computers. Before examining the sensors themselves in detail, it is necessary to understand something of the codes and the circuits used.

BINARY CODED SIGNALS

A digital signal has two states, on and off. Normal calculations operate in units of 10, but if an arithmetic is developed in units of two, then it is possible to represent these units by two-state devices. For example, switches are on or off; a light is present, or there is darkness; a voltage may be on or off. The presence of a signal is often referred to as the one state, and the absence of a signal as the zero state.

If one switch is available, it can be made to represent two conditions. If two switches are available, they may be made to represent four conditions. The sum total of choices for three switches is summarized in Table 7.1.

Table 7.1 Binary Representation of Three Switch Sequences

2^2	2^1	2^0	State	Represent	Decimal
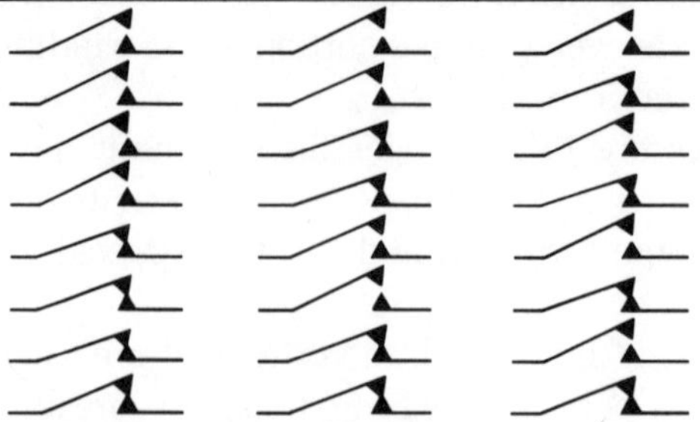			000	0	0
			001	2^0	1
			010	2^1	2
			011	2^1+2^0	3
			100	2^2	4
			101	2^2+2^0	5
			110	2^2+2^1	6
			111	$2^2+2^1+2^0$	7

Let each switch represent a power of two with the first switch on the right representing 2^0 which is decimal 1, the second 2^1 which is decimal 2, and the third representing 2^2 or decimal 4. With these three switches it is possible, therefore, to represent eight-decimal numbers or 2^3. This may be extended to four switches which may represent 2^4 or 16-decimal numbers, and so on.

Binary-Coded Decimal

It is possible with four switches to represent 16-decimal numbers. This may be deliberately restricted to ten, the remaining combinations being left unused. A group of four switches, therefore, could represent the numbers 0 to 9, a second group could similarly be used to represent the tens digits to 90, and a third group to represent the hundreds digits to 900. With 16 BInary digiTS or *bits* it is possible to represent decimal numbers in this way from 0000 to 9999. This method of coding is described as *Binary Coded Decimal*, usually abbreviated to BCD. If one of these groups is tabulated, it will be represented as shown in Table 7.2.

Table 7.2

BCD CODE				Decimal
0	0	0	0	0
0	0	0	1	1
0	0	1	0	2
0	0	1	1	3
0	1	0	0	4
0	1	0	1	5
0	1	1	0	6
0	1	1	1	7
1	0	0	0	8
1	0	0	1	9

The use of binary coded decimal is convenient when a digital display is required, such as with a digital voltmeter, which will be discussed in Chapter 16. When the output of this meter is to be read by a computer it is possible to connect the BCD lines directly to a digital computer. Figure 7.10 shows a schematic arrangement of the connections. A four-digit voltmeter would be connected to 16 bits of parallel binary input. Some type of amplifier is usually included to isolate the binary-coded decimal lines from external circuits. Where a BCD number is required to be displayed as a decimal, some type of *decoding circuit* is required.

The decoder illustrated is built up of circuits known as *logic gates*, originally made up of individual components, but now as integrated circuits.

BCD Decoders

Logic circuits are based on variations of two basic circuits, the AND gate and the OR gate (Figure 7.11a). The AND gate provides an output when a requisite number of signals are connected to the inputs. If three signals *A*, *B*, and *C* are at state 1, an output #1 will be obtained. If one of those signals is at state 0, the unit switches off, namely, the output turns to state 0. The OR gate switches on (state 1) if any, or all, of the signals *A*, *B*, or *C* is present (state 1).

Electronic circuits are more conveniently designed if gates produce an inversion. The first gate shown in Figure 7.11b provides an output with no signal input and changes to a zero state when *A*, *B*, and *C* are applied. This is known as a NOT

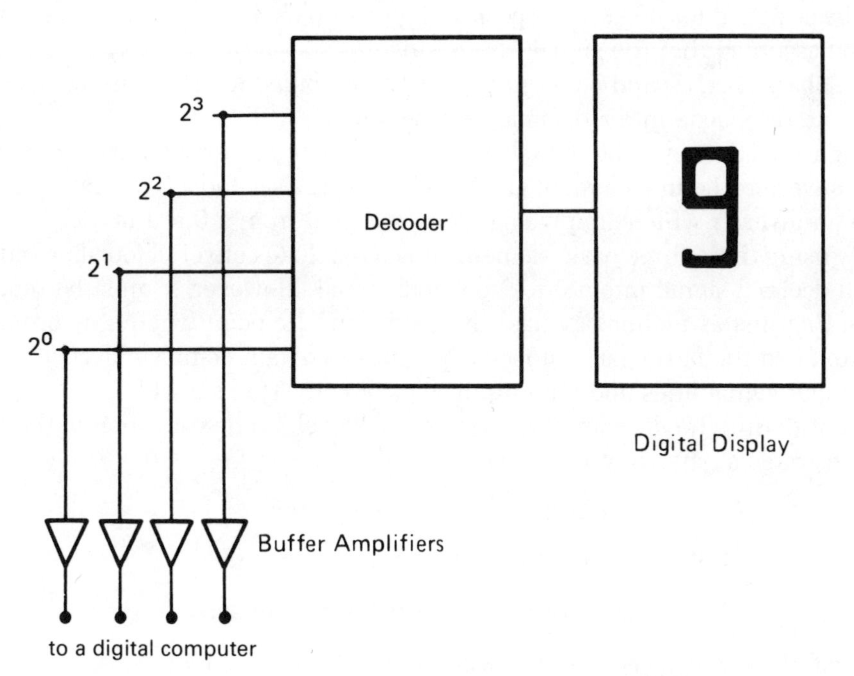

Figure 7.10
BDC Circuit Connection to a Digital Computer

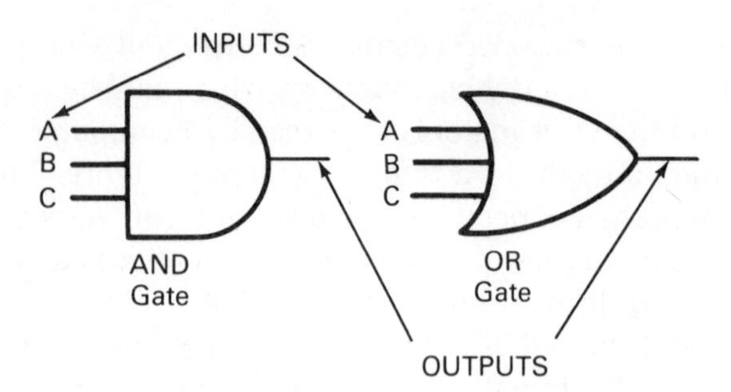

Figure 7.11a
Basic Logic AND and OR Gates

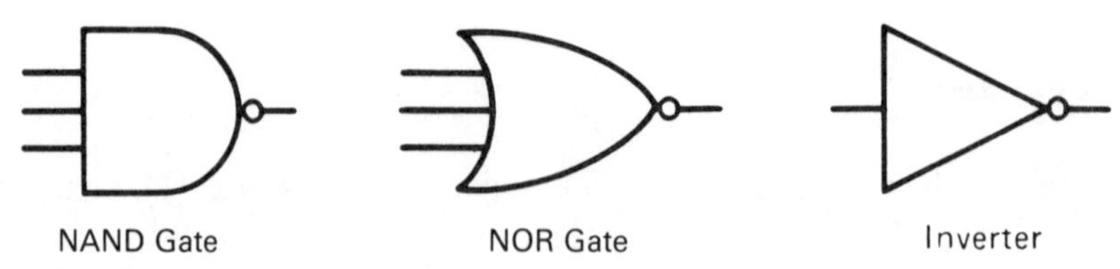

Figure 7.11b
Basic Logic Inverting Gates

AND or NAND gate and is distinguished from the AND gate by a circle on the output terminal. Similarly, the OR gate which is in a 1 state without a signal input and turns to a zero state when A, B, or C is applied, is known as a NOR gate. Gates may be designed for negative or positive input signals, but in these discussions it will be assumed that the signals are positive.

The basic NAND and NOR gates may be combined to build a decoder which turns a binary code into a decimal representation.

Figure 7.11b shows the basic logic elements required for a binary coded decimal converter; the first element is a NAND gate, the second a NOR gate, and the third an inverter which simply changes the signal from a 0 to 1 or vice versa.

By using these three basic elements it is possible to convert a four-line binary-coded decimal signal into a 0-to-9 decimal signal. However, it must be remembered that just as technology has advanced from the point where it is unnecessary to build the basic gates, it is also possible to obtain complete decoders with four input signal lines and ten output lines of converted signal.

Considering the decoder at the conceptual level, an investigation of the code demonstrates a pattern of repetitive states:

Decimal 0 is activated with no lines active.

Decimal 1, 2, 4, 8 are activated when one line is active.

Decimal 3, 5, 6, 9 are activated when two lines are active.

Decimal 7 is activated when three lines are active.

It follows that identical circuits may be used for decimal 1, 2, 4, and 8 which differ only in the way they are connected to the input lines. Similarly 3, 5, 6, and 9 will require identical circuits. Only 0 and 7 will be unique. Considering first decimal 0, the requirements are that there is an output from the circuit when no lines are active. The connection of a NOR gate across the four lines fulfills the basic requirements, and the circuit is shown in Figure 7.12. A table such as the one illustrated in Figure 7.12 demonstrates the logical operation of a circuit and is known as a *truth table.* With a decimal 1 circuit, line 1 must be activated when the other lines are zero, and activation of any other line must inhibit the circuit. The circuit in Figure 7.13 fulfills the necessary requirements.

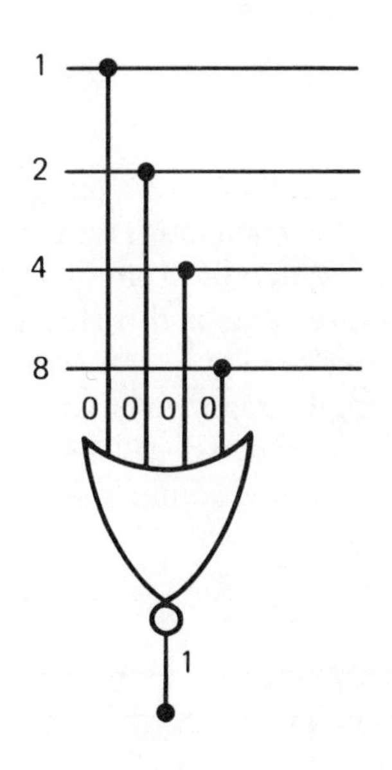

Truth table

Decimal	1	2	4	8	Output
0	0	0	0	0	1
1	1	0	0	0	0
2	0	1	0	0	0
3	1	1	0	0	0
4	0	0	1	0	0
5	1	0	1	0	0
6	0	1	1	0	0
7	1	1	1	0	0
8	0	0	0	1	0
9	1	0	0	1	0

Figure 7.12
Decimal 0 Decoder

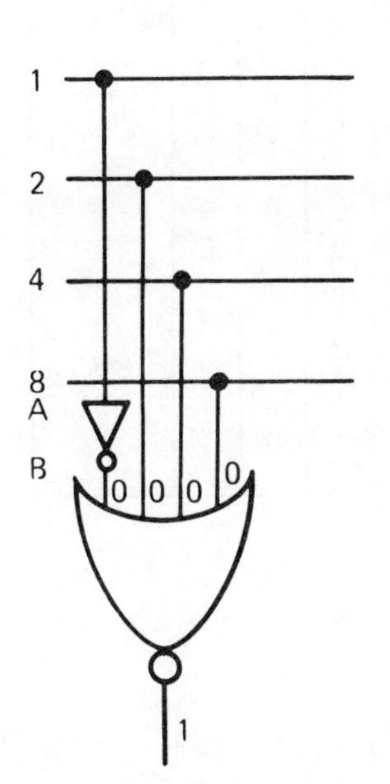

Truth table

Decimal	1		2	4	8	Output
	A	B				
0	0	1	0	0	0	0
1	1	0	0	0	0	1
2	0	1	1	0	0	0
3	1	0	1	0	0	0
4	0	1	0	1	0	0
5	1	0	0	1	0	0
6	0	1	1	1	0	0
7	1	0	1	1	0	0
8	0	1	0	0	1	0
9	1	0	0	0	1	0

Figure 7.13
Decimal 1 Decoder

If an inverter is inserted between the wire 1 (2°) and one of the NOR gate inputs, the output of the NOR gate remains in the 1 state when only that wire is at state 1. Similarly, for the remaining lines 4 and 8 the 1 state appears when the inverter is inserted in these lines and when a signal is applied. As far as decimal 3, 5, 6, and 9 are concerned, it is necessary to activate two lines and inhibit the output if either of the remaining two is active. Considering decimal 3 as typical, a circuit of the type shown in Figure 7.14 is required.

The truth table demonstrates an output only when the 1 and 2 lines are active. The figure decimal 7 is required to operate the 1, 2, and 4 lines and inhibit when 8 is active. The circuit and truth tables are derived in exactly the same way and shown in Figure 7.15. If these basic patterns are built up into a unit to give output from 0 to 9, the complete decoder takes the form shown in Figure 7.16

Serial BCD

Computers operate from a fixed number of input lines, 16 being common. This is convenient for the parallel connection of four BCD numbers. When BCD circuits are used, it is often economic both in computer input requirements and wiring requirements to switch each decimal digit into the computer serially.

Figure 7.17 illustrates an arrangement for obtaining signals from two binary to decimal decoders serially. In this arrangement, signals from two decades of binary coded decimals—units and terms—are connected to common output lines.

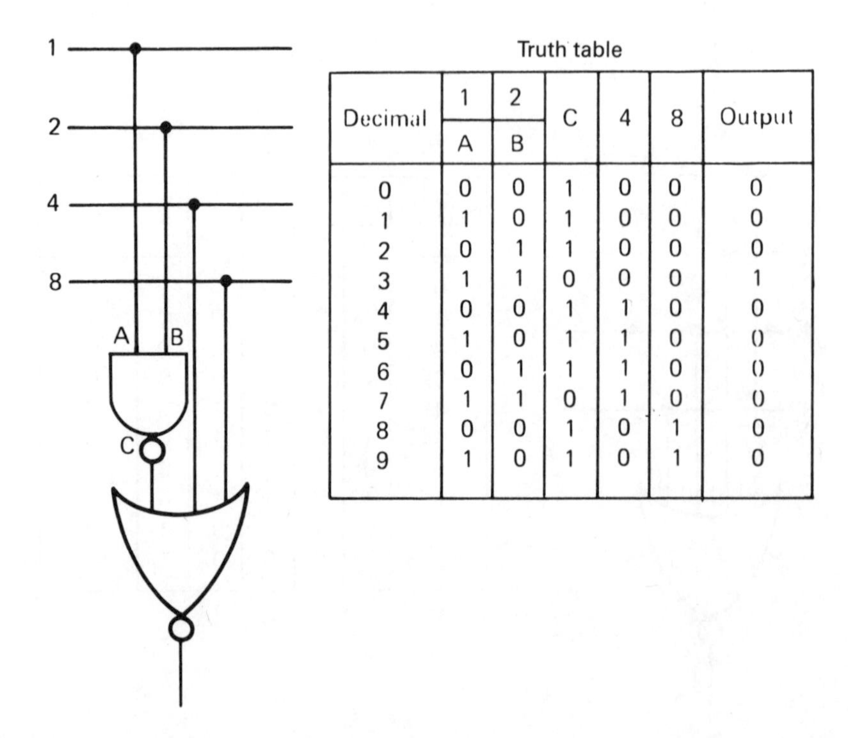

Truth table

Decimal	1 A	2 B	C	4	8	Output
0	0	0	1	0	0	0
1	1	0	1	0	0	0
2	0	1	1	0	0	0
3	1	1	0	0	0	1
4	0	0	1	1	0	0
5	1	0	1	1	0	0
6	0	1	1	1	0	0
7	1	1	0	1	0	0
8	0	0	1	0	1	0
9	1	0	1	0	1	0

Figure 7.14
Decimal 3 Decoder

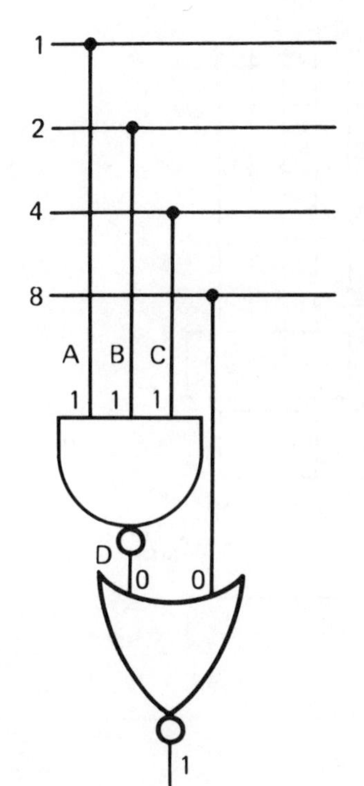

Truth table

Decimal	1	2	4	D	8	Output
	A	B	C			
0	0	0	0	1	0	0
1	1	0	0	1	0	0
2	0	1	0	1	0	0
3	1	1	0	1	0	0
4	0	0	1	1	0	0
5	1	0	1	1	0	0
6	0	1	1	1	0	0
7	1	1	1	0	0	1
8	0	0	0	1	1	0
9	1	0	0	1	1	0

Figure 7.15
Decimal 7 Decoder

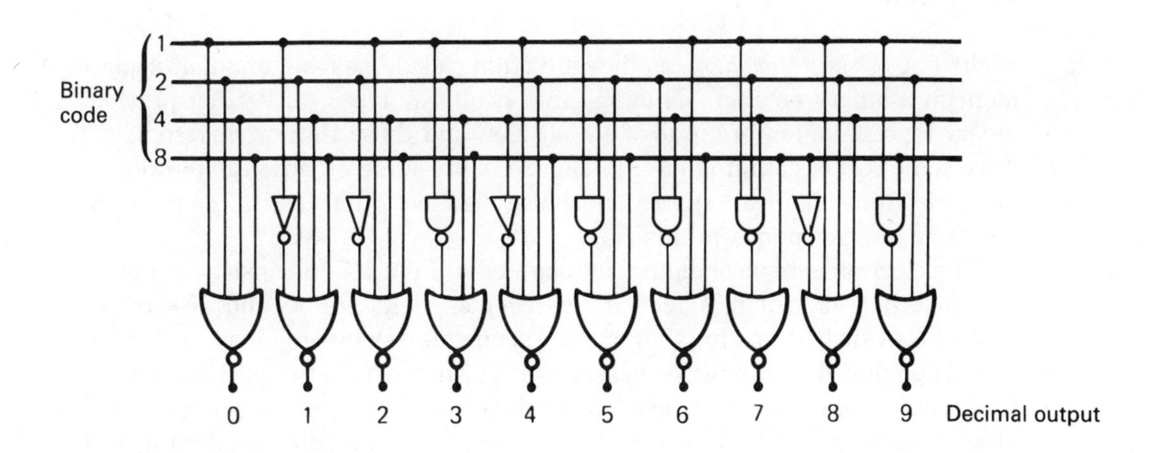

Figure 7.16
Binary to Decimal 0 to 9 Decoder

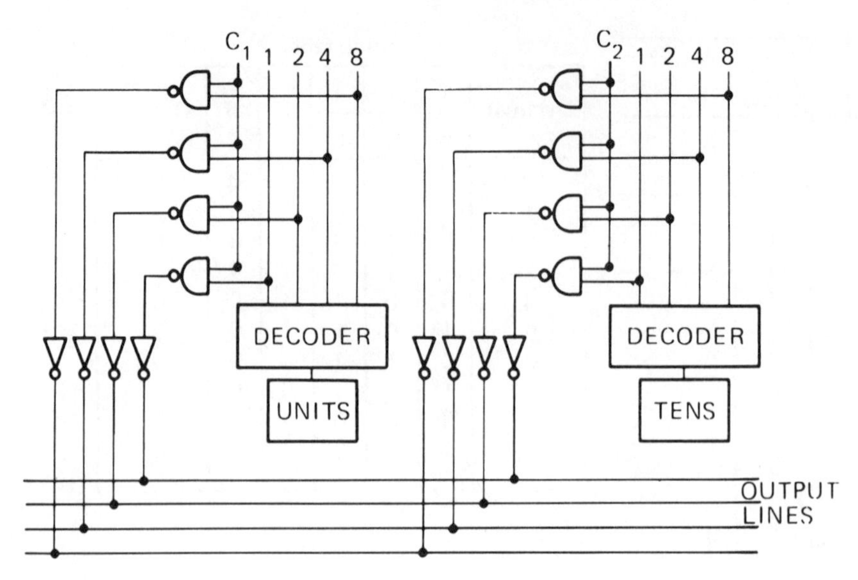

Figure 7.17
BCD Character Serializer

Considering first the units group: Regardless of the state of the binary-coded decimal input lines, no signal will appear on the output lines until a control signal is connected to C_1. With the input lines in their settled state, a signal on C_1 will reproduce the units on the output lines. If C_1 is removed and then a signal applied to C_2, the tens decade will appear on the output line. This method of switching allows economy of output lines but at the expense of increased reading time.

DIGITIZERS

Digitizers, or *shaft encoders*, as they are often called, measure angular displacement in a binary notation. They fall into two basic types, those that provide a binary representation of an absolute position and those that produce pulses to drive some sort of counting circuit. Digitizers providing an absolute position are often electromechanical devices and pulse digitzers often optical, or inductive, but there are exceptions to this rule.

The electromechanical digitizer illustrated in Figure 7.18 consists of a plastic disk, one face of which is "coded" with metal segments. As the disk rotates, contact is made between brushes and the segments, which produces a 1 state. A 0 state is produced as the brush passes over plastic areas between the segments. Early digitizers of this type used printed-circuit techniques to produce coded disks, which proved to be unsatisfactory. More recently, photoengraving techniques have produced much more reliable disks. The contact segments on these disks are gold-plated copper, usually mounted on melamine plastic, whose properties are carefully adjusted so that wear of the plastic and segments is matched.

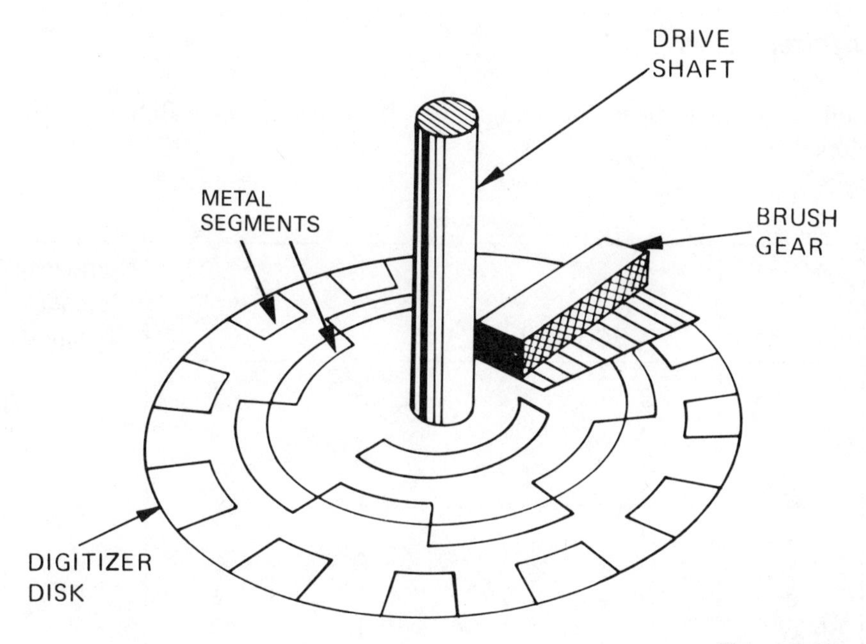

Figure 7.18
Electromechanical Digitizer

Brushes are usually made of precious metal alloys, with two brushes to a segment to overcome contact bounce problems.

It is possible, with disks of reasonable size, to produce codes from 000 to 999, but above this it is necessary to use multiple disks that are connected by gearing. The inclusion of gearing introduces backlash problems which can be overcome but, probably more significantly, in some circumstances increased torque is required to turn the disk.

The advantage of this type of digitizer is that its position indication is absolute. If the shaft is disconnected or if power supplies are lost, the original position can be set up again. The disadvantage, as with all electromechanical devices, is the finite life expectancy. This may be several million cycles, depending to a large extent on the speed at which they are operated. Most manufacturers provide a refurbishing service, and, provided spare units are available, mechanical digitizers may be used without undue problems.

One feature of the mechanical digitizer to be considered is the operational characteristic of the contacts. There is an obvious limit to the current that can be passed through these contacts, and in the short-circuit condition the current must be limited to that value. At the other extreme, deposits are liable to build up on the tracks over a period of time which will produce intermittent signals if low voltage and current are used. To counteract this the voltage might be raised to as high a level as possible, but the danger of shock hazards limits the desirable maximum to about 50 V. Some manufacturers recommend much higher voltages, but it is the author's experience that 50 V is usually quite adequate for reliable operation.

Digitizer Codes

If a binary code is used for a digitizer, the truth table in Table 7.3 may be developed.

Table 7.3

Decimal	BCD Code				Number of Contacts Changing
	8	*4*	*2*	*1*	
0	0	0	0	0	
1	0	0	0	1	1
2	0	0	1	0	2
3	0	0	1	1	1
4	0	1	0	0	3
5	0	1	0	1	1
6	0	1	1	0	2
7	0	1	1	1	1
8	1	0	0	0	4
9	1	0	0	1	1

The final column shows the total number of contacts that change between adjacent numbers. Between 1 and 2, for example, two contacts change state, and between 7 and 8 four contacts change. This is often not considered a desirable feature, as it may lead to ambiguity in reading. To offset multiple contact changes, various codes have been designed that change only one contact in moving from one decimal digit to the next. One of the early codes of this type was the *Petherick code*, which is similar to the better known *Gray code* shown in Table 7.4.

Table 7.4

Decimal	Gray Code				Number of Contacts Changing
	D	*C*	*B*	*A*	
0	0	0	1	0	1
1	0	1	1	0	1
2	0	1	1	1	1
3	0	1	0	1	1
4	0	1	0	0	1
5	1	1	0	0	1
6	1	1	0	1	1
7	1	1	1	1	1
8	1	1	1	0	1
9	1	0	1	0	1

Variations in codes are numerous, but when a binary code or its derivative is used, a compatible decoder is also required.

An argument that has merit states that it is better to allow multiple changes of contacts and use a decimal code, which eliminates the necessity for a decoder. The disadvantage lies in the fact that decimal contacts need more tracks on a digitizer's wheel than those of the binary representation. This disadvantage, however, is somewhat reduced by using a technique known as complementing. Consider the track shown in Figure 7.19. With this system two common tracks *A* and *B* are used, and the readout is split into two halves. The digits 0 to 4 use the first five steps, with a common return wire from *A*. As the brushes move from the 4 to the 5 digit, the brush on *A* runs off the end of the track, and *B* is picked up by an alternative common brush. This allows the digits 5 to 9 to use tracks in the same relative position as 0 to 4. Digitizers are often connected to numerical displays where digits 0–9 are made up of lamp filaments stacked one behind the other in a glass envelope. If this digitizer is connected to one of these displays, the wiring will be as shown in Figure 7.20.

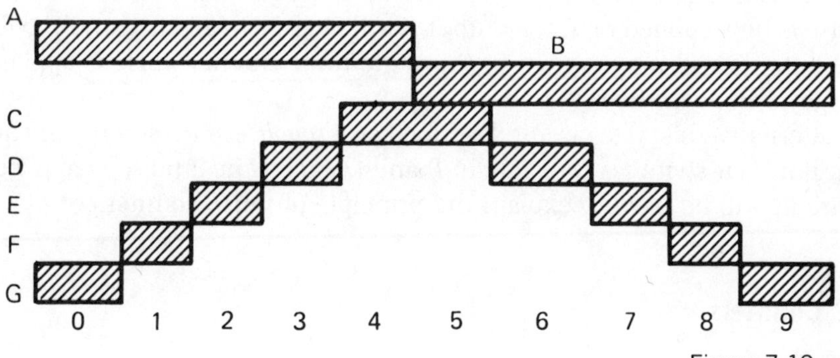

Figure 7.19
Decimal Digitizer Track Layout

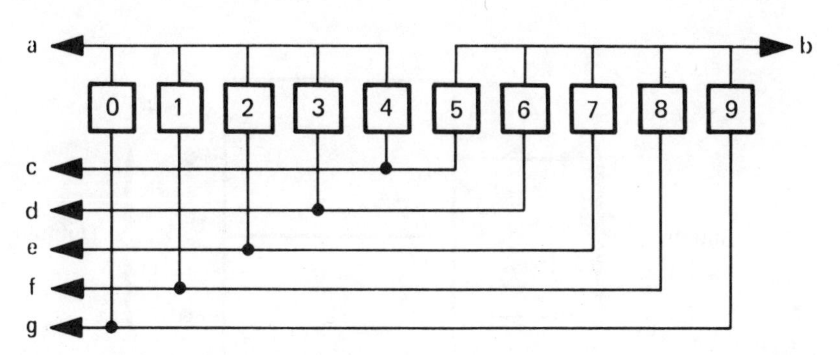

Figure 7.20
Decimal Digitizer Display

OPTICAL DIGITIZERS

The simplest optical digitizer is a toothed wheel with a photocell on one side and a light on the other. Each time the beam of light is broken by a tooth the photocell circuit changes from a 1 to a 0 state. These signals are trains of pulses and in no way reflect the digitizer position. It is necessary, therefore, to keep a count of the pulses from some arbitrary point and to be able to add to or subtract from the "count," depending on the direction of the disk. Such counters are known as UP/DOWN counters and may be purchased commercially to provide a variety of output codes, the most common being BCD, binary, or decimal.

Pulse Counters

The operation of pulse counters is based on a bistable circuit often known as a *flip-flop*. These are usually purchased as a complete integrated circuit. One type is shown schematically in Figure 7.21. A bistable circuit is one that has two stable states. When the circuit is first switched on, either *P* or *Q* can be in the 1 state but not both. The application of a pulse to *B* conditions the circuit so that the output *P* is in the 0 state and *Q* is in the 1 state. This input line *B* is known as the *reset line*. If a pulse is now applied to *A*, the output *P* will change to a 1 state and *Q* will go to *O*. The outputs will now remain in this state until another pulse is applied to *B* when they will again be reversed.

An alternative bistable circuit described as a *toggle circuit* is a modification of the flip-flop that switches the state of *P* and *Q* each time a pulse is applied to *A*. This circuit will be used to explain the principle of pulse counters.

UP Counters

A simple UP counter is shown in Figure 7.22 which converts serial pulses into binary code. Consider a pulse applied to the reset line. This sets all the toggles *T*1 to *T*4 so that the outputs *A*, *B*, *C*, and *D* are all set in the 0 state. If now a pulse is applied to *T*1 count line, *T*1 switches, and the output *A* turns to a 1 state. Simul-

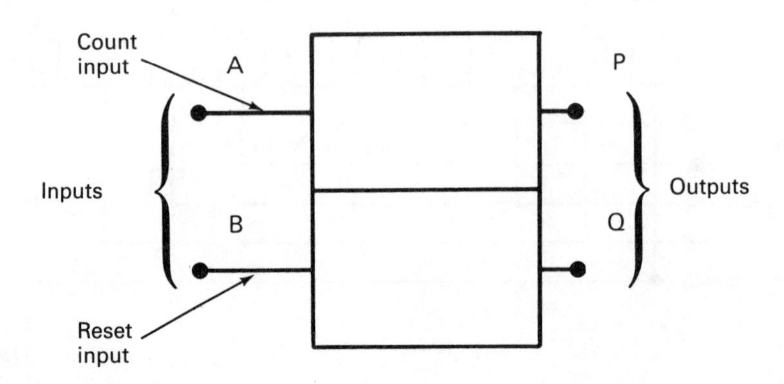

Figure 7.21
The Bistable circuit

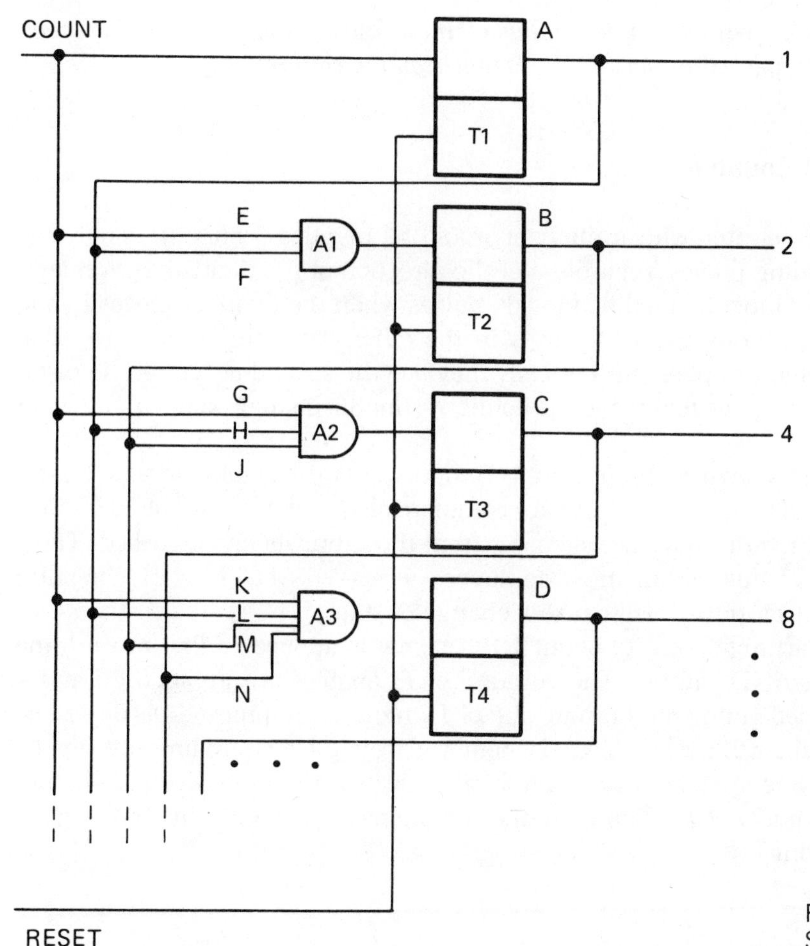

Figure 7.22
Synchronous UP Counter

taneously, the pulse being applied to $T1$ is also applied to the inputs E, G, and K on the AND gates, but because there is no other input, the remaining toggles are not switched. A second pulse applied to $T1$ will again change its state to 0, but because there is a delay between the input pulse arriving and a change taking place, AND gate $A1$, now briefly has two inputs applied, which turns $T2$ into a 1 state. The lack of signal on J and M inhibits the following toggles. The arrival of a third pulse will again turn $T1$ into a 1 state but the inherent delay in the circuit will leave the subsequent stages unchanged. After three pulses then, lines A and B are in a 1 state.

If a fourth pulse is applied, it will turn $T3$ into a 1 state, because signals are already present on H and J of $A2$. It will turn $T2$ into a 0 state because a signal is present on F, and $T1$ will be turned off.

If this process of reasoning is continued, it can be seen that the lines A, B, C, and D are switched in a binary sequence. The binary number may be extended by adding another toggle to the circuit and an AND gate with one more input than the previous one. Because the pulse is applied to all the toggles simultaneously, the counter is known as a *synchronous counter* and operates at a higher speed than its counterpart, the "*serial,*" or "*asynchronous counter.*"

UP-DOWN Counters

To use an UP counter with a digitizer would be pointless, since it would only respond by adding pulses, regardless of the direction of the digitizer movement. Such a counter must be modified to add pulses when the digitizer moves in one direction and subtract when it moves in the other. The circuits for reversible counters become complex, but basically they use the second output of the bistable and auxiliary gates to reverse the count. A simplified single stage is shown in Figure 7.23.

The toggles shown in Figure 7.23, comprising two bistable circuits *T1* and *T2* and two AND gates *A1* and *A2*, are commercially available integrated circuits that have a different input arrangement from those previously described. The *J* and *K* inputs of this circuit must be active—at state 1—before a trigger pulse called a CP (clock pulse) will cause a change of state.

With this arrangement, to count UP, a signal is applied to the forward line which enables AND gate *A*1. The output *P* of *T*1 enables or inhibits the *J* and *K* inputs of *T*2, depending on the state of *P* of *T*1 and a count proceeds as described for the UP counter. If now the signal is removed from the forward line and applied to the reverse line, *A*2 takes over control and activates or inhibits *K* of *T*2, depending on the output *Q* of *T*1. These are the conditions required for the first stage of reverse counting.

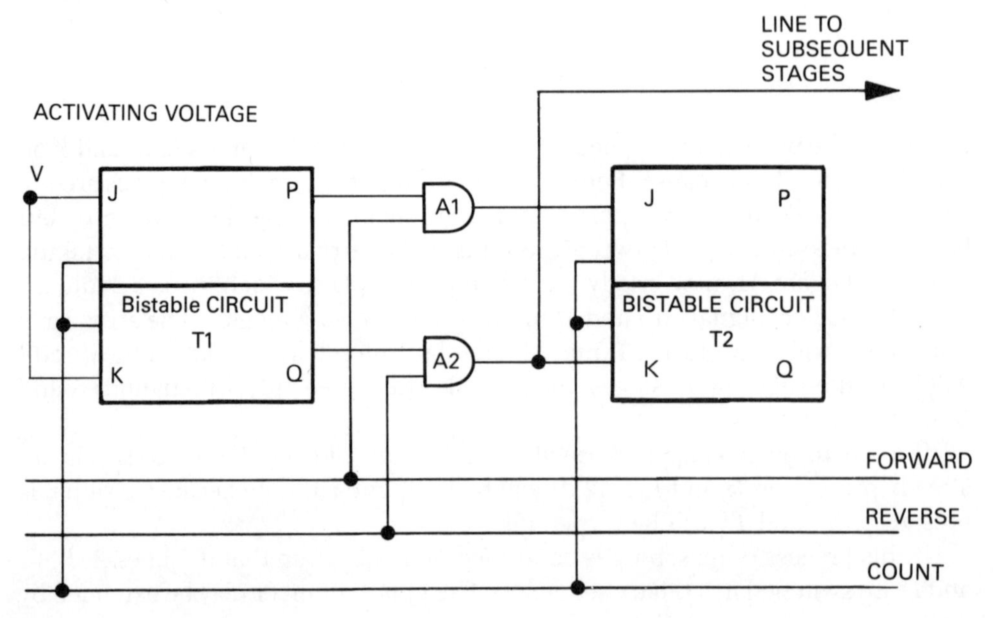

Figure 7.23
UP-DOWN or Forward-Reverse Counter

Direction Sensitive Optical Digitizers

If UP-DOWN counters are to be applied to digitizers, the digitizer must not only produce count pulses, it must provide a means of activating the forward or reverse lines. There are a number of methods of obtaining this signal: one provides the digitizer with two tracks that are separated by a 90° phase relationship.

Suppose a digitizer disk made of glass contains a pattern of dark patches, around its perimeter as shown in Figure 7.24. As these dark patches pass between the photo cells P_1 and P_2 and a light source, they will cause a pulse to be produced. If the disk is moving in the direction of the arrow, light will always be cut off from P_1 before P_2. In the reverse direction the light will be cut off from P_2 before P_1. Electronic circuits are available that operate a bistable, the output of which depends on this phase relationship.

Figure 7.25 is self-explanatory, but it should be pointed out that the electronics are, in fact, more intricate than have been so far implied. The outputs of P_1 and P_2, for example are not sharp ON/OFF signals, but vary in a way that more nearly approaches a sinusoidal variation. Consequently, these crude signals must be conditioned and turned into sharp rise pulses by what are known as *squaring circuits*.

It is possible with a single glass disk to obtain finer resolution than with an electromechanical digitizer, and this resolution can be improved several orders by

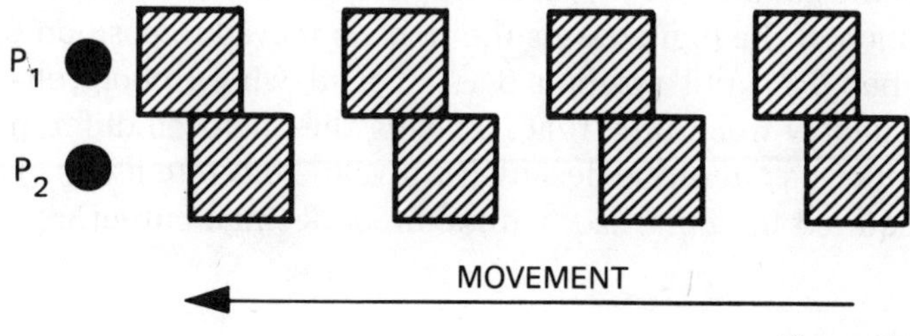

Figure 7.24
Phase-Sensitive Optical Digitizer

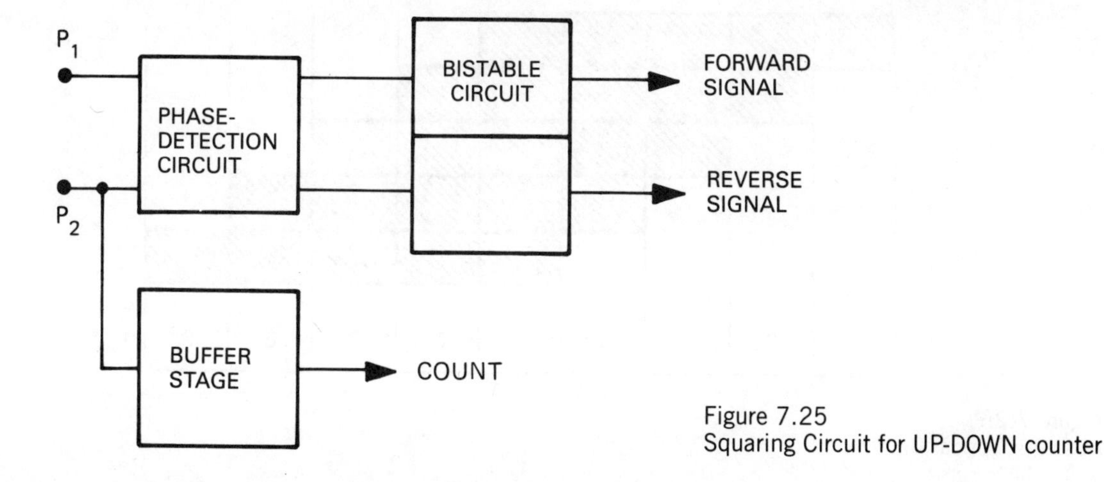

Figure 7.25
Squaring Circuit for UP-DOWN counter

using *optical diffraction gratings* on two separate digitizer disks. When two such gratings are suitably arranged, interference patterns called *moire fringes* are produced of a wavelength equivalent to the pitch of the grating. It is possible to make gratings of very fine pitch and the resolution of a digitizer is usually restricted by mechanical constraints rather than the grating. Optical digitizers of this type measure relative movement, and if there is a failure of the associated circuits which maintain the measurement of absolute position, a setting-up procedure must be operated. The mechanical digitizers described measure absolute movement and maintain their position indication even though associated circuits may fail. To their advantage, these optical digitizers have longer life expectancy than their mechanical counterparts, and they may be driven at much higher speeds. They can be also made with much higher resolution.

Absolute Optical Digitizers

Because of the limitations of pulse-counting circuitry, attempts have been made to combine the advantages of the absolute measurement of the mechanical digitizer with the reliability of the optical. Suppose a track shown in Figure 7.26 is laid down on a glass substrate and suppose the output of the photocells P_1 to P_5 is electronically summed. As the track moves in the direction of the arrow, first P_1 is covered, then P_2 and so on, until in position 5 all are hidden from the light source. The output of the adding circuit, therefore, decreases in a series of discrete steps, with the position of the digitizer. As the track is moved to position 6, P_1 is again uncovered, then P_2, until position 0 is reached where photocells are all uncovered. By suitably weighting, that is, giving the photocell different output levels, the output of its circuit be designed with nine discrete levels. These are the conditions required for a digitizer with a direct decimal output.

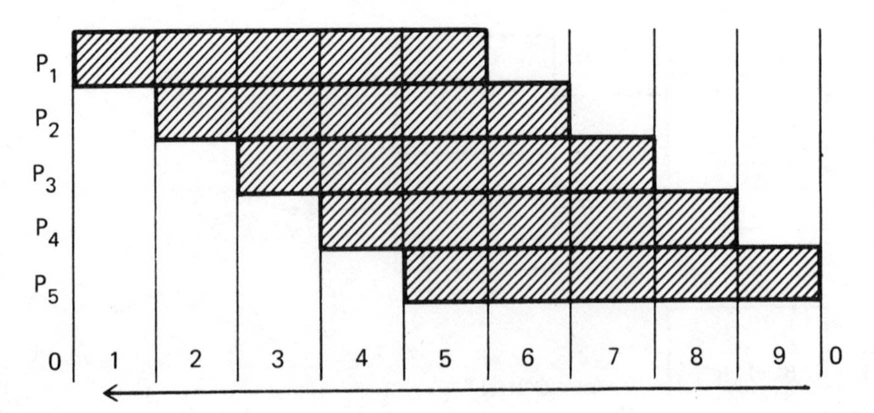

Figure 7.26
Absolute Optical Digitizer

INDUCTIVE DIGITIZERS

Inductive digitizers are not a very common form of digitizer, but they are useful in severe environmental conditions. Ferrous tracks laid down on a substrate are used to change the inductance of pick-off coils. The resolution is not very high, but within their limitations they will survive conditions of temperature which destroy the other types.

LINEAR DISPLACEMENT DIGITIZERS

The discussion of digitizers up to this point has emphasized their use to measure rotary motion. They can, of course, be used equally well to measure linear displacement. Mechanical digitizers with their discrete-track requirement become bulky if fine resolution is required for a long scale and are, therefore, usually restricted to linear displacements of no more than 10 inches (0.25 meter). Provided the effects of count loss, in the event of power failure, can be accepted, there is no apparent restriction on the length of an optical digitizer, other than the physical strength of the substrate and the problems involved in mounting.

If counting techniques are to be employed and if resolution is restricted, it is possible to use inductive devices with a pickup coil and a toothed armature (Figure 7.27).

As the armature moves relative to the pickup coil, the inductance of the circuit fluctuates and may be converted into an electrical signal which may be used to drive a counter. A second coil may be used to provide the direction signal.

Toothed armatures are expensive to manufacture because each slot requires milling. It is possible, however, to use a threaded rod as an alternative, and these are less expensive to produce.

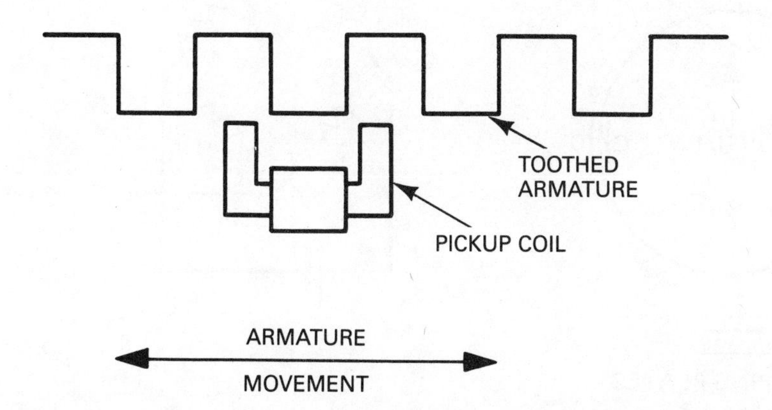

Figure 7.27
Inductive Linear Digitizer

MAGNETIC TAPE DIGITIZERS

A magnetic recording tape imprinted with a series of equally spaced 1 and 0 signals and a standard playback head form the basis of a useful digitizer. The flexibility of recording tape and the ease with which it may be glued to a contour make it adaptable to applications in which other types of digitizers may be difficult to apply. Resolution of such a device is high and may discriminate about 0.002 inch (0.05 mm). The tape and the head, however, are somewhat fragile for an industrial environment, and if used, must be protected from contaminants and magnetic fields.

HOOP PICKUPS

A very simple device for measuring displacements up to about 0.1 inch (2.5 mm) is the *hoop pickup*. These are very easily made for measurements where space is no problem and may be produced in considerable quantities at short notice. Strips of aluminum alloy or steel about 20 inches (0.5 meter) long and about 0.5 inch (12 mm) wide are drilled at each end and in the center with a clearance hole to accept a small bolt. The strips are then formed into hoops, with the ends bolted where the holes overlap. A bolt through the center hole provides a second mounting fixture. Strain gages are used to sense the deformation of the hoop and are wired as shown in Figure 7.28. The hoop will obviously introduce some spring loading to the parts producing the movement, and this should be considered in the design. An alternative to the hoop, which may be more appropriate in certain instances, is a strain-gaged cantilever, which provides a reasonably linear output for small deflections.

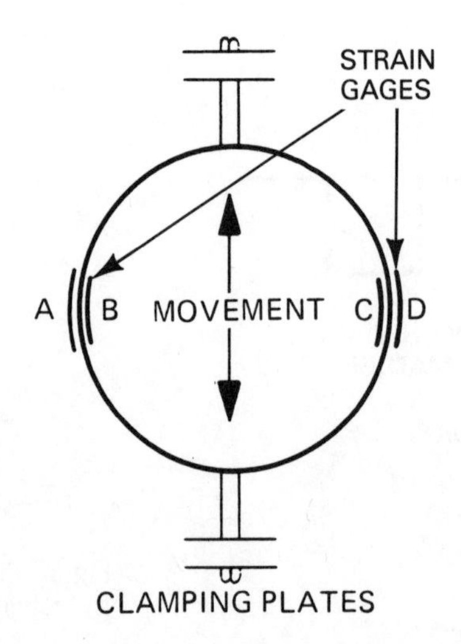

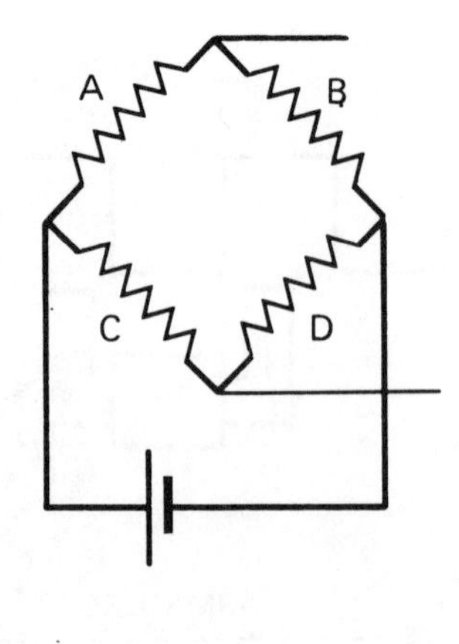

Figure 7.28
Hoop Pickup

CHAPTER 8

LOAD MEASUREMENT

The variety of load sensors, many of which are custom built to meet particular applications is almost unlimited. These range from modified strain-gaged pressure sensors, with ranges of a fraction of a gram, to load cells, with ranges of many tons. They may be conventional weight scales with sensors added to give electrical outputs, or they may be the supports of a silo strain-gaged to turn it into a self-contained weight scale. This chapter, at best, can only contain a representative cross section and provide the briefest outline of their application.

LOAD CELLS

Strain gages play a substantial role in load measurement. The small changes in dimensions of a structure necessary to give a working signal make gages very attractive for this application. High signal levels are available with semiconductor gages, and deposited gages offer a line of development with the prospect of measurements of even higher accuracy and stability. Load measuring devices using strain gages as secondary transducers are often described as *load cells*.

Cantilever Load Cells

There are two common designs of load cell, those that employ strain-gaged cantilevers and those that employ strain-gaged tensile or compressive members.

Considering each in turn, the cantilever is probably of most use in restricted spaces and may be designed to measure very small loads, while still being capable of accepting gross overload. Figure 8.1 illustrates the basic design of such a device.

Load is applied at one end of a simple cantilever that has been clamped firmly at its opposite end. Strain gages in a four-arm bending configuration allow maximum sensitivity to be obtained from the bridge, and damage is prevented by an overload stop. The range of the sensor is determined by the stiffness of the cantilever; with thin-shim steel the range would be a few grams; with a substantial beam, a range of several tons is feasible. Both extremes of range have their problems. Very thin cantilevers are stiffened locally by the attachment of gages, and in these situations deposited gages may be used to advantage. Even then care must be taken to ensure that leads do not interfere with the natural operation of the beam. At the other end of the range, high coupling moments at the fixed end of the beam are experienced which tend to distort the case appreciably.

Tensile Load Cells

Tensile load cells, as illustrated in Figure 8.2, are widely used with lifting machinery. They are expected to receive rough treatment and are designed accordingly.

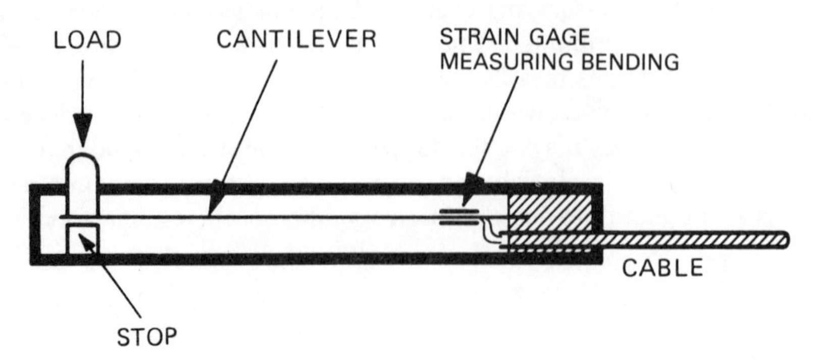

Figure 8.1
Simple Cantilever Load Cell

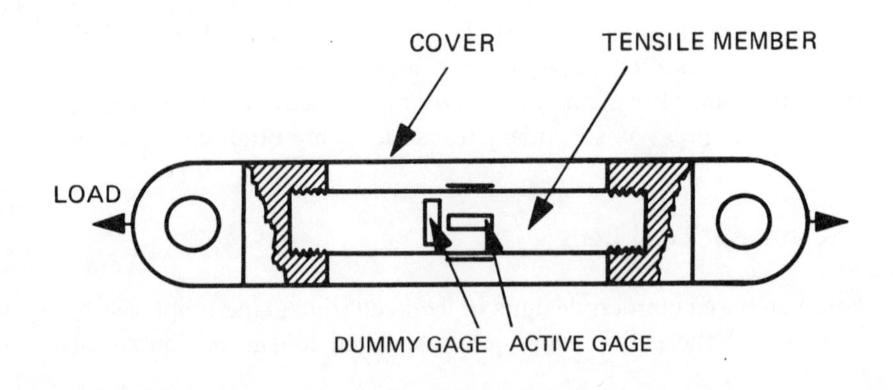

Figure 8.2
Tensile Load Cell

Load is transmitted to the tensile member by two end fittings and to increase its sensitivity the tensile member is sometimes made with a tubular cross section. An end-load-measuring strain-gaged bridge is employed, and usually the dummy gages are bonded to the specimen at right angles to the line of principal strain. Dummies fitted in this fashion provide good temperature compensation and are known as *transverse dummies*. Most gages have some cross sensitivity and, in addition, the diameter of the tube will be reduced under load because of Poisson's ratio effects. Transverse dummies, therefore, are partially active gages. This feature is usually of little consequence providing the sensor will be calibrated before use.

The tensile member is protected by a tubular cover fastened to the end fittings. To some extent this too becomes a load-bearing member, but its dimensions usually are such as to make little difference to the sensitivity of the whole unit.

Up to this point the description of the sensor has been one of a rugged device that can withstand mechanical shock and overload admirably. It has one weak point: The strain gages must be wired to equipment, possibly in a crane driver's cab. This cable is very exposed and liable to damage. Basically, the designer has two choices. The cable can be made changeable on site. In this case a waterproof plug and socket or terminal block must be provided. Alternatively it is possible to connect the cable directly into the case of the cell through sealed glands. If this course is followed, it is necessary to maintain an adequate supply of spare load cells.

Compressive Load Cells

Compressive load cells are often used in weighing platforms and structures which may be supported by a compressive member. The one illustrated in Figure 8.3 is a typical design. These compressive load cells employ a flexible diaphragm

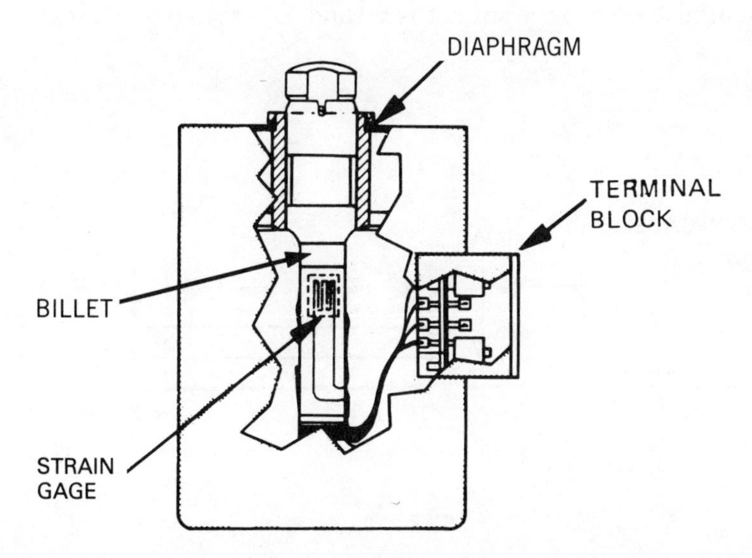

Figure 8.3
Load Cell. (BLH Electronics, a Bofors Company)

restrained by a compressive member usually described as a *billet*. The billet strain gaged to measure end load is usually fitted with transverse dummies. Cases are substantial and the cell is a very rugged device. Accuracies claimed for strain-gaged load cells are often surprisingly high, but it should be noted that in some designs these accuracies only apply so long as the load is acting directly along the line of the billet.

LOAD CELL DESIGN

Design of Cantilever-Type Load Cells

The calculation of the size of a cantilever to produce a reasonable working strain is based on the formula:

$$\frac{M}{I} = \frac{\nu}{y} \tag{1}$$

where

M = the bending moment in the beam

I = the second moment of area round the neutral axis

ν = the surface stress at the fixed end

y = the distance of the neutral axis from the surface.

Consider the cantilever described in Figure 8.4 whose length from the point of application of the load to the center of the strain-gaged element is d. If the load L is applied to this cantilever, the bending moment at the gage element centers will be:

$$M = Ld$$

The distance of the neutral axis from the surface is $t/2$ and for a rectangular cross section:

$$I = \frac{wt^3}{12}$$

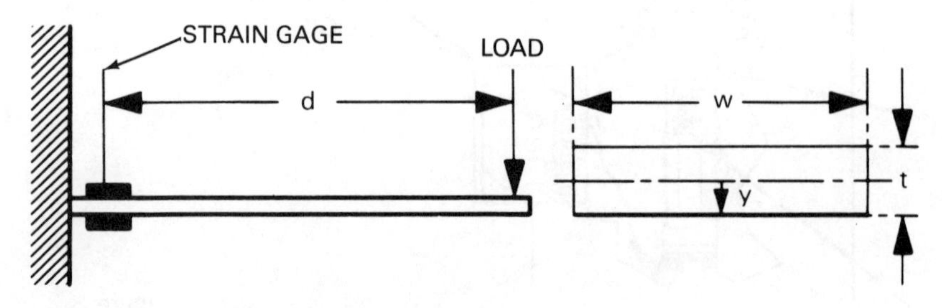

Figure 8.4
Cantilever Load Cell Design

where

w is the width and t is the thickness of the beam.

The surface stress of the cantilever due to the load is E multiplied by the surface strain, where E is Young's modulus for the material.

Substituting in (1):

$$\text{Surface strain} = \frac{My}{\epsilon I}$$

As with many engineering problems, the design now becomes a series of successive approximations. Presumably the maximum load is fixed and the maximum surface strain experienced by the gages is decided by the manufacturer's recommendations, but d, w and t are all variable to some degree. The final design will depend on the geometry of the load cell and the allowable deflections of the load end of the cantilever. This is given by the formula:

$$\delta = \frac{Ld^3}{3EI}$$

where δ is the deflection at the load end of the beam. The bridge configuration to provide maximum sensitivity will be a four-arm closed bridge.

Design of Sensing Elements for Compressive Load Cells

If cells use tensile and compressive members to carry all the load in the billet, then the stress-strain relationship applicable is:

$$\epsilon = \frac{\text{stress}}{\text{strain}}$$

$$\epsilon \text{ for steel} \simeq 30 \times 10^6 \text{ lb/in}^2 \ (2.1 \times 10^6 \text{ kg/cm}^2)$$

$$\text{and stress} = \frac{\text{load}}{\text{cross-sectional area of billet}}$$

If a maximum working strain is assumed, a billet of suitable dimensions can be calculated.

LOAD CELLS WITH DIFFERENTIAL TRANSFORMER SENSORS

The use of the LVDT as a secondary transducer can be extended to load-measuring devices. Simplest of these is where the load deflects a spring and the LVDT is used to measure the deflection. There are a number of variations on this basic design, one of which is shown in Figure 8.5. In this cell the load is balanced by a coil spring selected to match the range requirements. The weighing platform is constrained vertically and horizontally by leaf springs, thus eliminating bearings and pivots with their accompanying frictional problems. Deflection of the load

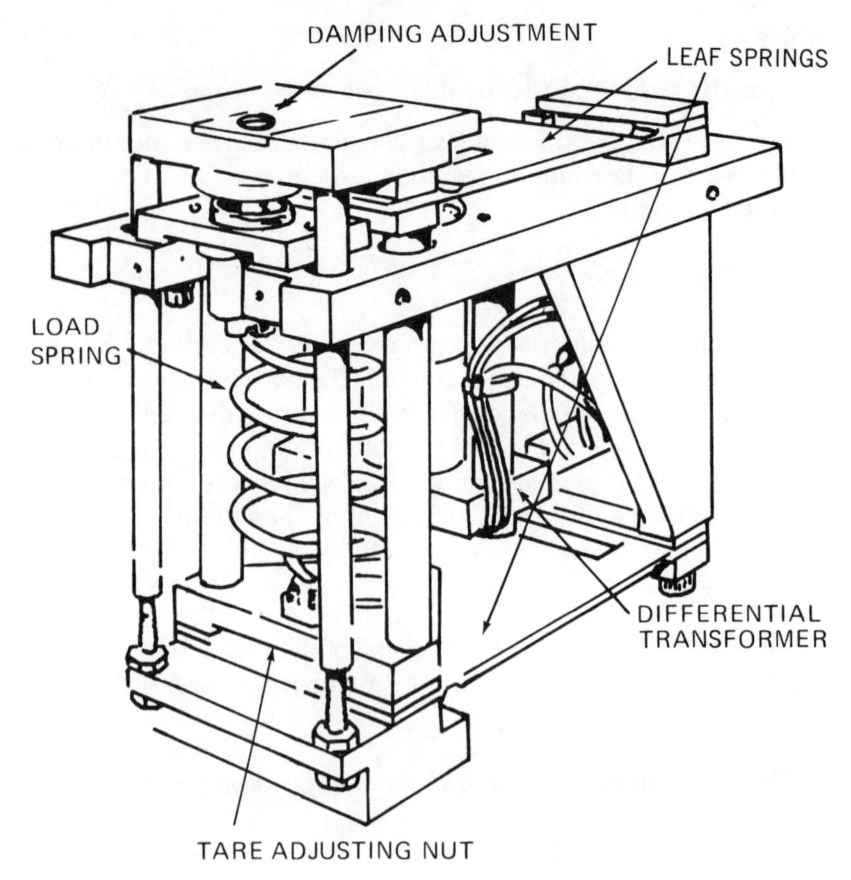

Figure 8.5
Weighing Cell.
(Hunting Electro Controls)

spring is sensed by a LVDT, the output of which is proportional to the load applied. Load cells of this type are designed for lower load ranges and may have ranges as low as one ounce (28 g). In devices using low-stiffness springs, sustained vibration becomes a problem, and it must be eliminated if an accurate measurement is to be made. Also when using springs, care must be taken in the selection of material. Some suffer *strain hysteresis* analogous to magnetic hysteresis. If this is high, the resolution of the weigher will be restricted accordingly.

Spring Vibration

It is a well-known feature of spring systems that if they are disturbed, they will continue to oscillates for some time. If the spring system shown in Figure 8.6 is extended by an amount $-X$ and then released, the mass will accelerate back to its zero datum and then decelerate until it reaches a point X above the datum. The mass will then begin to move downward again, reaching a point somewhat short

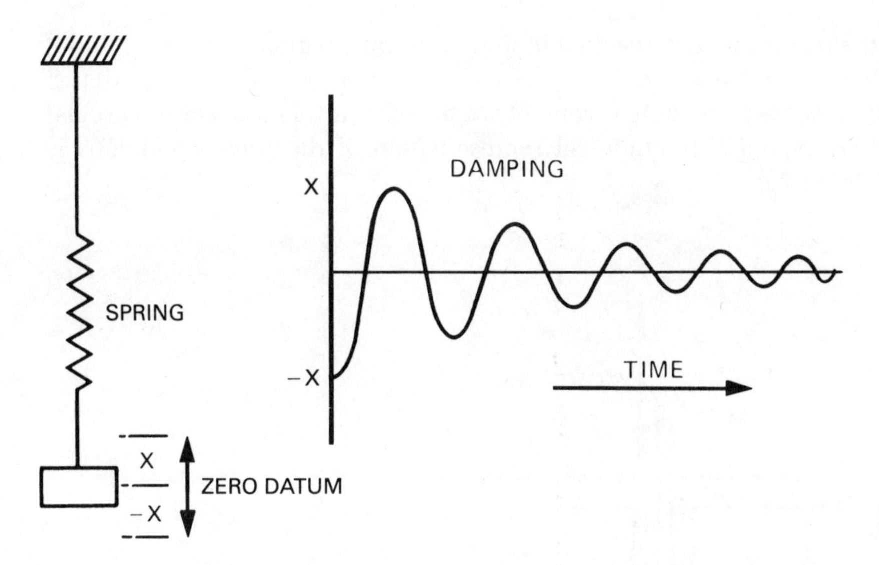

Figure 8.6
Vibration of a Spring

of $-X$ before again reversing. Each subsequent cycle will be smaller in amplitude, but the time for a complete oscillation will remain constant. Oscillating motion of this nature is called *simple harmonic motion*, and the fact that the oscillations are decreasing causes it to be described as *damped simple harmonic motion*. The rate at which the amplitude decreases is a measure of *damping*, and the number of oscillations per second is known as the *natural frequency* of the system. The amplitude of an oscillating spring decreases for two reasons. The mass is moving against the resistance of the air surrounding it and this absorbs some of the energy. In addition, structural distortions inside the spring material itself absorb energy. If this damping is small, the application of very small disturbing forces each time the mass reaches its extremity of travel will maintain the oscillation.

Because of its importance, the subject of vibration is discussed at length in Chapter 13.

Vibration Dampers

To eliminate spurious oscillation of spring balances, some form of artificial damping that will resist the oscillatory motions but will not affect its rest position must be introduced. Such devices are called *dampers* or *dashpots*. A simple version is shown in Figure 8.7.

The cylinder of the damper illustrated is filled with oil and the balance is connected to an immersed piston. If the piston changes its vertical position, oil must be displaced either round the edges of the piston or through holes drilled in its face. The effective area of the holes can be varied, thus varying the effective resistance of the oil to movement of the piston. When the effective piston area is insufficient to resist an oscillation, the system is said to be *underdamped*. In the optimum, the rest position is reached in minimum time without overshoot and it is then said to be *critically damped.* When the piston area is so large that it prevents the rest position being reached for a period equivalent to several oscillations, it is *overdamped.* A graphical representation of the three conditions is shown in Figure 8.8.

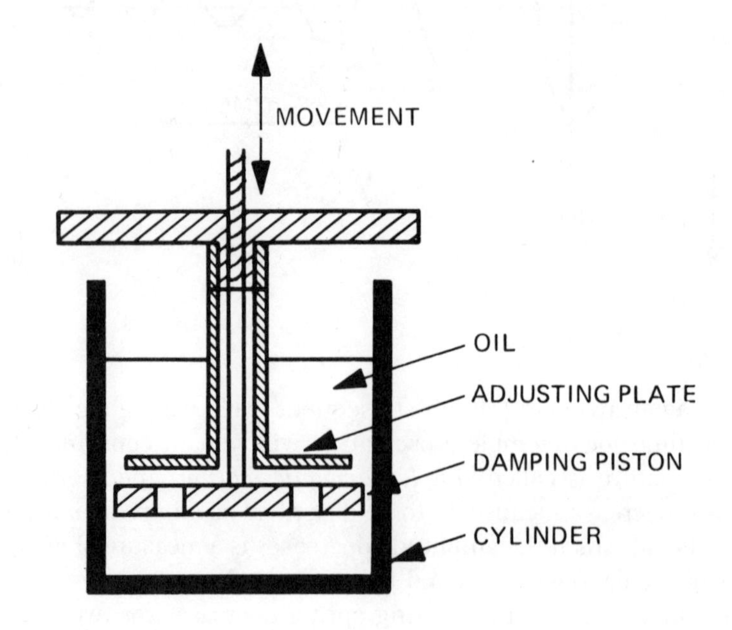

Figure 8.7
Vibration Damper

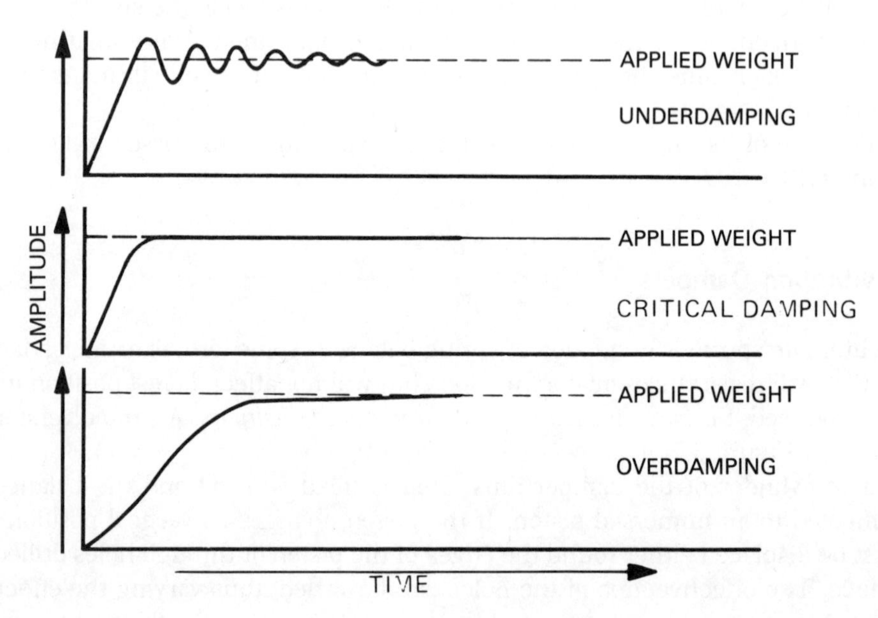

Figure 8.8
Conditions of Damping

Regardless of damper adjustment, a weighing platform will eventually reach its true loaded position but when it is either under- or overdamped, some time is lost before a measurement can be made. The difference in time between a load being applied and when a meaningful measurement can be made is described as the *settling time* of the balance. When individual loads are being measured infrequently, the degree of damping is unimportant. The common chemistry balance, for example, is practically undamped, whereas many scales in stores are overdamped. In fact, a degree of overdamping in some situations is a protection against shock loading. In a production line, where individual items must be weighed in the minimum time, critical damping must be achieved if the load cell is to be operated at its optimum speed.

Unfortunately, the adjustment of damping is not quite as straightforward as may be at first imagined, but depends upon the type of product being weighed. Consider the weighing system in Figure 8.9a and suppose the weighing platform is pressed down and released. When the damper is set to an underdamped adjustment, the system will oscillate for some time. The damper may be adjusted by a series of repeated operations until critical damping is achieved. Suppose now an item of substantial weight is dropped onto the platform; this weight combines with the weight of the weighing mechanism to lower the natural frequency of the system. Consequently, the system damping will be no longer critical. Figure 8.9b illustrates a powder such as flour being weighed. In this instance the platform moves downward under the first impact, but its return movement is resisted by the stream of flour still being added. In this situation the flour itself is acting as a damper. A scale that has already been set for critical damping in a static condition will now be overdamped. When a weighing system is to be used "in-line" the damper must be adjusted in the dynamic situation with the materials it will be used to weigh.

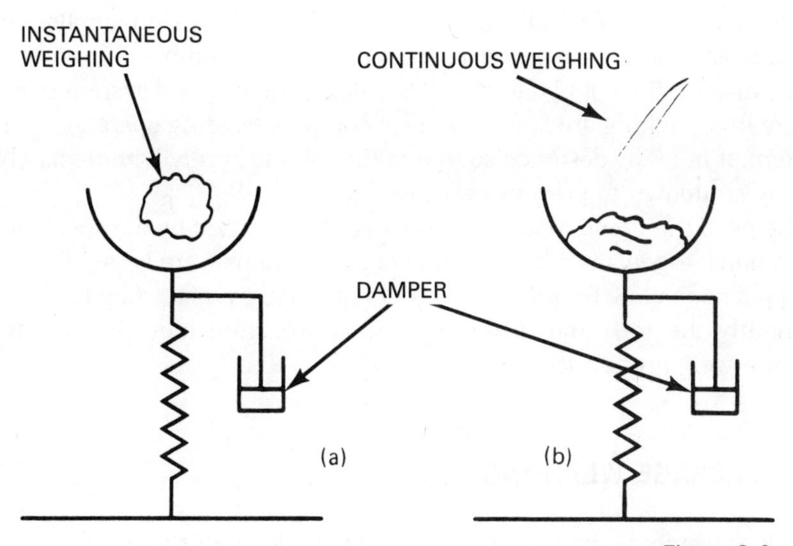

Figure 8.9
Effect of Product on Damping

IN-LINE WEIGHING

If high-speed weighing is defined as that speed above which manual reading is not possible, weighers are divided into three categories. The first category includes those weighers with low natural frequencies and a pointer and dial readout. These weighers, although slow, are usually very accurate, have good repeatability, and are capable of withstanding considerable abuse without damage. Scales of this type can be designed with ranges of many tons but their mechanisms can be extremely bulky and must often be accommodated in large pits.

A second category of weighers have stiff balances with natural frequencies in the hundreds, if not thousands, of cycles per second. These often feature high internal damping. Load cells described at the beginning of this chapter fall into this category. Load cells with accuracies commensurate with their mechanical counterparts are common but since these accuracies are dependent on amplifier and power-supply stability, they are somewhat less easy to maintain. In situations in which the weighing cycle is less than about 0.25 sec, however, there is no alternative but to use this type of device.

Between the conventional dial scale and the load cell is a whole range of weighters combining the principles of the spring balance with electrical output. These weighers have the advantage of comparatively large movements to operate secondary transducers. They also introduce their own problems associated with spring-mass systems.

The problems of in-line weighing are not confined to those involved in setting up dampers. The range of weighers in current use gives some indication of the numerous attempts to solve specific problems.

A load cell, or weigher, in a dynamic system may not be considered in isolation. It must function as part of a complete *feed*, *weigh*, and *disposal* system. The feed system is made up of conveyors, hoppers, and gates which bring the product to the weigher platform and dispense it as demanded by the weighing cycle. The form the feeder takes depends on the product. It can be a simple roller conveyor that carries slabs of metal over a load cell, or it can be a complex vibrator conveyor to create an even flow of potato chips. Similarly, the disposal system may simply terminate in a stacking area, or it can be a complex bagging operation. Whatever the system, it must be designed so that failure of any part will immediately cause an orderly shutdown to prevent overflow.

In the past, the production cycle was mainly governed by the speed of weighing operations, especially when weights of a few grams were being dispensed. In many applications low frequency weighers are being replaced by load cells, and consequently the feed and disposal systems are often more important to the timing of a weighing cycle.

IN-LINE PACKAGE WEIGHING

With in-line weighing systems the arrival of the load on the weighing platform produces a shock load to the weigher, followed by an unstable period as the package runs onto the platform. This causes the weigher to oscillate. Following this, a stable period occurs when the package is completely on the platform and a measurement is possible. The size of the package, its velocity, and the natural frequency of the weigher all determine the rate at which packages may be weighed. A typical roller conveyor and weighing platform is shown in Figure 8.10, and the load cell response to a typical weighing cycle is shown in Figure 8.11.

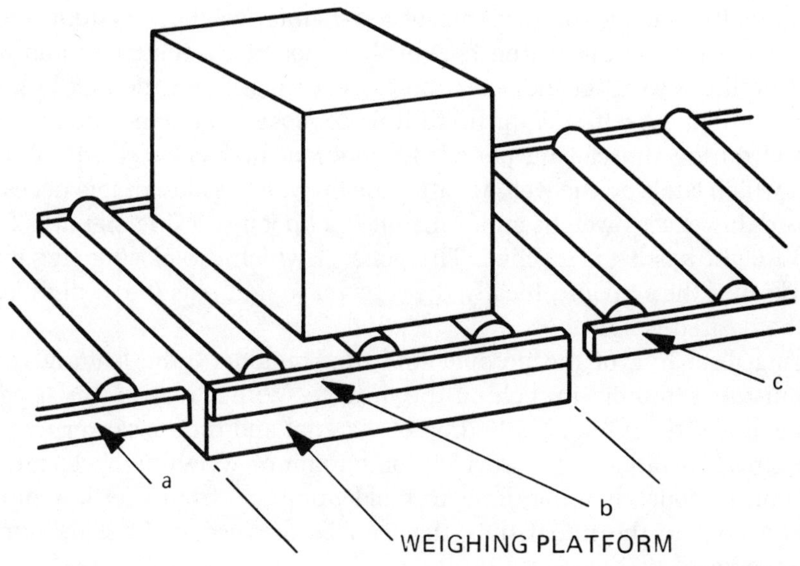

Figure 8.10
In-line Conveyor Weighing System

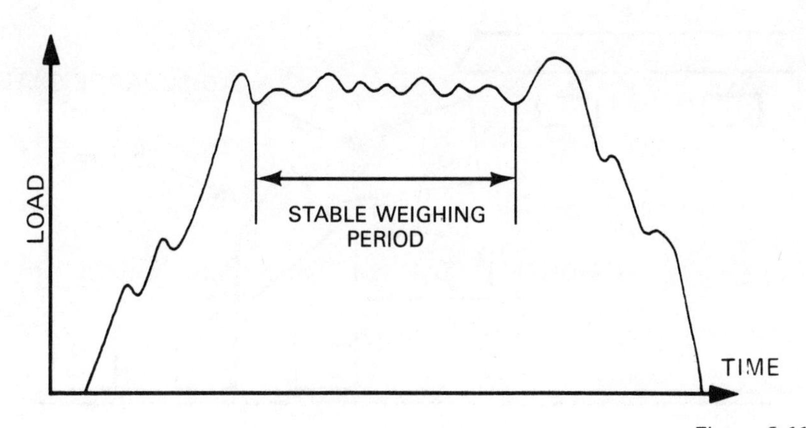

Figure 8.11
Load Distribution of In-line Package Weigher

IN-LINE PRODUCT WEIGHING

In-line weighing of products such as sugar or flour presents a different set of problems from those encountered in package weighing. Figure 8.12 is a sketch of a typical weighing system.

The product is brought to the weigher by a conveyor and discharged into the hopper through a lip gate. When the correct amount of product has been collected in the hopper, the lip gate is closed and the discharge gate is opened to fill one of the containers. As a further charge is being loaded into the hopper, a conveyor moves the containers forward, so that the next one is in position.

The problem facing the designer of a dynamic weighing system is how to achieve correct weighing in the hopper. It is not just a matter of pouring the product until the weigher indicates the correct weight and then closing the lip gate. The lip gate itself takes a finite time to close, and some product will be dispensed during the closing period. In addition, there is a steam of *in-flight* product which lands on the weigher after the lip gate has closed. It is necessary to anticipate this excess weight and transmit the lip-gate closing signal before the desired weight has been reached. The point at which the closing signal is produced is called the *setpoint*. Ideally, the setpoint weight plus the in-flight product equal the desired weight.

During discharge of the product into the containers, the loading conveyor causes material to build up behind the lip gate. Within reason this is advantageous, as it allows an initial bulk drop of material and then a slower topping up feed. Such an arrangement provides for minimum weighing-cycle times, but there is an obvious allowable limit to build up. Apart from overflow problems, there is a limit to the initial bulk that can be dropped without upsetting the measurement.

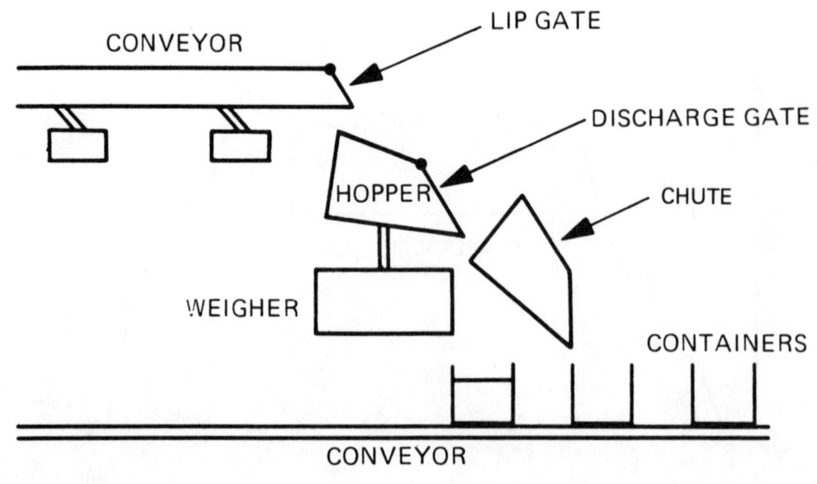

Figure 8.12
In-line Product Weigher

The graph in Figure 8.13 illustrates the buildup of weight that might be expected in the hopper of a typical product weigher. When the lip gate is opened an initial fall of product will cause peak *A*, the dynamic value of which will be higher than the static weight delivered. It is undesirable to allow this shock load to exceed the setpoint since "chattering" of the lip gate may result. This limits the amount of material that can be built up behind the gate before the conveyor is stopped.

After bulk delivery, product will continue to be dispensed until the setpoint is reached, and then the lip gate will be closed. Addition of in-flight material to the dispensed weight will then, it is hoped, bring the total weight in the hopper to the desired weight. Even with an accurately positioned setpoint and a critically damped system, the motion of the weigher will carry it over the desired value to point *B* before it eventually settles at its true value *C*. This is not important unless each dispensation is being automatically check weighed, in which case a delay must be built into the weighing system.

Certain products, such as salt and sugar, are homogeneous in nature. Given a consistent feed system and a weigher that has good repeatability, it can be expected to achieve repeated weighing close to the desired value. After the weighing hopper has emptied, a consistent tare value results. Under these circumstances it is possible to allow the weigher to rest on stops until a weight is reached just below the setpoint. This provides two advantages; it allows shock loadings to be carried by stops and it allows a weigher to be designed with a comparatively large movement from the unstopped position to the setpoint. In principle, therefore, dispensed weights very close to the desired weight ought to be possible.

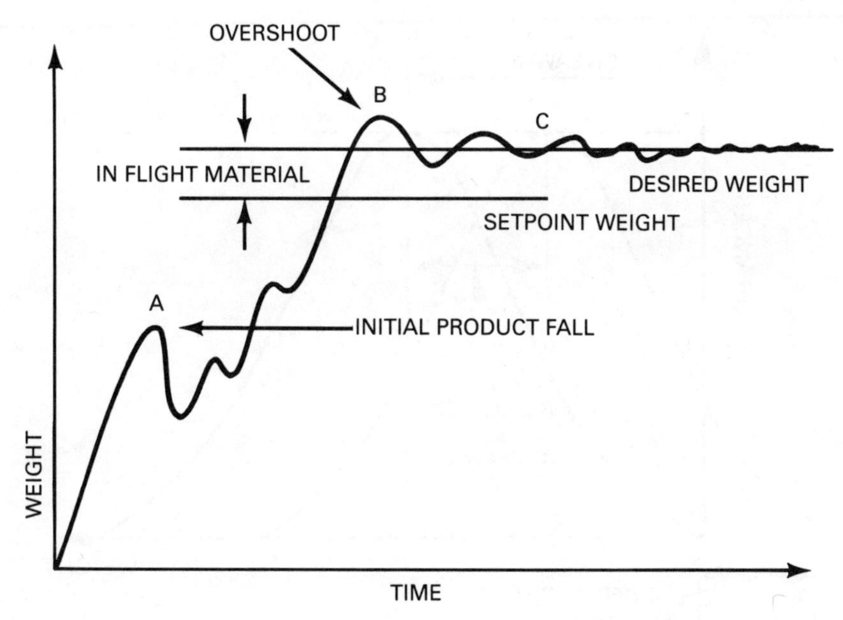

Figure 8.13
Buildup of Weights in Hopper of In-line Weigher

Unfortunately, it is not an ideal world; mechanical feed systems are inconsistent, and product dust can get into scale mechanisms and change their characteristics. Since it is illegal for manufacturers to sell less weight of product than advertised on the package, weighing systems are always adjusted to provide a surplus of product. This overweight is known as *product giveaway*.

Weighing homogeneous "clean" products presents problems, but these are small compared with the manufactures of heterogeneous fatty products. Potato chips are a classic example of such a range. Potato chips vary considerably in individual size and weight, depending upon the size of the potato in a batch and upon cooking conditions. In-flight product during a weighing cycle will vary considerably, and for his own protection the manufacturer must allow a large giveaway. The nature of the potato chip is such that small fragments will stick to the scale hopper and gradually cause a buildup, producing an increase in tare weight. Increases in tare weight effectively change the zero position of the weigher and triggers the setpoint with a lower dispensed weight. Confronted by these variables, it is not surprising that the manufacturer of fatty products is particularly careful about check weighing operations and accepts the necessity for considerable product giveaway.

Regardless of the product being weighed, the manufacturer must be prepared to allow some giveaway in each package. This giveaway represents lost profit and, when the setpoint is adjusted manually, is entirely dependent on the skill of the setter.

If a large number of samples of a product are check weighed, a distribution curve may be plotted, as shown in Figure 8.14. The shape of the distribution

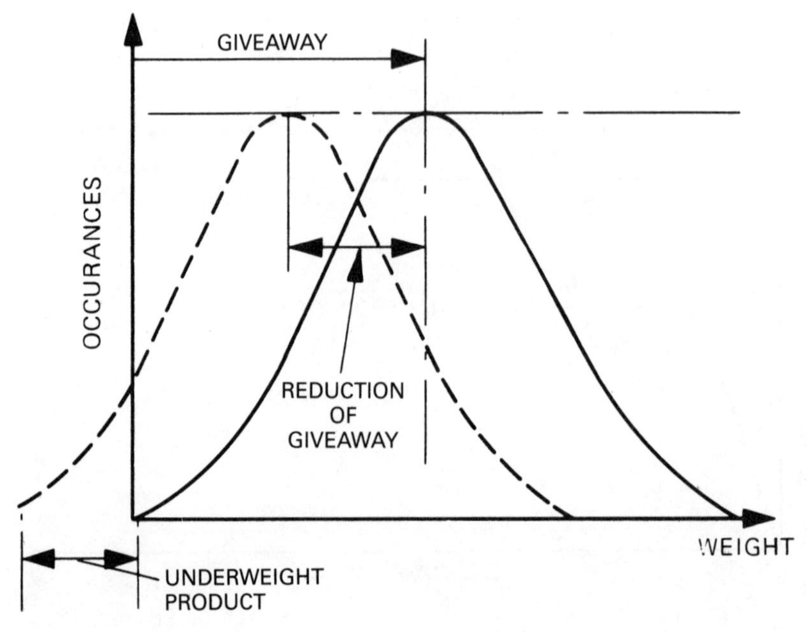

Figure 8.14
Product Giveaway Distribution Curve

curve will depend on the product. Sugar, for example, will give a much smaller spread than potato chips. Any reduction in giveaway involves moving the distribution curve to the left, but if the setpoint has already been adjusted to cope with the maximum spread, this must involve producing some underweight dispensations. This is illustrated in Figure 8.14 by the dotted curve.

Because it is illegal to sell underweight products, some means of eliminating underweight packages from the production line must be provided and this requires that every package be weighed. However, provided underweight packages can be eliminated in some way, the savings resulting from reduction in giveaway are considerable and, if achievable, are worth substantial capital investment to obtain. One method uses weighers in subsequent conveyor systems to operate gates which divert underweight packages. An alternative check-weighs each dispensation while it is still in the hopper, and if underweight, *dribble feeds* a fixed additional weight of product. With such a procedure a few packages may end up appreciably overweight, but this will be insignificant compared with the total savings. Although the principle of check-weighing in the hopper is simple, in practice it involves complications to the scale. If it is assumed that the product will be overweight, all that is required of a scale is that it leave its stop at a weight just below the setpoint and hit the other stop as the weight in the hopper reaches its setpoint value. This second stop is an electrical contact used to operate the lip gate. A simplified version of such a scale is shown in Figure 8.15.

At one end of the beam is a hopper, and at the other a counterbalance weight. The balance weight performs two functions: It balances the weight of the hopper and its mechanisms, and it also provides additional weight to balance the weight of the product.

With clean products it is reasonable to allow the balance to rest on stops in its unloaded condition and periodically check that the setpoint has not changed. With "sticky" products it must be accepted that the tare weight will change and

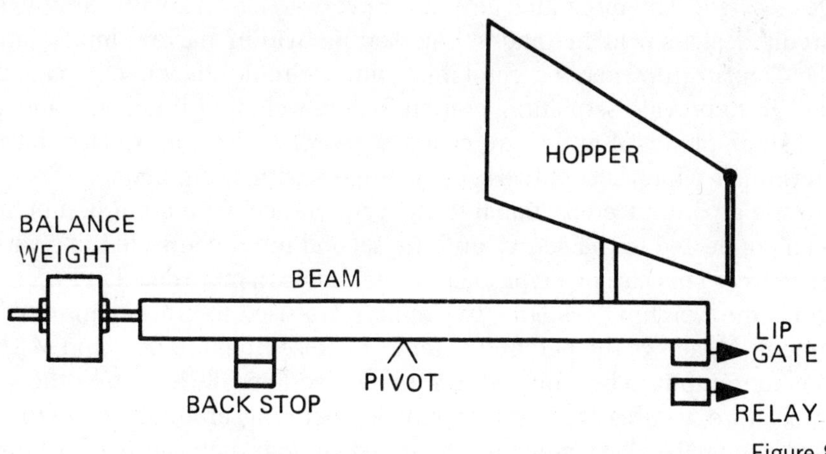

Figure 8.15
Simple Beam Balance

this must be detected by frequent check-weighing, or alternatively corrected automatically. A self-correcting system to be used for automatic weighing must:

1. Provide a signal proportional to the weight being measured (gross weight).
2. Provide a signal proportional to the zero position of the balance (tare weight).
3. Subtract the tare signal from the gross.
4. Provide an adjustable setpoint signal to operate the lip gate.
5. Maintain long-term accuracy and repeatability.

Electrical Setpoint Control

If a weighing system is to trigger the setpoint mechanism automatically, it must be provided with some type of switch to cut off the supply of product. One method providing good repeatability uses a photocell and a light beam that is interrupted by the movement of the scale, another method uses a magnet and reed switch. In addition, there are inductive devices that perform the same function. Sensors of this classification are called *proximity switches* and operate without physical contact between the moving parts.

Electronic Setpoint Control

The use of a proximity switch to control the setpoint does not allow tare correction which requires an analog measurement. If analog sensors and electronics must be provided for this function they may as well be used to provide the setpoint signal. Unfortunately analog signals are not ideal for operating switching circuits directly and some provision must be made to provide consistent operation of the lip gate. Among the range of integrated circuits available is a device called a *comparator*, the output of which is quiescent until a set input is reached when it changes its state. The integrated circuits are very stable and, with careful ancilliary circuit designs will maintain their setting within narrow limits for long periods. *Comparators* may be combined into a circuit shown schematically in Figure 8.16 to provide setpoint operation, check-weighing functions, and alarm limits. The amplifiers A and C are called *differential amplifiers*. They have two input terminals which accept two separate inputs with a common reference point and provide an output proportional to their difference. One input of amplifier A has been connected to the sensor and the second input connected to a variable-voltage source. This part of a typical circuit is shown in more detail in Figure 8.17.

With some weighing systems, two meters are used to provide an initial zero adjustment. M_1 allows the potentiometer P_1 to be set to midrange, and M_2 allows the amplifier output to be adjusted to zero by mechanically adjusting the sensor. This procedure assures that amplifier A, is operating around the center of its designed range. If M_2 is suitably calibrated, it may now be used to give an indication of the offset required to P_1 to provide a desired setpoint. Where no

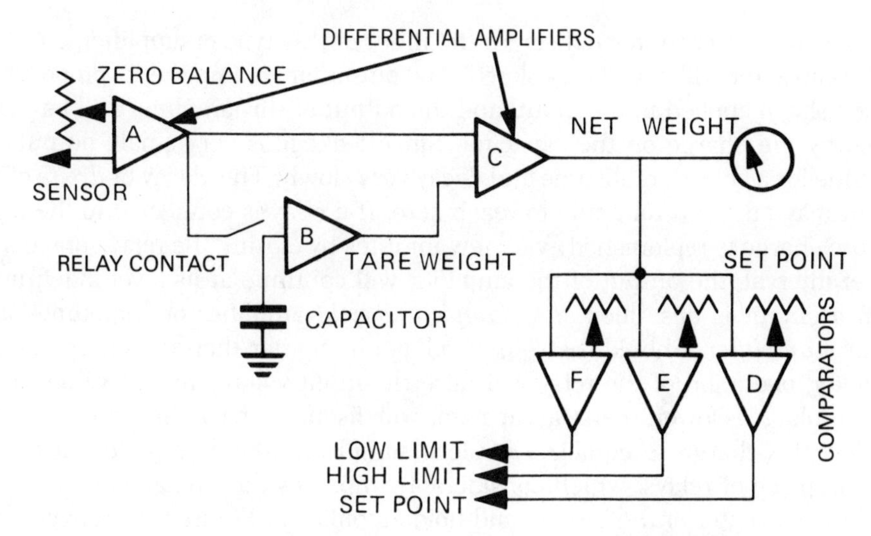

Figure 8.16
Automatic Setpoint and Tare Correction Circuit

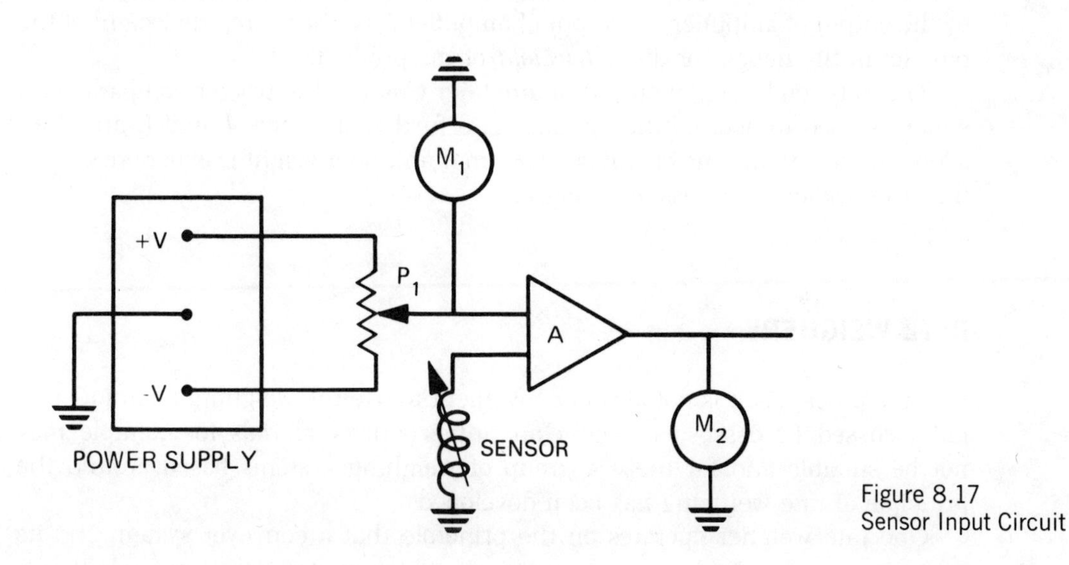

Figure 8.17
Sensor Input Circuit

change in tare weight is anticipated the output of amplifier *A* can be used to operate the weigher lip gate, but where changes in tare occur some type of correction circuit must be introduced.

Automatic Tare Correction

Tare compensation in the circuit illustrated in Figure 8.16 is accomplished by the introduction of a second amplifier *B*. This is a rather different device from amplifiers previously discussed; it has an extremely high input impedance made possible by the use of transistors called *field-effect-transistors* (FET).

If a capacitor is connected across the input of this type of amplifier, a charge on the capacitor will decay very slowly. The output of an amplifier is proportional to the voltage applied to the input, and the output of this amplifier will be maintained by the charge on the capacitor. Since leakage is very small, output will continue for a considerable time and decay very slowly. This decay or *droop* of the output may take several hours to reach zero. If a relay is connected to the input and the charge is replenished every few minutes by closing the relay contacts for a brief interval, the output of the amplifier will continue at its level indefinitely. Such a circuit is described as a *sample-and-hold* amplifier or sometimes as a *memory amplifier*. Should the source voltage be higher than the charge on the capacitor, operation of the relay will raise the input voltage to that value. If the source voltage is lower, then the capacitor will discharge back through the source to allow the charge to equalize. Modern sample-and-hold amplifiers use FET gates in place of relays, which considerably improves their reliability.

If the capacitor of the sample-and-hold amplifier in Figure 8.16 is exposed to the output of *A* while the weighing hopper is empty, a signal proportional to the tare weight is applied to one input of amplifier *C* for the remainder of the weighing cycle. This tare weight signal subtracts from the gross weight signal produced by the output of amplifier *A*. Output of amplifier *C* is, therefore, the weight of the product in the hopper or the *net weight* of the product.

The net-weight signal output of amplifier *C* is used to trigger comparator *D*, which is used to operate the lip gate. Two further detectors *F* and *E* are often added to the circuit; one operates an alarm if an underweight is experienced and the other indicates excess overweight.

RATE WEIGHERS

In some processes it is not always convenient to use the weighing techniques so far discussed. Processes using large quantities of raw materials, for example, may not be suitable and for these a group of weighing systems based around the principle of rate weighing has been developed.

One rate weigher operates on the principle that a conveyor system and its contents may be weighed, and the speed of the motor adjusted such that a constant feed rate is maintained. The product of feed rate and time provides a measurement of the amount delivered. An alternative system weighs the product in an auxiliary hopper, and the rate at which the hopper loses weight is a measure of the feed rate. The system illustrated in Figure 8.18 employs a conveyor which is weighed by a load cell and its speed is measured by a coil and magnet tachometer illustrated in Figure 8.19. The tachometer uses a toothed wheel on the drive pulley shaft to vary the magnetic field round a coil and magnet sensor. In the process, a series of voltage pulses are induced in the coil, and these are used to drive the tachometer. The tachometer itself is essentially a pulse counter which counts pulses over a fixed time period and converts the value into a measurement of speed.

With this system it is normal to calculate the weight of material discharged by summing the weights discharged over a continuous series of sampling periods. During a sampling period the rate of discharge is calculated from measurements of the weight of material per unit length of belt and belt speed. The total discharge is equal to the rate of discharge multiplied by the sampling period.

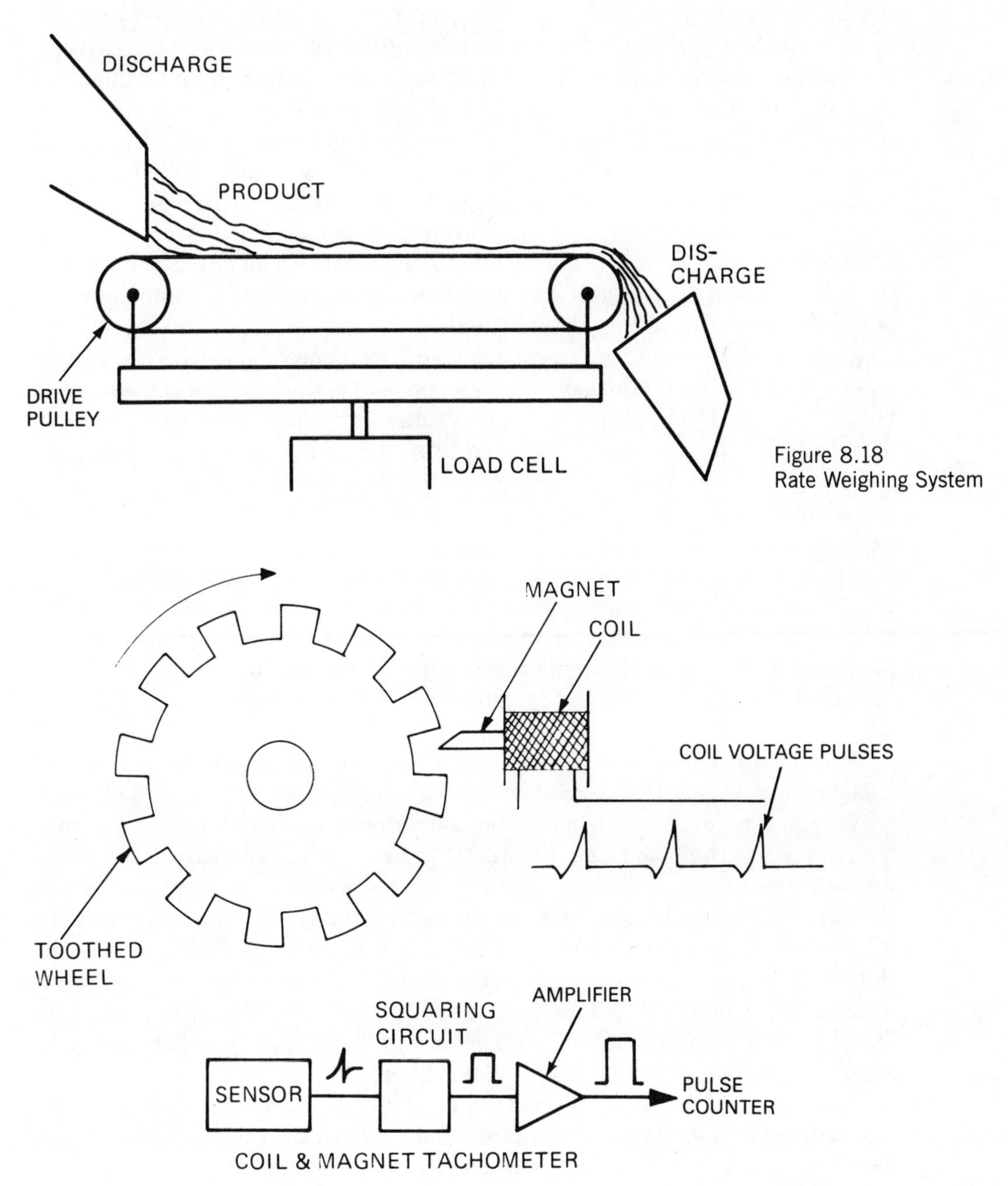

Figure 8.18
Rate Weighing System

Figure 8.19
Coil and Magnet Tachometer

LOAD MEASUREMENT IN THE STEEL INDUSTRY

Of all the environments to which sensors are subjected, those found in the steel-manufacturing industry must rank among the worst. In these conditions variable inductance and differential transformer devices are invaluable, and many weighing systems have been built round them. One load measurement that presents considerable difficulties is that of a ladle of molten steel on a crane hoist. The simple expedient is to lower the empty ladle onto a scale, weigh it, then carry it to the pouring point, add the melt, and then weigh again. Apart from the handling costs of such a procedure, which are high, the melt is losing heat during the weighing process, and this heat often has to be replaced at appreciable cost. In addition, it may be necessary to share the contents of a single ladle between a number of furnaces, and know the weight delivered to each furnace. In this case, returning to a scale after each pour is likely to be completely unacceptable. The alternative use of small ladles introduces considerable handling costs. Another method of *in situ* weighing, is a more practical proposition and eliminates many of the disadvantages of the previous methods.

There are two reasonably accessible points on a crane hoist where load measurement is possible: at the ladle lifting points and at the bearings of the hoisting pulley. The ladle lifting points are not very satisfactory and if used, cables must be detached each time a ladle is removed from the lifting hook. Considering the alternative, sensors fitted to bearings on the hoisting pulleys are subjected to considerable radiant heat but even so, the pulleys usually offer the best locations for load measurement.

Apart from performance considerations of any device fitted to the hoisting gear, there is also the consideration of safety. Sensors will be subjected to frequent shock loadings and must not only survive electrically but maintain the same safety factors as the remainder of the crane structure. One interesting device which contains all these features is illustrated in Figures 8.20, 8.21, and 8.22.

The system being considered in Fig. 8.20 requires the installation of load cells at points *A*, *B*, C, and *D*. Several designs of load cells may be used, most of which are based on inductive principles. The one described is based on the Pressductor®. Suppose a laminated core, Figure 8.21a, has four holes drilled through it and through these holes are wound primary and secondary windings that cross at right angles. In the unloaded state an alternating current in the primary winding will not affect the secondary (Figure 8.21b) but if a force is applied to the core (Fig. 8.21c) , the permeability of the material is reduced in the direction of the force. This phenomenon is known as the *magnetoelastic effect*. Changes in the flux pattern around the coils induce a voltage in the secondary, which within the limits of operation is proportional to the load applied.

The sensing element in this load cell is mounted on a flexure arrangement so that only vertical forces are transmitted to the element. In the unlikely event of flexure failure the pulley shaft is restrained by the main body of the cell, and because operating gaps are small, hoist operation is not impaired. This arrange-

® Registered trade-name of the ASEA Corporation

ment provides a very rugged and reliable sensor capable of surviving the most severe environments.

Installation of sensors to crane hoisting gear introduces problems of electrical connection. Provision of power to operate the sensor and connection of signal lines to a location where they may be useful can be difficult. Most overhead cranes have three motions. The load must be moved vertically; it must be moved across the width of the crane bay, and it must be moved along the length of the gantry. Each motion requires an electric motor drive and power is often supplied

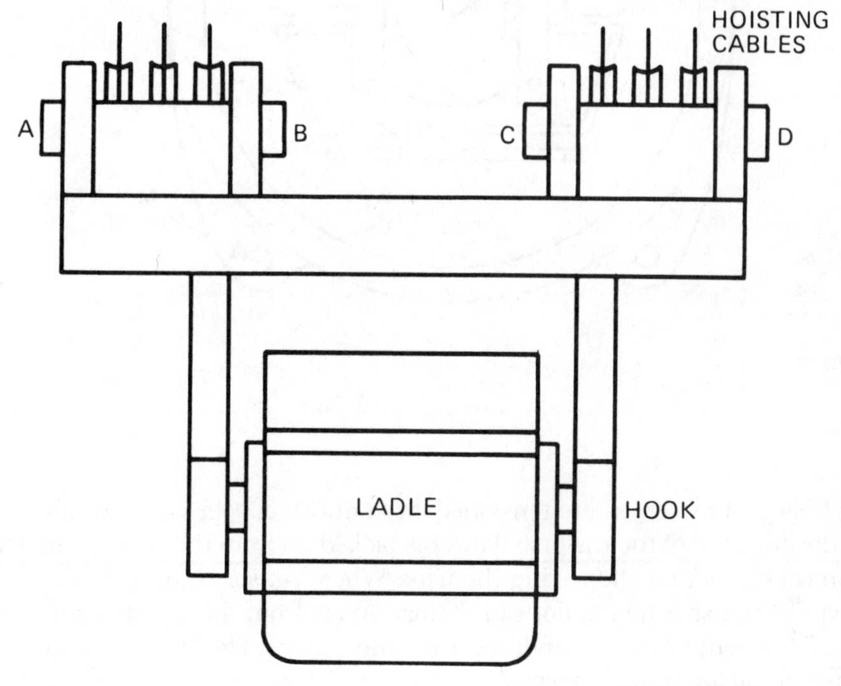

Figure 8.20
Load Measurement Points on Crane Hoist Pulley

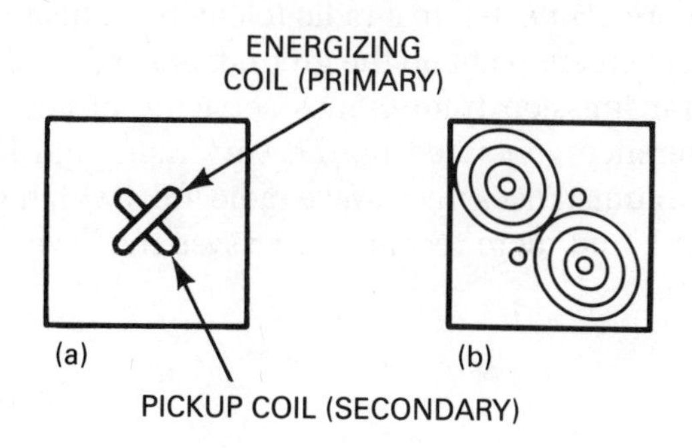

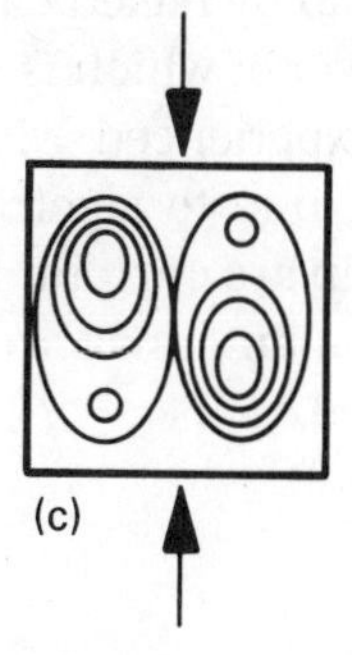

Figure 8.21
Principle of a Magnetoelastic Effect Load Cell

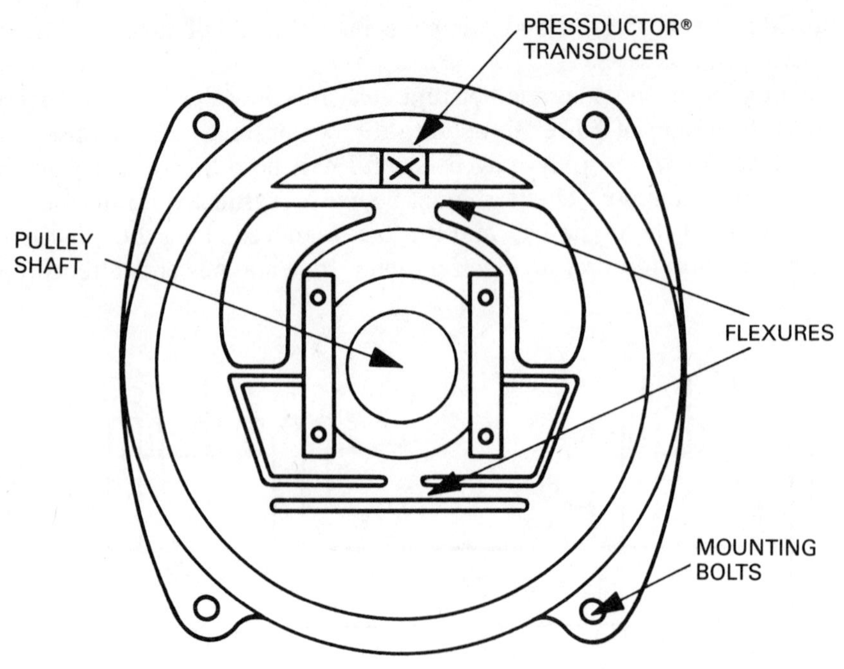

Figure 8.22
Load Cell (ASEA)

along trolley wires. These are tensioned, hard-drawn copper wires running parallel to the direction of the traverse. Power is picked up from them by spring-loaded carbon collectors that slide along the wire. Where sensor signals are concerned, this type of transmission is not satisfactory. Apart from the variations in contact resistance, electrical noise is induced from the motor trolley wires. It is fairly easy to connect cables from a hoist to a cross-travel carriage by using a cable that winds into a spring-loaded drum. These are in common use for supplying power to motors and they are satisfactory to return a signal to the driver's cab. When the signal must be transmitted to the ground, the problem is of different magnitude. One solution to this problem uses radiotelemetry. Signals may be returned to the driver's cab by reeled cable and there connected to a radio telemetry transmitter, the receiver of which is at some convenient point on the ground. Severe problems can be experienced with radio transmission from cranes servicing electric arc furnaces, due to electrical interference generated by the very rapid and large changes in arc current. In these circumstances microwave radio links with highly directional antennas are often used, and more recently laser systems have been introduced.

CHAPTER 9

FLOW MEASUREMENT

FLOW METERS

Flow measurement is a very early science practiced by the Romans who used the *notched-weir* flow meter. This was a device which used a V-shaped notch in the top edge of a weir board. As the weir was lowered, the aperture below the surface level of the water increased, from zero at the bottom of the V to a maximum at the top. This provided a flow rate, which was a function of the area of the notch below surface level and which could be calibrated. Since then, many types of flow meters have been developed, most of which stem from the early nineteenth century onward. In general, they may be divided into two basic types: *quantity meters* that measure the quantity of fluid directly, and *rate-of-flow meters* whose output must be integrated to provide a measurement of total quantity delivered.

Mass Flow Meters

Fluids may be divided into two groups, gases and liquids, both of which may be monitored by quantity or rate-of-flow meters. In either case the masses of fluid delivered will depend on its specific gravity, which is in turn usually dependent on its temperature. The measurement of volume flow is fairly straightforward, but the measurement of mass flow because of temperature and specific gravity variations is more complex. The simplest mass-flow meter is operated by the weight of a liquid. One type uses a container counterbalanced by a weight. When a predetermined mass of fluid is poured into the container, it rotates and empties its contents, then, being lightened, rotates back to its filling position. The number of emptying operations is counted and gives a measurement of the total mass delivered. Such a device is crude and not suitable for many liquids being monitored today. Modern instruments commonly measure the specific gravity and the volume flow and calculate the mass flow with an electronic circuit, often described as a *mass-flow computer*.

Volumetric-Flow Meters

By far the simplest volumetric-flow meter is the tank that is filled to a predetermined level and then emptied. The number of times the tank is emptied gives a measurement of the total volume delivered.

POSITIVE-DISPLACEMENT METERS

Although the simple tank flow meter in itself is of little value, it is the basis of a number of *positive-displacement meters* which rely on delivering a fixed volume of liquid into a chamber and then expelling it into a delivery pipe. The reciprocating piston flow meter illustrated in Figure 9.1 is typical of such meters.

This type of meter operates in the same way as the cylinder of the double-acting piston of a steam engine. Fluid enters cylinder *A* through the inlet port and drives the piston to its extremity. At this point the piston of the slide valve is changed, cutting off the supply from cylinder *A* and opening up the entry to cylinder *B*. Simultaneously, the outlet from cylinder *B* is closed and opened to cylinder *A*. As fluid expands into cylinder *B*, the piston is driven down the length of its bore, expelling the liquid in *A* through the outlet. At the opposite extremity of the stroke, the action is again reversed, allowing cylinder *B* to discharge. The number of piston strokes is a measurement of the total volume delivered.

There are a number of variations on this basic type. The slide valve, for example, may be replaced by nonreturn valves and the piston by an elastic diaphragm. All positive displacement meters drive a piston stroke counter of some type. Where electrical outputs are required proximity sensors may be used. Reed switches and magnets provide one method and inductive sensors another. In hazardous areas where an explosion risk is present the use of photoelectric devices with fiberoptic-isolation is a valuable technique.

RESTRICTION FLOW METERS

By far the largest proportion of flow meters measure rate of flow, and the total volume is calculated by integrating the rate of flow over time. Of the rate-of-flow meters, those that monitor flow by measuring pressure drop across a restriction in

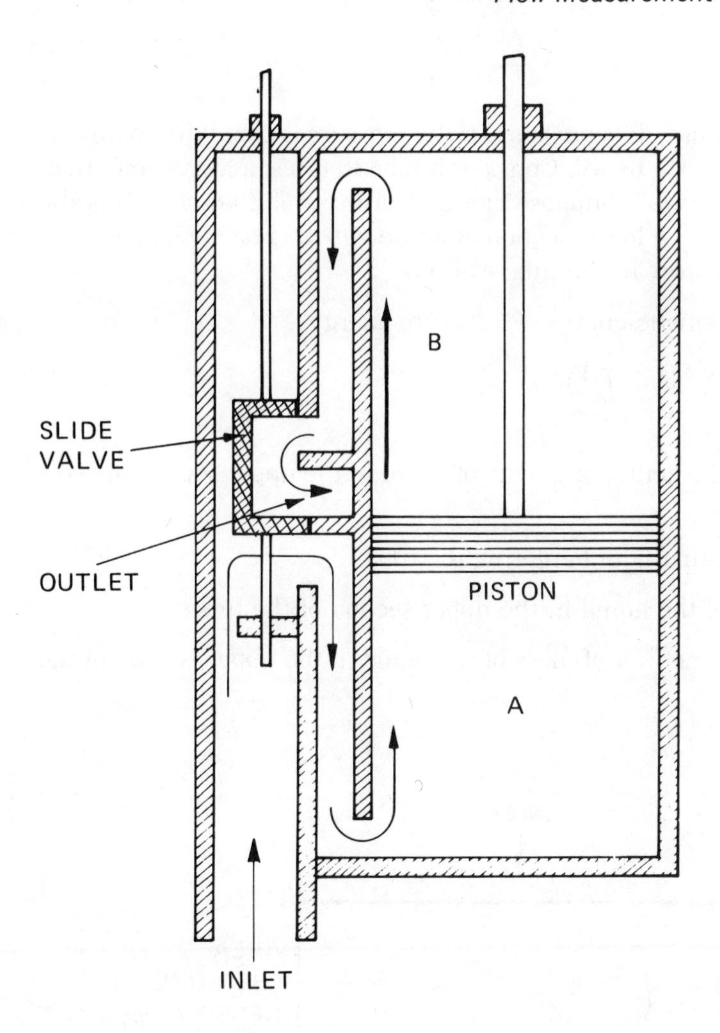

Figure 9.1
Reciprocating Piston Flow Meter

the pipe are probably the most common. Restriction may take several forms, but the *orifice plate* restriction is the most easily produced.

To understand the theory of orifice-flow meters it is necessary to be aware of some of the basics of energy theory.

Bernoulli's Theorem

This states that energy cannot be created or destroyed; it can only be changed in form.

Potential Energy:

Suppose a mass m is at the height h above a reference plane. The potential energy of the mass is mgh where g is the acceleration due to gravity.

Kinetic Energy

If a mass m is in motion, it has a kinetic energy of $\frac{1}{2} mv^2$ where v is the velocity.

Pressure Energy

A liquid has energy because of its pressure. If the pressure is p and the volume V, then its pressure energy will be pV. Consider a tube that has a larger inlet than outlet as shown in Figure 9.2. Suppose one unit of mass of a liquid enters the upper section and leaves the lower section at a rate which creates no change in the level H. The total energy in the tube will be:

potential energy + kinetic energy + pressure energy

$$= 1 \times h_1 g + \tfrac{1}{2}(1 \times v_1^2) + p_1 V_1 \quad (1)$$

where

h_1 = the height of the center of gravity of liquid in the upper section above a reference plane

v_1 = the velocity of that liquid entering the tube

p_1 = the pressure of the liquid in the upper section of the tube

V_1 = the volume of one unit of mass of the liquid in the upper section of the tube

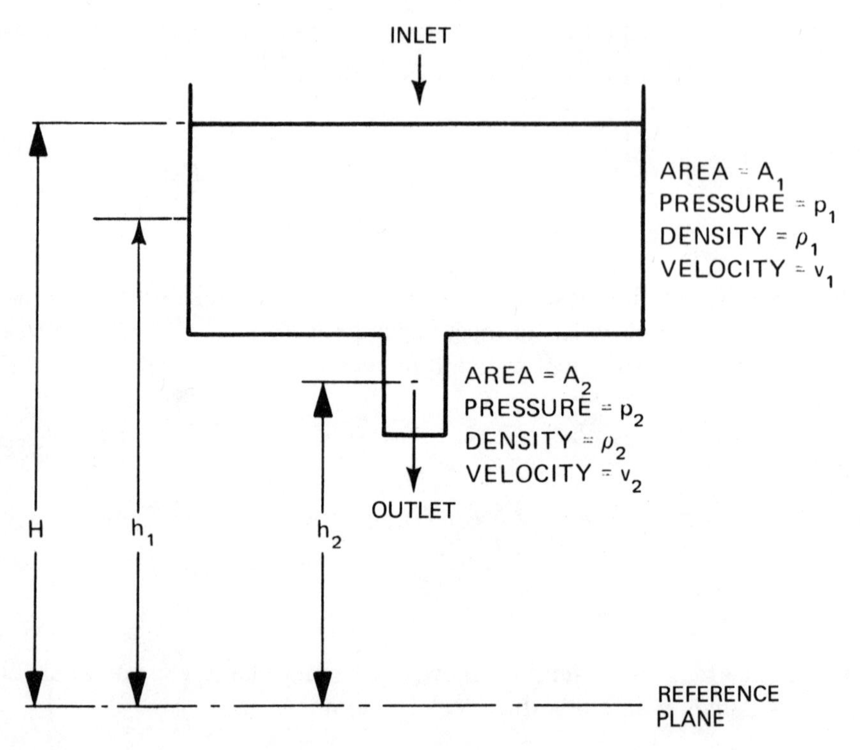

Figure 9.2
Flow Through and Restriction

The total energy of the liquid leaving the tube will be:

$$1 \times h_2 g + ½(1 \times v_2^2) + p_2 V_2 \tag{2}$$

where

h_2 = the height of the center of gravity of the orifice

v_2 = the velocity of the liquid leaving the lower section of the tube

p_2 = the pressure of the liquid leaving the tube

V_2 = the volume of one unit of mass of the liquid in the lower section of the tube

Liquids may be considered as incompressible, and therefore their volume per unit mass and their density will not change:

$$V_1 = V_2 = \frac{1}{\rho}$$

where

ρ = the density of the liquid.

According to Bernoulli's theorem, energy cannot be created or destroyed so that:

Equation (1) = Equation (2)

$$h_1 g + \frac{v_1^2}{2} + p_1 V_1 = h_2 g + \frac{v_2^2}{2} + p_2 V_2$$

and substituting $1/\rho$ for V_1 and V_2

$$h_1 g + \frac{v_1^2}{2} + \frac{p_1}{\rho} = h_2 g + \frac{v_2^2}{2} + \frac{p_2}{\rho}$$

This equation may be rewritten:

$$\left(\frac{p_1}{\rho g} + h_1\right) - \left(\frac{p_2}{\rho g} + h_2\right) = \frac{v_2^2}{2g} - \frac{v_1^2}{2g} \tag{3}$$

Suppose a sealed manometer with a vacuum above the liquid is connected to the tube at a point h_1 above the reference plane and another connected at a point h_2 above the reference plane. Suppose that both manometers contain the same liquid as the pipe. The liquid level will rise in the manometers so that the pressure due to the liquid in the tube balances the static pressure.

The height of the liquid above h_1 will be $p_1/\rho g$, and the height of the liquid above h_2 will be P_2/ρ_g.

Pressure difference between the two points expressed in the difference in heights of the two liquids therefore, will be:

$$\left(\frac{p_1}{\rho g} + h_1\right) - \left(\frac{p_2}{\rho g} + h_2\right)$$

and this is known as the *differential head h*. Substituting in (3):

$$h = \frac{v_2^2}{2g} - \frac{v_1^2}{2g} \tag{4}$$

If the differential head h of the liquid is to remain constant, the volume flowing into the tube per second must equal the volume leaving. Let Q be the volume of liquid flowing per second.

$$Q = \text{velocity} \times \text{area} = vA \tag{5}$$

with v denoting the velocity and A the tube area.

For an incompressible liquid, the velocities of liquid entering and leaving the tube v_1 and v_2, and the tube areas at the entering and leaving points A_1 and A_2, obey the equation

$$v_1 = \frac{v_2 A_2}{A_1}$$

Substituting in (4) for v_1:

$$h = \frac{v_2^2}{2g} - \frac{v_2^2}{2g}\left(\frac{A_2}{A_1}\right)^2$$

$$2gh = v_2^2\left[1 - \left(\frac{A_2}{A_1}\right)^2\right]$$

or

$$v_2 = \frac{\sqrt{2gh}}{\sqrt{1 - \left(\frac{A_2}{A_1}\right)^2}}$$

If the factor

$$\frac{1}{\sqrt{1 - \left(\frac{A_2}{A_1}\right)^2}}$$

is represented by a term K, the expression may be simplified to:

$$v_2 = K\sqrt{2gh}$$

Applying this, along with equation (5) to the liquid leaving the tube, we obtain

$$Q = KA_2\sqrt{2gh}$$

The mass of liquid flowing per second equals ρQ.

Also when $h_1 = h_2$, the differential head h becomes

$$h = \frac{\delta p}{\rho g}$$

where δp is the differential pressure $p_1 - p_2$. From (6)

$$Q = KA_2 \sqrt{\frac{2g\delta p}{\rho g}}$$

so that mass flow

$$\text{mass flow} = \rho Q = KA_2 \sqrt{2 \rho \delta p}$$

This equation applies to liquids only and assumes streamline flow. Usually the flow is turbulent, and requires a greater pressure difference across the plate to generate a given rate of flow than is required in the ideal case. To take this into account the equation is modified by a factor called the *discharge coefficient* defined as the ratio of actual volume flow/theoretical volume flow. For an orifice plate this value is about 0.6.

Practical Restriction Flow Meters

The orifice plate provides the basis of a reasonably accurate, simple and inexpensive flow meter (Figure 9.3). When flow of fluid begins to wear or corrode the orifice, it is easily replaced.

Pressure drop is measured upstream and downstream of the orifice with a differential-pressure sensor. The maximum pressure drop across a square-edge orifice is obtained by placing the upstream tap close to the orifice plate and the downstream tap at a position equal to half the pipe diameter.

The disadvantage of the orifice meter lies in the large pressure drop, across the plate. This is improved considerably by replacing the orifice with a *venturi.* (See Figure 9.4).

The venturi consists of an accurately machined upstream parallel section that carries the upstream sensor tap. This parallel section runs into a conical section to

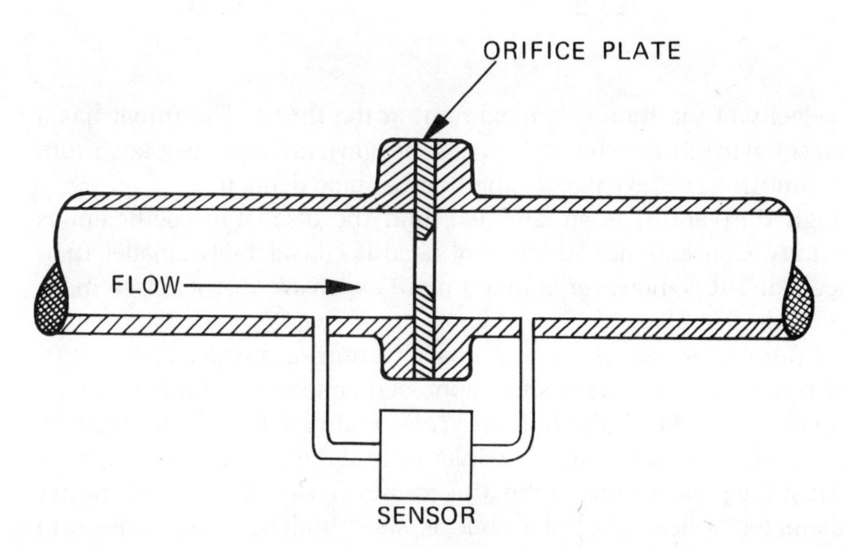

Figure 9.3
The Orifice Flow Meter

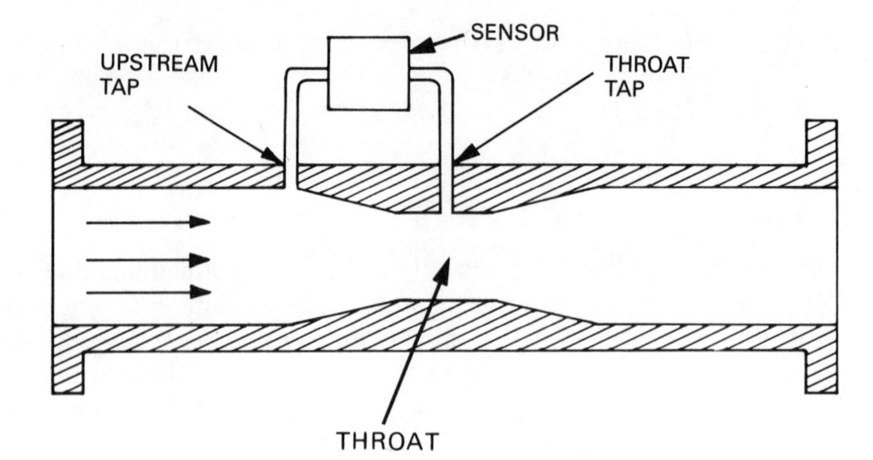

Figure 9.4a
Venturi Flow Meter

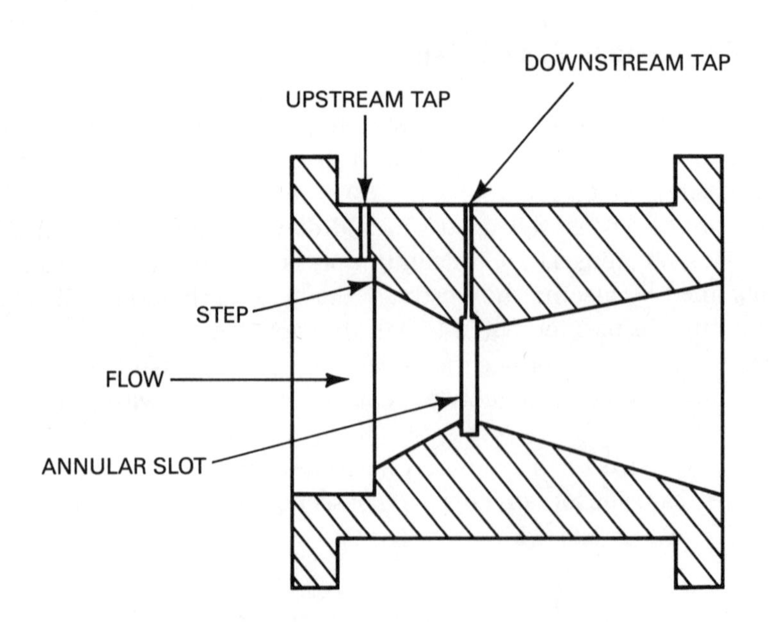

Figure 9.4b
Dall Flow Tube

increase the velocity of the fluid to a maximum at the throat. The throat has a parallel section equal to its diameter and carries the downstream sensor tap. From the throat the venturi again expands to the original pipe diameter.

Flow through the venturi is streamlined, and the discharge coefficient is much nearer unity. Consequently, the loss of head is considerably smaller than with the orifice plate. It is, however, a much more expensive restriction to manufacture than the orifice plate.

There are a number of variations of the basic venturi principle. The venturi nozzle type of restriction has a very short approach cone but still maintains its long expansion cone. Another is the *Dall flow tube*, Figure 9.4b, a device halfway between an orifice plate and a venturi. The inlet cone of the Dall tube is of smaller cross section than the pipe, the end of the tube forming a step that behaves rather like a large-diameter orifice. The inlet cone is short and has steep sides and guides the fluid into a throat that contains an annular slot. From here the fluid expands through a short expansion cone and then into a larger diameter pipe. The pressure differential of this arrangement is higher than an equivalent venturi, while the loss of head is considerably lower than in a conventional orifice.

Sensors for Restriction-Type Flow Meters

Pressure differential across a restriction is usually very small compared with the working pressure in the pipe. When differential-pressure sensors are used, this is of little direct consequence. Under normal operating conditions, a thin diaphragm may be used to provide a working signal.

But if the orifice or one of the sensor taps becomes obstructed, the diaphragm will experience full working pressure. Unless adequate overload protection is provided, this will rupture the diaphragm. In addition to diaphragm considerations, it must be remembered that the sensor case is subjected to line pressures. If these pressures are high, or if corrosive fluid is being used, case failure could have serious consequences.

THE TURBINE FLOW METER

The turbine-flow meter illustrated in Figure 9.5 is widely used where reliability and high accuracy are required, coupled with low head losses. It is also used for

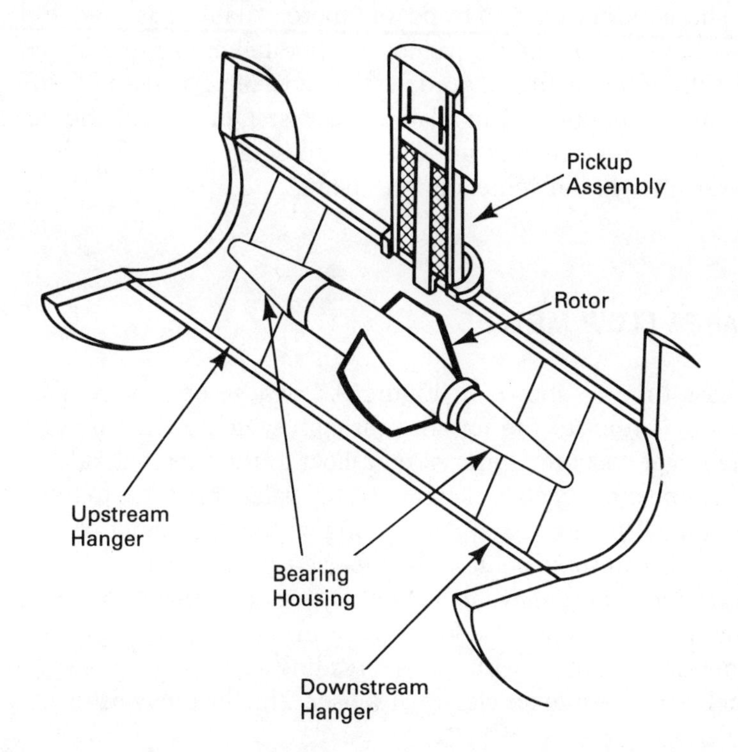

Figure 9.5
Turbine Flow Meter.
(Meterflow Ltd.)

measurements where low flow rates are encountered. Turbine meters may be used in applications where half-inch (12-mm) pipes are used and with flow rates of 0.1 gal/min (0.5 1/min), or in large pipes with flow rates of 3000 gals/min (15,000 1/min). Accuracies with these meters when flow straighteners are used in the pipe is about ±0.25% FS with a flow range of 10:1.

The flow meter consists of a *rotor* or *impeller* suspended in the flow stream by *hangers* which may have jeweled bearings. As the impeller turns, the blades induce pulses in a coil and magnet sensor, the pulse rate of which is a measurement of flow rate. Apart from cost, the major disadvantage of turbine meters is that they suffer damage when exposed to dirty fluids.

ELECTROMAGNETIC FLOW METER

Electromagnetic flow meters (Figure 9.6) have no moving parts and offer no restriction to the fluid flow. This makes them ideal for the measurement of slurries. For satisfactory operation they require that the fluid being measured have appreciable electrical conductivity. This makes them unsuitable for gases and oil-derived products.

The flow-meter principle is based on Faraday's law which states that the voltage induced in a conductor moving in a magnetic field is proportional to the velocity of the conductor. In this instance the conductor is the fluid itself, and the potential measured across the diameter is proportional to the fluid's velocity through an insulated section. The magnetic field is generated across the section by two coils. These may be excited by dc, but more usually ac is used. Fluid flows at right angles to the field, and the voltages generated are sensed by small electrodes in the walls of the insulated section. The most likely cause of failure in this type of meter is the buildup of insulating encrustations round the electrodes. Consequently, some flow meters are fitted with scrapers so that they may be cleaned periodically without removing the meter.

VARIABLE AREA FLOW METERS

Consider a glass tube as shown in Figure 9.7, whose cross-sectional area increases from the bottom to the top and through which a fluid flows from the smaller to the larger section. Upthrust on a float in the tube will depend on the flow rate and on the cross-sectional area, and the balancing force to this upthrust will be the weight of the float itself. Depending on the flow meter characteristics the float will stabilize at some point in the tube at a given flow rate. An increase in flow rate causes the float to move to a point of greater cross section, and, a lower flow rate causes it to stabilize at a lower point in the tube. This provides a very simple flow-measuring device with a direct reading scale. There are several ways of turning such a device into an electrical sensor. The float may be used as a core

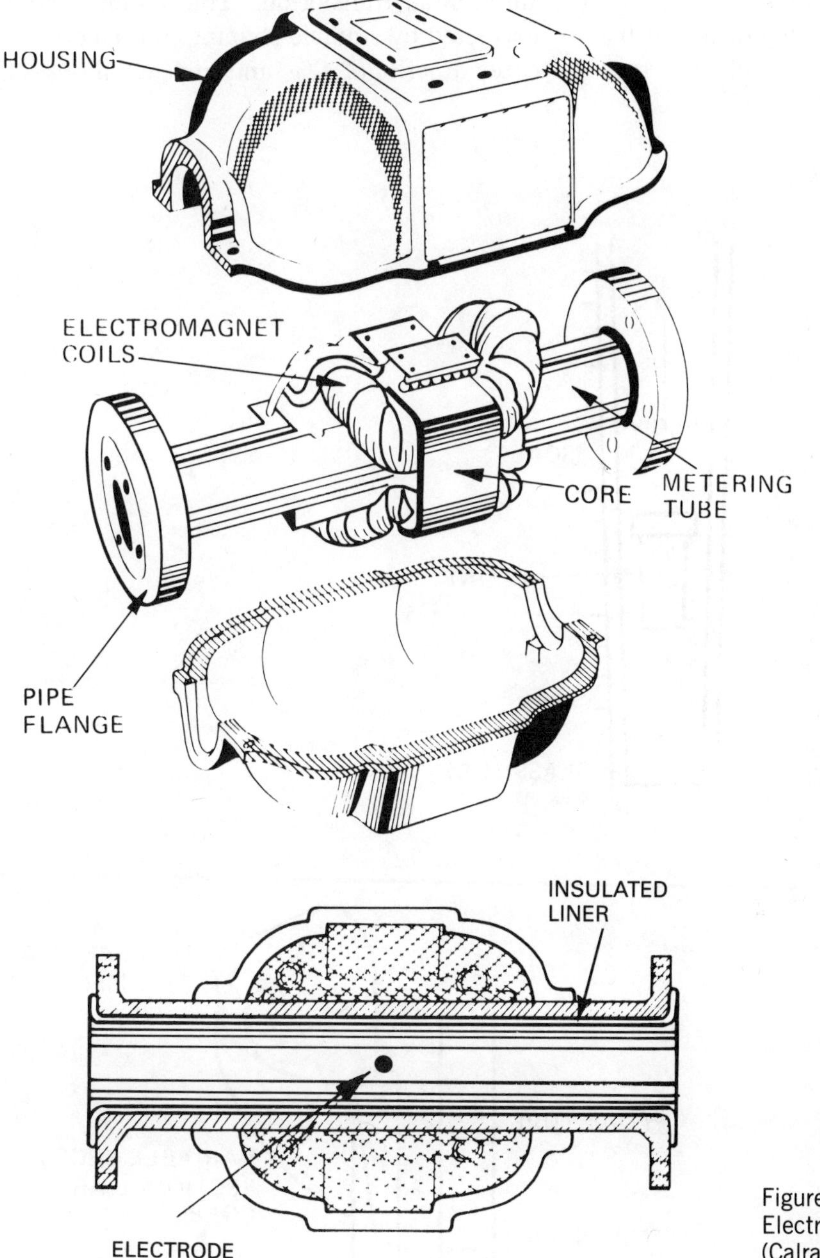

Figure 9.6
Electromagnetic Flow Meter.
(Calray Co. Ltd.)

of a differential transformer and the coils moved to a null position by a servo system. Self-balancing mechanisms tend to be expensive; a much simpler type of transducer is shown in Figure 9.8. In this device the float contains a magnet that attracts a ferrous strip wound helically on a pivoted drum. As the float moves, so it attracts the opposite portion of the helix. This causes the drum to rotate, moving the core relative to the coils of a differential transformer. The system would normally be nonlinear, but may be corrected by suitable shaping of the core.

Flow meters of this type have accuracies of 2% and repeatabilities of 0.25% FS.

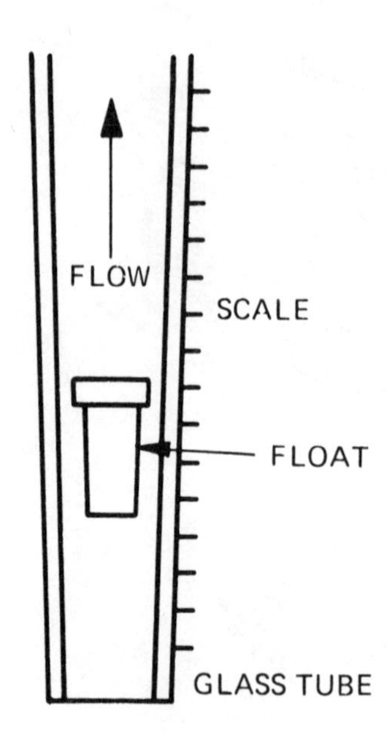

Figure 9.7
Variable Area Flow Meter

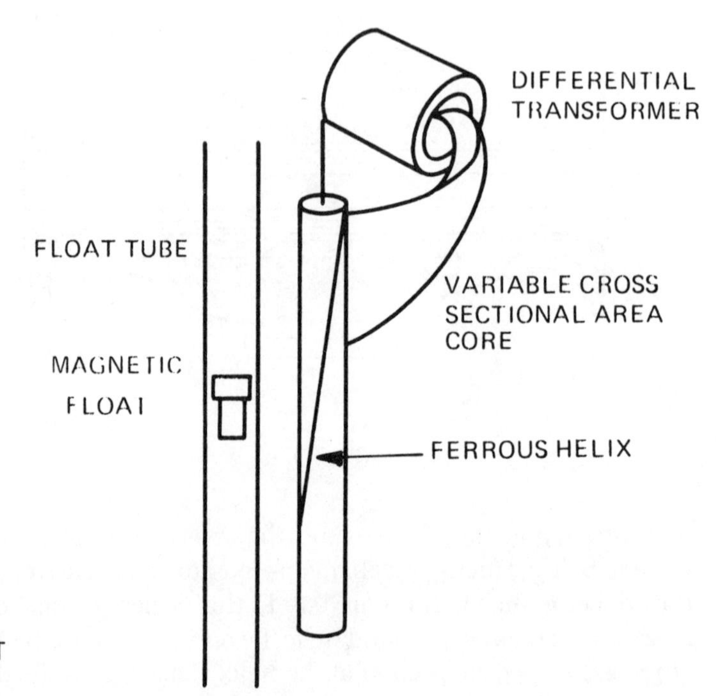

Figure 9.8
Variable Area Flow Meter with LVDT Output. (Brooks Instruments)

ULTRASONIC FLOW METERS

Another type of flow meter uses an *ultrasonic transmitter* mounted on the outside of the pipe with two ultrasonic sensors on the opposite side of the pipe, spaced downstream and upstream to the transmitter. With the fluid stationary, both sensors receive equal amounts of ultrasonic energy and, therefore, produce equal outputs. As the fluid moves in the pipe, the ultrasonic field is distorted proportionally to the rate of flow, and a differential signal is produced between the two sensors.

ANEMOMETERS

A number of *anemometers* are in use for measuring gas flow, an appreciable proportion being designed for specialist application. Two common types are described in this section.

The propeller anemometer usually consists of a multibladed propeller that rotates at a speed proportional to a gas flow. If good accuracy is to be achieved, the propeller must be light and frictional losses low. Such devices are usually fitted with jeweled bearings and are made from light-gage aluminum. This type of anemometer is widely used by meteorologists for the measurement of wind speed.

Another method of gas-flow measurement is the *hot-wire anemometer*. This requires that a wire carrying an electric current be placed in a gas stream. Heat is conducted away from the wire by the gas flow and, consequently, changes its resistance, the change being a function of the flow rate. There are two ways of measuring this change; either the probe can be connected to a Wheatstone bridge and the change in bridge output measured; or a servo system may be used that maintains the probe at constant temperature. In the latter case, power supplied to the probe provides a measurement of flow rate. Flow meters of this type are useful for measuring very low flow rates in large-diameter pipes.

CHAPTER 10

TORQUE MEASUREMENT

When an angular force is applied to a shaft it twists and produces shear stresses in the shaft material. The twisting moment that produces these stresses is described as *torque*, and it is numerically equal to the product of a tangential force and the distance at which it is applied from the center of the shaft.

TORQUE MEASUREMENT WITH STRAIN GAGES

It can be shown that the shear stresses in a solid shaft resulting from the application of a torque T, is governed by the relationship:

$$T = \frac{\pi \nu_s D^3}{16} \tag{1}$$

where ν_s is the shear stress and D is the diameter of the shaft.

Similarly, for a hollow shaft:

$$T = \frac{\pi \nu_s (D_1^4 - D_2^4)}{16 D_1} \tag{2}$$

where D_1 and D_2 are the respective outside and inside diameters of the shaft.

Considering the case of the hollow shaft:

$$G = \frac{\text{shear stress}}{\text{shear strain}} \tag{3}$$

where G is the modulus of rigidity of the shaft material.

But ν_s can also be expressed in terms of the shear strain ϕ as

$$\nu_s = G\phi$$

and substituting in (2)

$$T = \frac{\pi G\phi(D_1^4 - D_2^4)}{16D_1} \tag{4}$$

It is possible to measure ϕ with strain gages by measuring the complementary tensile and compressive strains, using four gages G_1, G_2, G_3, G_4 (see Figure 10.1). The output of these gages will provide a measurement of torque.

Suppose the total change in resistance of the bridge is δR and the gage resistance is R. Since there are four active arms, the strain measured by the bridge is:

$$\frac{\delta R}{4\ R\ G_F} \tag{5}$$

where G_F is the gage factor.

But these are the complementary tensile and compressive strains and equal to half the shear strain (see Chapter 4). ϕ therefore will be:

$$\frac{\delta R}{2\ R\ G_F} \tag{6}$$

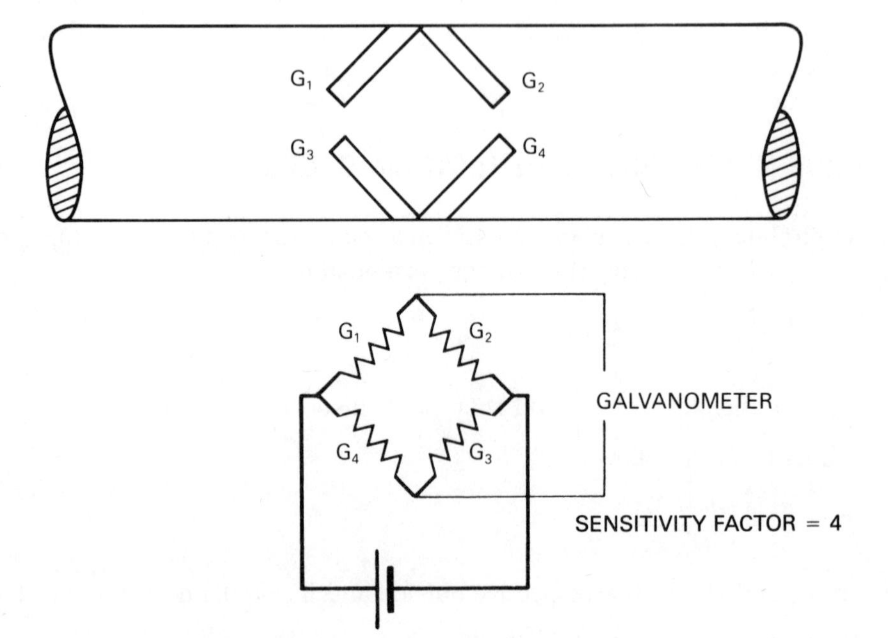

Figure 10.1
Torque Measurement with Strain Gages

and the torque equation will become:

$$T = \frac{\pi(D_1^4 - D_2^4)\ \delta RG}{32D_1RG_F} \qquad (7)$$

In practice, this equation is of limited value, as it is usually more convenient to measure the output of the bridge in volts rather than values of δR. In this case a modified formula must be used. Suppose a calibration resistance when switched across one arm of the bridge produces a change of 0.1% in the bridge output and suppose the corresponding output of the bridge is m_1 millivolts. If the output of the bridge when torque is applied is m_2 millivolts, then the value of $\delta R/R$ when torque is applied $= \frac{0.001m_2}{m_1}$ and from equation (7):

$$T = \frac{0.001\ \pi(D_1^4 - D_2^4)\ G\ m_2}{32D_1G_Fm_1} \qquad (8)$$

This method of determining torque provides a means of either calculating its value from a bridge output or alternatively producing a meter scale calibrated in torque.

TORQUE MEASUREMENT WITH STRAIN GAGE ON ROTATING SHAFTS

Occasionally it is required to measure torque in a stationary shaft. More often the shaft is rotating, and then a problem is experienced in obtaining signals from the bridge and also in supplying it with an excitation current. Slip-rings are the traditional solution, and in many cases are still the only solution in the available space.

Slip-rings

When designing slip-rings for a strain-gaged shaft it must be remembered that signals are of a very low order, often only a few millivolts, and slip-rings are notorious generators of electrical noise. It follows, therefore, that slip-ring manufacture and maintenance will be comparatively expensive.

Slip-ring assemblies that fit onto the ends of shafts are commercially available, but often the shaft ends are inaccessible, and it is necessary to design slip-rings that suit the application. In this situation several criteria must be met:

- The contact resistance between the brushes and slip-rings must be as low as possible.
- The contact resistance must not vary as the shaft rotates.
- The surface speed of the rings must be kept as low as possible.

- The resistance between rings must be as high as possible, as must the resistance between the rings and the shaft.
- The surface of the rings must not produce high-resistance oxides during use.
- The rings must be strong enough to survive the hoop strains imposed during rotation.

Contact-resistance problems suggest that the rings should be made from a noble metal or at least electroplated with a noble metal. Plating presents several problems. It is difficult to plate a ring evenly around its circumference and to the thickness required for a reasonable life. After plating, the ring must be turned to give the required surface finish, and this operation removes a large portion of the plating which has laboriously been laid. In addition, high plating currents may not be used, because the resulting plating tends to contain pinholes and to have a brittle structure.

An alternative to plating is flame spraying. The ring is first turned to produce a very fine screw thread, and then the surface is flame sprayed to a thickness of about 0.1 inch (2.5 mm). This allows eccentricity to be turned out and a fine surface to be obtained while still leaving a layer thick enough to provide the strength and wearing properties required. Flame spraying with a noble metal such as silver is extremely expensive in materials, much of which is wasted. However, this may be offset against the cost of labor for alternative designs.

The silver-wire slip-ring is an alternative which has proved very successful in practice. This is illustrated in Figure 10.2. A shaft to be fitted with slip-rings is first coated along part of its length with epoxy resin which when cured is turned to the internal diameter of insulated bases manufactured from resin-impregnated laminated sheet. These bases, designed as two half-collars, are fitted around the shaft to provide insulation between rings and act as a form for the ring wire. Manufacture of the bases is most easily accomplished by turning from the solid and then splitting the resulting ring for fitting. Slots milled across the underside of the bases allow the passage of slip-ring wires to the strain-gage end of the assembly, and a further slot milled in the edge of the base allows the free end of the wire to be recessed and soldered to the end that has been passed under the rings.

Formed silver wire about 0.1 inch square is used for the rings, and if it is wound carefully, only the slightest machining is required to provide a good ring surface. The width of the base is often designed to accommodate about five turns of wire, but this may be decreased when space is restricted.

The first base carries a single slot and the second two, one to carry the wire from the previous ring and the second to carry its own wire. Before fitting the second base, it is prudent to paint the face of the previous ring with epoxy resin. When cured, the two rings then virtually become one solid assembly, which resists the ingress of moisture and dirt.

It is possible to build a three-ring assembly and use a half bridge on the shaft to measure torque, but because of sensitivity and noise problems, it is usual to produce a four-ring assembly and use a four-arm bridge.

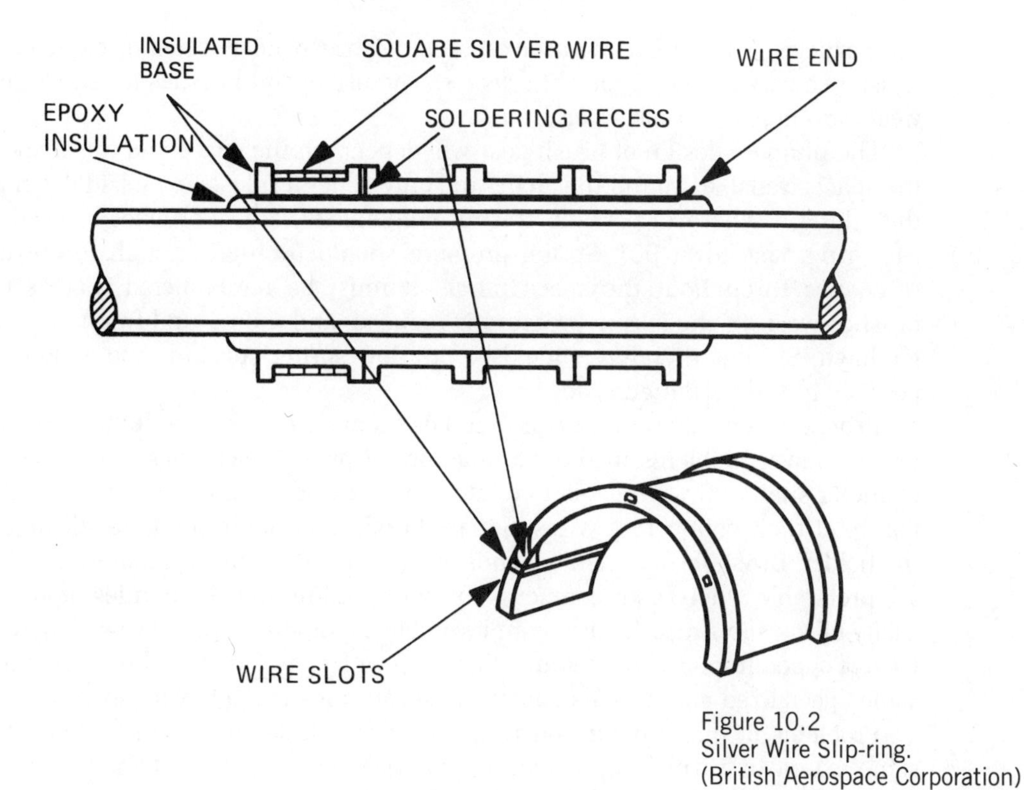

Figure 10.2
Silver Wire Slip-ring.
(British Aerospace Corporation)

Brush Assembly Considerations

When using slip-rings, the mistake is sometimes made in concentrating on slip-ring design to the point where design of good brush assemblies is forgotten. Yet the success of every installation depends as much on one as the other. In designing brush assemblies the following criteria must be considered:

- The brushes must not leave the rings at any speed below below the maximum. This effect described as *brush bounce* is usually caused by eccentricity of the rings or shaft generating a vertical movement of the brushes against the springs, which at some speed becomes resonant.
- The electrical resistance of the brushes must be as low as possible.
- The brushes must remain free to slide in the brush holders during service.
- The brush pressure must be as light as possible to reduce wear.
- The electrical resistance between adjacent brush holders must be as high as possible and remain constant.

Some of these criteria are in contradiction, and any solution must be a compromise. To fullfil the first three criteria, it seems logical to use a high brush pressure. However, high brush pressure will cause ring wear and brush wear, and the particles will spread themselves over the brush gear, causing changes in resistance between the adjacent brushes. The use of soft brushes to reduce ring wear can increase brush-wear problems.

The ultimate design of brush gear will depend on the space and the shape of the space available around the shaft, and only general rules can be laid down for their design. Wherever possible, it is recommended that two brushes be used on each ring, spaced at 90°. Spring pressure should be high enough to prevent resonance throughout the speed range. It must be remembered that as the brushes wear, so the spring pressure is reduced, and the natural frequency can go down. Springs therefore should be of a low spring constant and as long as possible in the permitted space.

Brushes themselves may be of round or square cross section. Round brushes present fewer problems in the manufacture of brush assemblies, but on large-diameter shafts they are liable to rotate, causing uncertain contact and possible failure of their connecting wires. Square brushes are more prone to sticking in the holder. Brush formulation depends on circumstance. Silver graphite brushes are preferable where poor contact is experienced due to irregularities or eccentricity of the slip-rings. Under more favorable conditions carbon brushes may be used. Copper graphite brushes are often used with stainless steel slip-rings and in some specialized applications contact with the rings is made with mercury.

Problems of dust buildup on rings and brush assemblies can be helped to some extent by varnishing insulating materials with a polyurethane-based varnish, and wherever possible providing shoulders that lengthen the resistance path. In the final analysis, there is no substitute for adequate maintenance and frequent cleaning.

Brushless Transmission

Space permitting, an alternative method of transmitting strain-gaged signals to the recording equipment is available that eliminates the need for slip-rings. A battery version of such a system is shown in Figures 10.3 and 10.4.

The system consists of a *frequency-modulated* (FM) transmitter mounted on the strain-gaged shaft and signals are transmitted from a rotating transmission coil to a stationary pickup coil. When torque is applied to the shaft, the bridge is off balanced and provides a signal that controls a *frequency modulator*. Variations in the strain-gage signal cause variations in frequency proportional to the applied torque. Output of the modulator is applied to the transmission coil which is inductively coupled to the pickup coil. A *bandpass filter* removes extraneous noise from the transmitted signal, which is then discriminated and amplified.

Several variations of the design exist, one of which uses a second transmission coil to provide ac power, in place of the battery.

The use of FM transmission has considerable advantages over slip-rings. Maintenance problems associated with slip-rings are eliminated, and the equipment is much less affected by adverse environments. In addition, the system provides preamplification of the signal, and combined with the fact that brush noise is eliminated, excellent signal-to-noise ratios are achievable. Older systems present space problems, especially where a number of strain-gage circuits are accommodated on one shaft. However, as electronics are reduced in size and power requirements become smaller the problems become less significant.

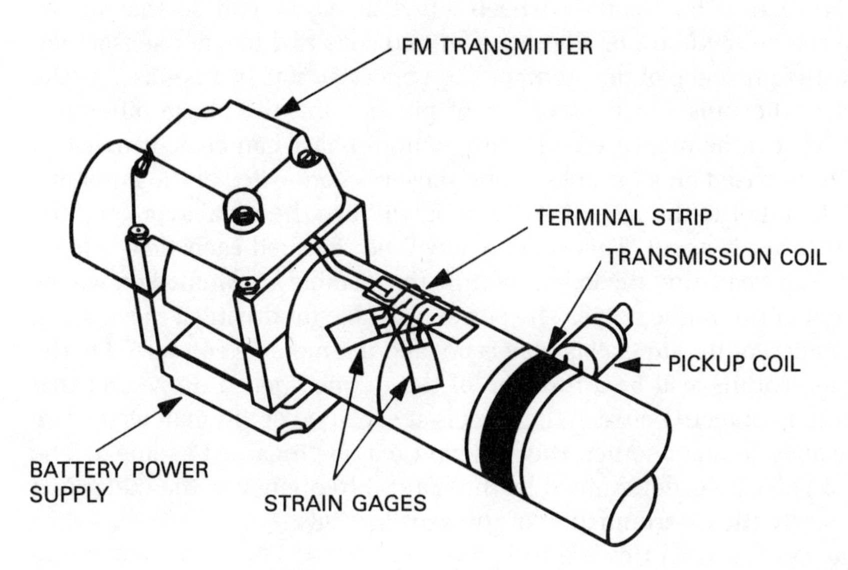

Figure 10.3
FM Wireless Transmission Signal

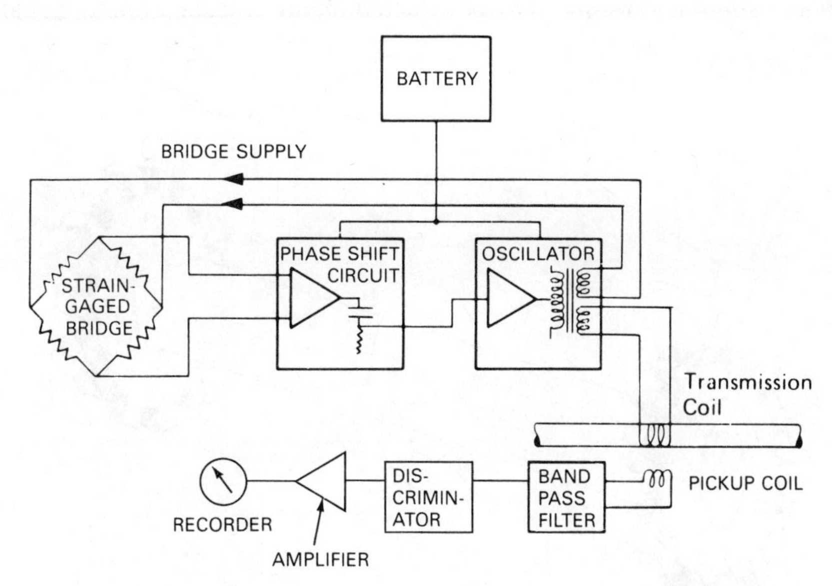

Figure 10.4
Schematic Diagram of Wireless Transmission System

TORQUE MEASUREMENT BY PHASE MEASUREMENT

When torque is applied to a shaft, it twists, and the angle of twist is proportional to the torque. Measurement of this angle gives a measurement of torque. One way of measuring angle of twist is to use toothed wheels, one at each end of the shaft, and then detect the change in relative position of the teeth. Suppose the shaft shown in Figure 10.5 has toothed wheels fitted at either end so that in the untorqued state the teeth are in alignment. If two coils and magnet sensors are used to sense the presence of the teeth, pulses generated will be in phase. As the shaft is loaded, the pulses will move out of phrase, and the phase difference between the two can be measured. Measurement of phase can be accomplished by using a circuit based on a bistable. If one sensor is connected to the input of a bistable and the other to the reset line, the first will turn the bistable on, and the second will turn it off again. This sequence will be repeated each time a tooth passes the pickup head, and the length of time the bistable is switched on will be a measurement of the phase angle. The power available for driving a moving coil meter is a function of the time a bistable is on, and if a meter is connected to the output line, its reading will be a function of the torque applied. Providing that there is no overlap of teeth between the wheels at each end of the shaft as the two ends twist relative to one another, this function may be regarded as linear. The power limit of linearity is determined by the natural frequency of the meter and the upper limit by the discrimination of the sensor heads.

Torque meters based on this principle are very robust. The problem of using them is one of calibration. Coil and magnet pickups only provide an output when the shaft is rotating. Consequently, the calibration must be dynamic. If one sensor is mounted on an adjustable bracket that is calibrated in degrees of offset to the other, a simulated torque may be effected on an unloaded rotating shaft.

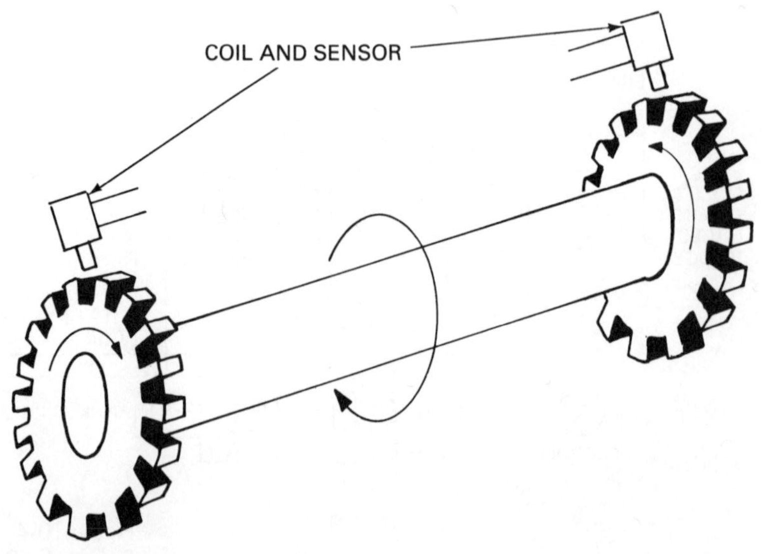

Figure 10.5
Phase Change Torque Meter

CHAPTER 11

LEVEL MEASUREMENT

SIGHT GLASSES

By far the simplest method of measuring the level of liquid in a tank employs a *sight glass* tube mounted vertically beside the tank, the ends of the tube being connected to the bottom and the top of the tank as shown in Figure 11.1. Liquid in

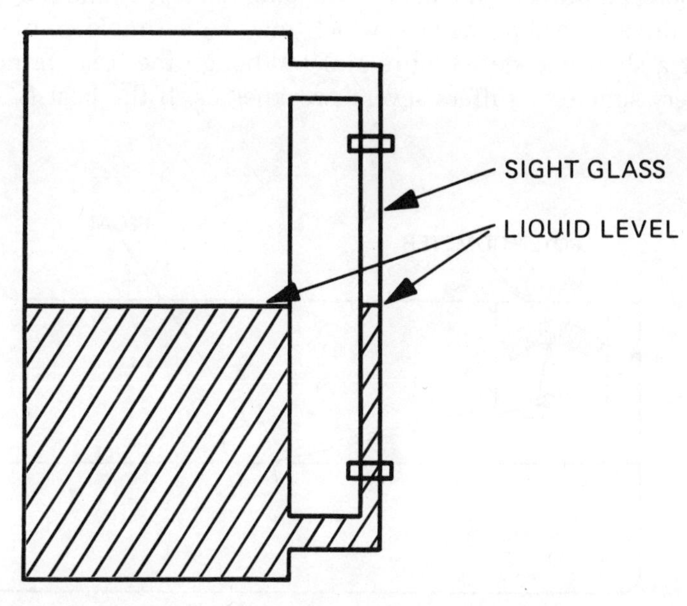

Figure 11.1
The Sight Glass

the tube and in the tank take up the same level thereby providing a direct visual measurement. Sight glasses have several obvious limitations. Tanks are often inaccessible and viewing a sight glass can be difficult. In addition, the glass is liable to be broken which can result in considerable spillage before the inlet can be sealed off. If hot or corrosive liquids are being stored, the consequence of an accident can be serious. In addition there is also a limit to the length of glass tube that can be mounted in this way.

To overcome the limitations of the sight glass two traditional methods of level measurement are used. The first uses floats of various types and the second is an indirect method, in which the pressure at the bottom of the tank is used as a measurement of level. More recently, ultrasonic and laser techniques have been developed.

FLOAT SYSTEMS

When tanks containing liquids have changes in level of less than about 12 inches (30 cm) a float with a pivoted arm may be used. The classic example of this is the fuel gage for the gasoline tank of an automobile (Figure 11.2).

This device consists of a float attached to the end of a pivoted arm which often operates a potentiometer. As the gasoline level varies, the voltage ouput from the potentiometer changes. This is often a nonlinear output and is sometimes left that way, to provide better resolution at the bottom end of the scale.

When deep tanks are employed it is not possible to use this method and *reeled tape* systems are often used as an alternative. The arm is replaced by perforated tape that carries the float at one end, while the other end is attached to a spring-loaded drum. As the level falls, the weight of the float is sufficient to overcome the spring tension and unreel the tape. By wrapping the tape round the sprocket it is possible to drive a reading device, which may be a simple counter, a potentiometer, or a shaft encoder (Figure 11.3). Although the basic principle of this design is very simple, it suffers several weaknesses. If the float is allowed free

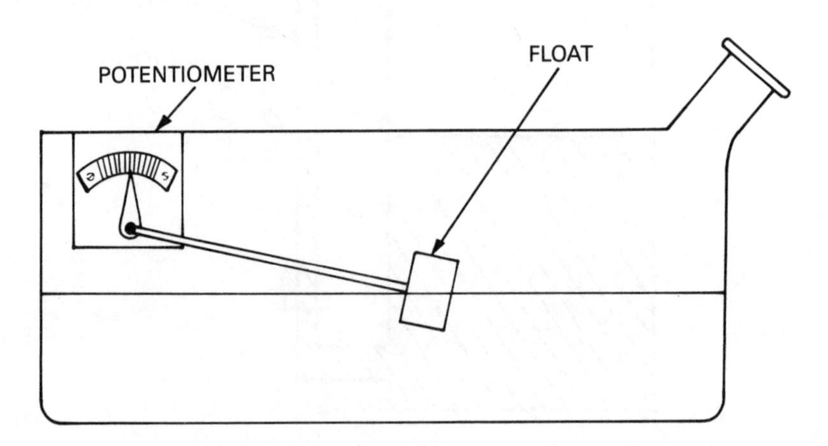

Figure 11.2
Motor Car Fuel Tank

movement in the tank, it will be constrained by the tape in one plane only. This will allow it to move across the surface of the liquid and give erroneous readings. In addition, during filling operations there is usually considerable turbulence, which can damage or cause the tape action to jam. Both of these problems may be overcome by introducing a *still pipe* into the tank. This is a tube that runs vertically from the tank top to a few inches from its bottom. The float is confined by this tube, which constrains and protects it.

A second weakness of the basic system is that the tape must pass over two pulleys if the reading equipment is to be placed near the bottom of the tank. Friction in these pulleys can cause errors in the float position. One way to overcome this is to vibrate the tape with some type of vibration exciter; often a small electric motor fitted with an eccentric.

A further source of error is to be found in the changing weight of the float. When the level of liquid is low, the float not only supports its own weight, but also the weight of the tape above it. At high liquid levels the tape is considerably shorter and the effective weight of the float that much less. One method of correcting this employs a spring system whose characteristics have been designed to maintain constant tension in the tape as the level drops, effectively counterbalancing the additional weight of the tape.

There are a number of variations of the float type of level indicator. It is possible with electrical outputs to mount reeling equipment at the top of the tank and carry signals to a remote reading point by cable. One method uses devices called *synchros*.

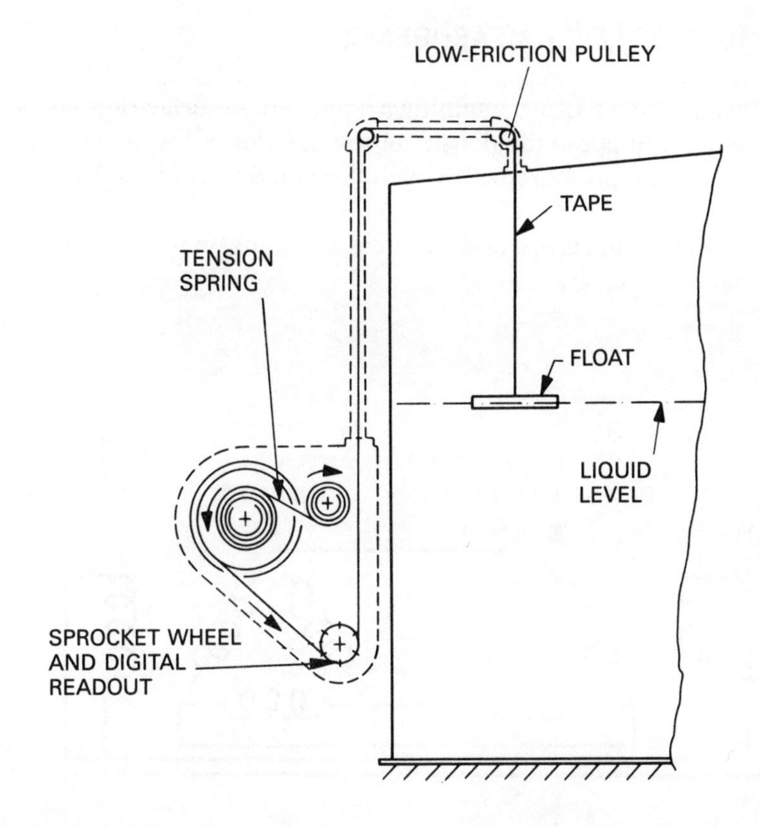

Figure 11.3
Level Gage. (Whessoe Ltd.)

The synchro may be considered as a transformer/motor having a three-phase stator and a single-phase rotor. If two such devices are coupled, as shown in Figure 11.4, the rotor of synchro 1 (a slave) will follow the movement of the master rotor of synchro 2.

Consider first the case in which rotor 1 is connected to an ac supply and rotor 2 is open circuit. Because of the transformer coupling between the rotor and stator a voltage is induced in stator 1 which is reproduced in stator 2. Stator 2 will now generate a magnetic field which induces a voltage in rotor 2. This voltage is proportional to the sine of the difference angle between the two rotors and is zero when the two are in alignment. If the two rotors are not in alignment, the application of a common power supply to both will create a distortion of the magnetic field, which in turn will produce a torsional couple on both rotors. If one rotor is restrained, the other will move to the point at which the couple is zero, and this point will coincide with the alignment of the two. Any movement of one rotor will cause the other to follow faithfully, within the design limits of the synchro. Synchros are rugged devices requiring practically no servicing. They are capable of providing adequate torque for driving sensors, gear trains, and pointers. One feature, however should not be overlooked. There are three alignment positions the rotor may take up. If for some reason power is switched off or the driven unit jams, it is possible that the rotor may take up a new alignment 120° from its set value. This could provide a gross error in calibration.

PRESSURE METHODS OF LEVEL MEASUREMENT

Consider a tank shown in Figure 11.5, containing a liquid whose surface level is at a height h above the base. Suppose the density of the liquid is ρ, the area at the base of the tank is a, and the pressure with which the liquid acts on the base of the tank is P.

The mass of liquid in the tank is $a\rho h$ and the pressure acting on the base of the tank due to the liquid is:

$$P = \frac{ah\rho}{a}$$

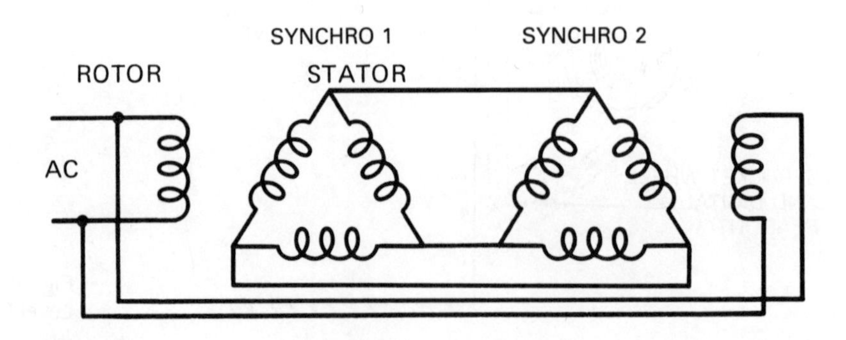

Figure 11.4
Slaved Synchros

or

$$h = P/\rho$$

In addition to the pressure exerted by the liquid the atmosphere also exerts pressure. If a gage pressure sensor is used, the atmospheric pressure acting on the back of the diaphragm cancels the effects of atmosphere on the liquid and the output of the sensor is proportional to h. If the tank is sealed and pressurized, this relationship no longer holds, and it becomes necessary to use a differential sensor with the reference port connected to the space above the liquid.

Connecting a sensor to the base of a tank may not always be feasible. The base may be inaccessible, or it may contain liquids that are incompatible with the sensor materials, or it may not be possible to change sensors without first emptying the tank. In this case an alternative system is used called *the purge system* illustrated in Figure 11.6. With this method, an air pipe is inserted into the tank,

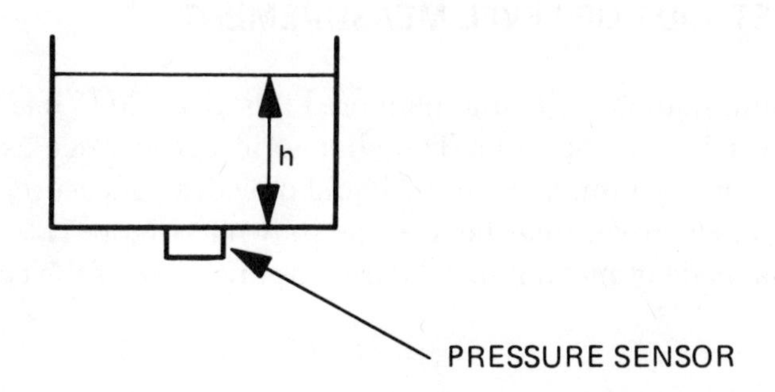

Figure 11.5
Pressure Methods of Level Measurement

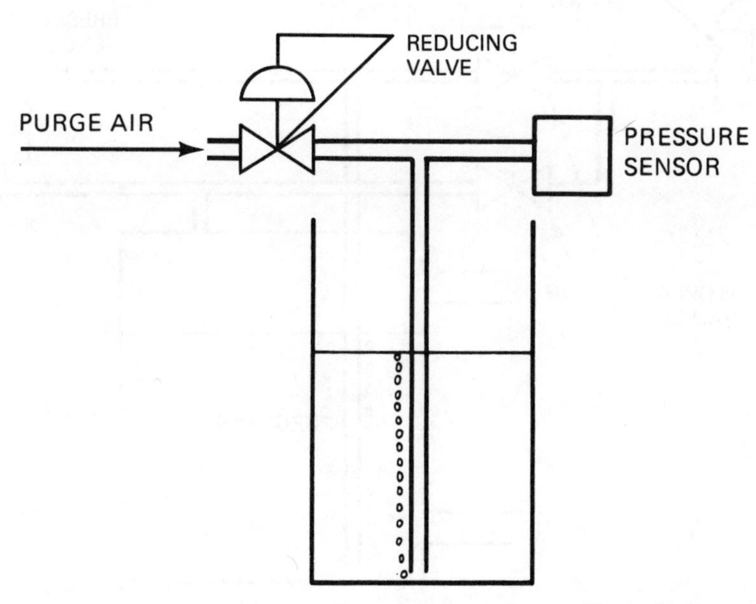

Figure 11.6
Purge System of Level Measurement

with its open end close to the base. At the top of the pipe is a tee, one side of which is connected to a sensor. The other side is connected through a valve to an air supply. Air pressure builds up in the pipe, forcing the liquid level down against the liquid pressure until air bubbles emerge from the open end of the pipe. At this point, pressure in the pipe becomes constant, equal to the pressure at the base of the tank, and the pressure sensor provides a reading proportional to liquid level.

Purge systems may be modified to operate in sealed tanks by using a differential sensor and a reference purge supply, as shown in Figure 11.7.

A reducing valve is set at a slightly higher pressure than the pressure in the tank, and the flow adjusting valves are set so that a stream of bubbles issues from the open end of the stand pipe. The level is sensed by a differential pressure sensor. To facilitate setting the flow valves sight glasses are often used as illustrated in Figure 11.8.

CAPACITANCE METHODS OF LEVEL MEASUREMENT

When nonconducting liquids are being monitored it is possible to use the liquid as a dielectric material for a capacitor. The change in capacitance as the electrodes become more or less immersed in the liquid provides a measurement of its level. Several types of electrodes may be used, as shown in Figure 11.9a, b, and c. When a tank wall is made of metal it may be used as one plate of the capacitor. A

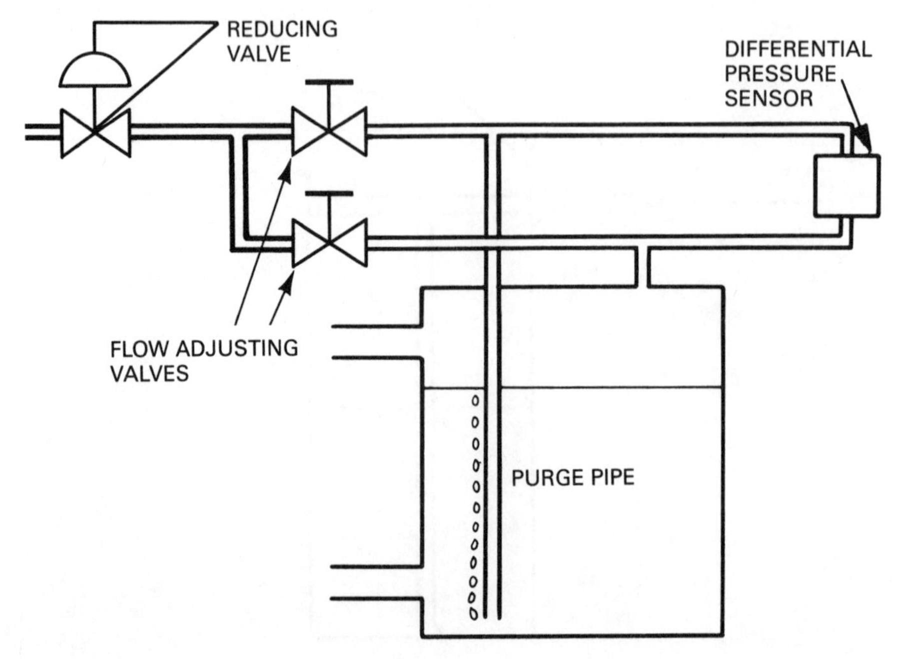

Figure 11.7
Purge System for Sealed Tanks

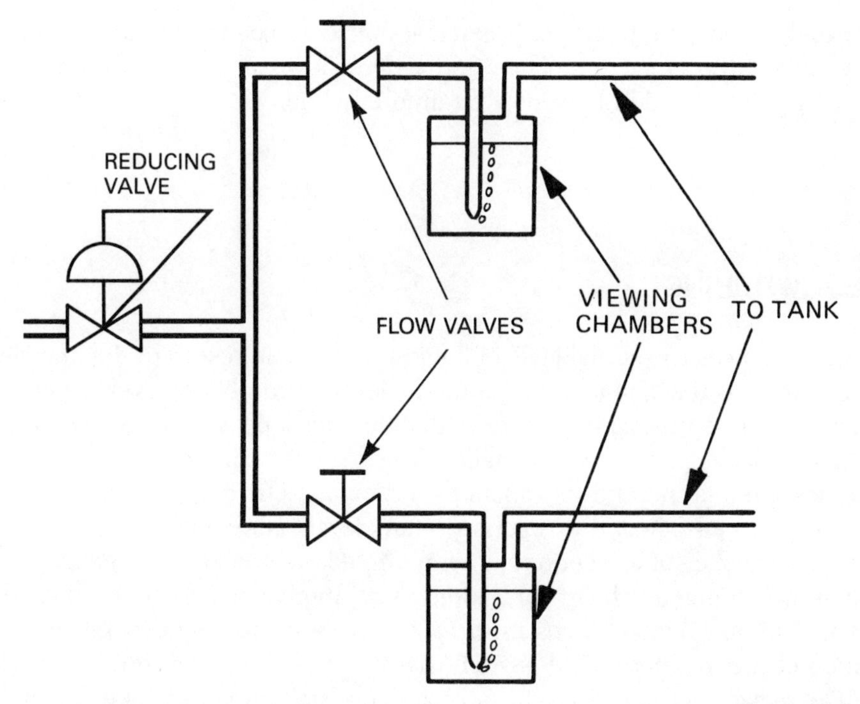

Figure 11.8
Bubble Viewing Chambers

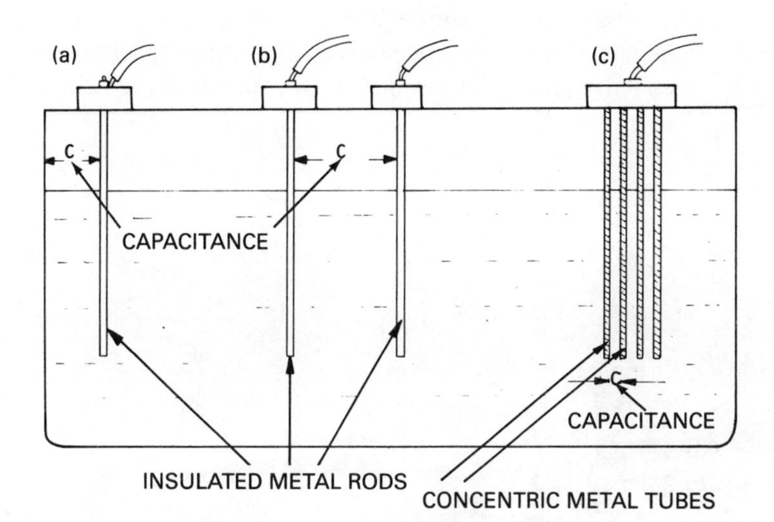

Figure 11.9
Capacitance Level Sensor

single electrode running parallel to the wall but insulated from it provides the other plate (a). If the tank is made of insulating material two parallel electrodes may be employed with the same effect (b). An alternative to the rod electrodes uses two concentric cylinders of metal as plates of the capacitor (c). Generally speaking, the changes in capacitance of a concentric cylinder sensor are considerably higher than the rod type. It is usual to use inexpensive materials such as mild steel for the capacitor plates and where necessary protect them with a coating of polyvinylchloride or polytetrafluorethyline.

LEVEL SWITCHING

In many instances the actual level of liquid is of less interest than the knowledge that the tank has reached a critical contents level. A number of level switches are available for these functions. Figure 11.10 illustrates a float-operated switch operated magnetically. This design makes possible the complete isolation of the switch from the liquid and its vapors by a nonmagnetic diaphragm.

Movement of the float due to a rise in tank level causes magnet m_1 to move in an arc past the face of a second magnet mounted on a rocker. The poles of these magnets are arranged to be in opposition. As m_1 approaches m_2, m_2 is repelled but restrained by mechanical stops until m_1 has passed the center point. Here the direction of the magnetic field is such that it drives the rocker onto its opposite stop. The rocker therefore has a bistable action which imparts a snap operation to the switch.

Pressure switches are also used to detect tank levels. The one illustrated in Figure 11.11 makes use of the buildup in pressure of a trapped air column as the liquid level rises around it. Sensors are mounted at the top of a tank, with their ports connected to stand pipes called *risers*. The open ends of these risers terminate at a distance below the alarm points determined by the operating pressure of the switches.

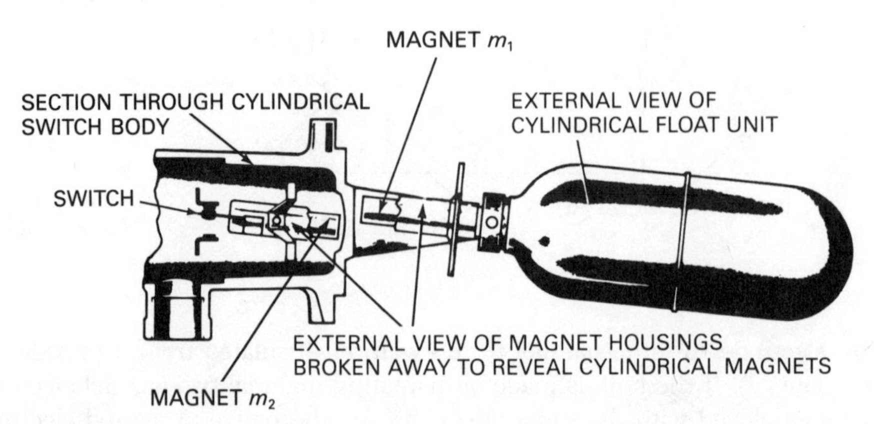

Figure 11.10
Magnetic Level Switch.
(Bestobell Mobrey Ltd.)

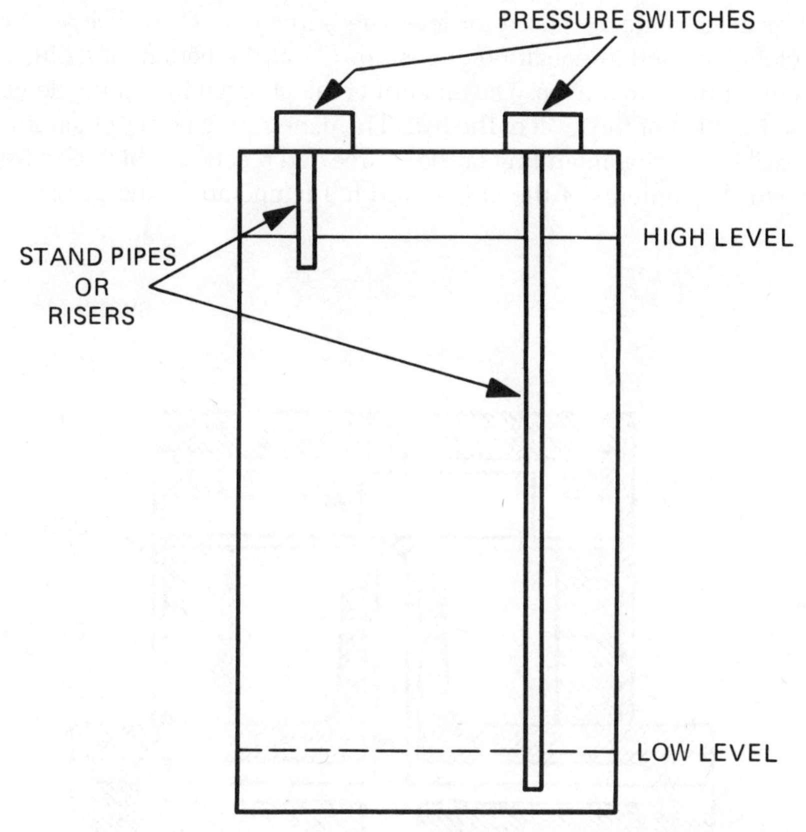

Figure 11.11
Pressure-Operated Level Switches

MEASUREMENT OF THE LEVEL OF SOLIDS IN TANKS

This discussion of solid-level measurement assumes that the solids are in the form of powders, grains, or crystals and are free flowing. Such materials may often be measured in the same way as liquids. It is possible, for example, to use a capacitance-level measurement, or it is possible to use a diaphragm-pressure switch (Figure 11.12). Pressure switches designed for use with solids usually have large diaphragms which may be 10 inches (25 cm) or more in diameter.

An alternative principle employs a paddle wheel driven by a low-torque synchronous motor through a spring coupling. If the level of solids is below the level of the paddle, the motor drives the paddle. As the level reaches the vanes of the paddle, the motor stalls, and in doing so winds up the spring to operate a switch.

Another device uses a pendant cone. This device relies on the tendency of solids to "heap" as they are being poured into a container. As the solids level reaches the required level the cone is tilted by the heaped material, causing a tilt switch to operate.

Nuclear devices are also used for level measurement. These use a source of gamma radiation such as cobalt-60 or cesium-137 at the bottom of a bin, with a Geiger counter fitted to the top. The amount of radiation falling on the detector is an inverse function of the level of the bed. The penetrating power of gamma rays makes possible external mounting of the source and receiver, which is extremely useful when the contents of the bin are at high temperature and pressure.

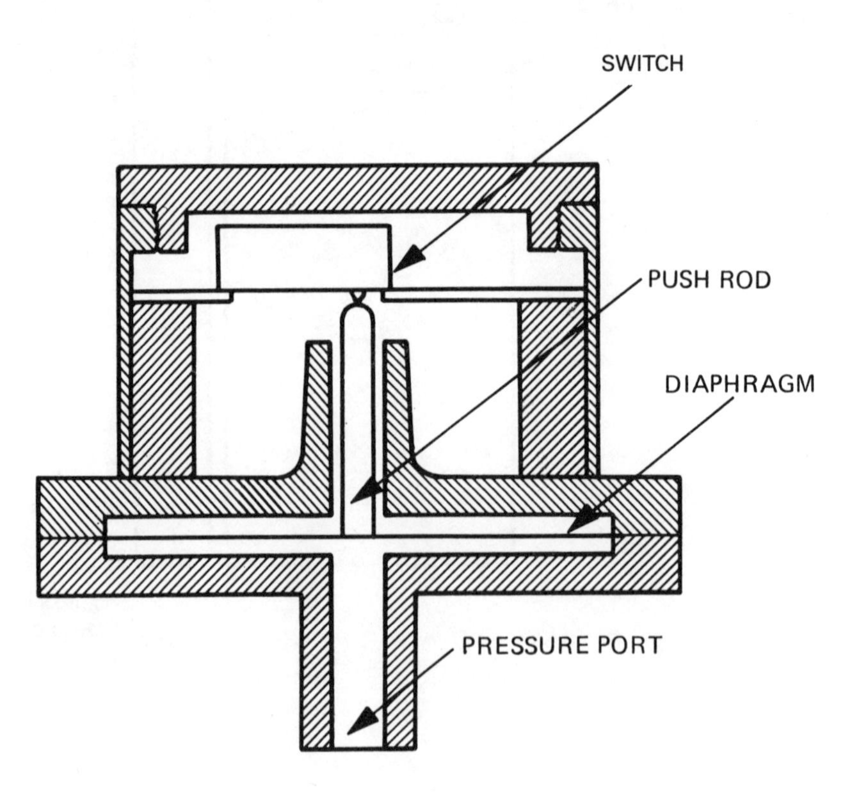

Figure 11.12
Pressure Level Switch

CHAPTER 12

MISCELLANEOUS PROPERTIES OF MATERIALS

Previous chapters have attempted to cover the more common measurements and techniques which are encountered by instrumentation engineers. A considerable number of measurements, although of less general interest, are equally important to those working in specific fields. In this chapter some of these applications are considered.

HUMIDITY MEASUREMENT

Relative humidity (RH) is defined as the ratio of water-vapor pressure present to the saturated-water-vapor pressure at the temperature at which the measurement is being made. High atmospheric relative humidity coupled with a high ambient temperature is detrimental to health, and its control is very important in some industrial processes.

Hygrometers

One traditional method of measuring relative humidity is the *wet-and-dry-bulb hygrometer*. This device employs two thermometers using liquid in a glass. One thermometer is exposed to the ambient air and is said to have a *dry bulb*. The *wet bulb* thermometer is one with a wetted porous wick wrapped round its bulb. The end of this wick is immersed in a reservoir of water. Evaporation from the wick causes a depression in the thermometer reading, and the difference in readings between the two thermometers may be used in conjunction with *psychometric charts* to determine relative humidity. To obtain accurate readings of humidity, air must flow freely round the thermometers, and to facilitate this they are often mounted in open-ended transparent enclosures with airflow maintained by a fan. Liquid-in-glass thermometers do not produce electrical output. When this is necessary, it is usual to replace them with either resistance thermometers or thermocouples.

Resistivity Sensor

An alternative to wet-and-dry-bulb methods is the *resistivity sensor*. This sensor has electrically conducting surface layers integral with a nonconducting substrate (Figure 12.1). Changes in relative humidity cause the resistance of the gap between the conductors to vary, and this change may be measured in a bridge.

Organic Material Hygrometers

Many organic substances change their length with moisture content. Human hair, for example, increases its length with increased humidity; cotton decreases its length. It is possible to use these materials to operate a pointer or to provide an electrical output from a coupled electrical transducer.

In many cases it is not necessary to measure humidity, only to maintain it around a predetermined value. For this purpose the change in length of organic substances may be used to operate switches. These devices are called *humidistats*. The humidistat shown in Figure 12.2 uses cotton threads as a

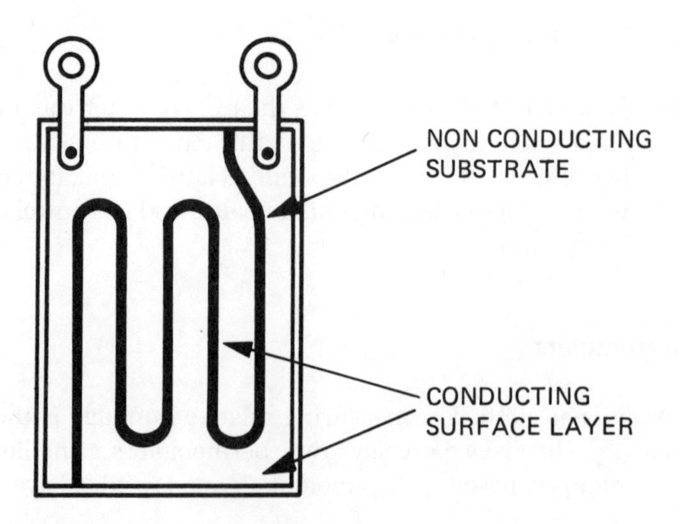

Figure 12.1
Resistivity-Type RH Sensor

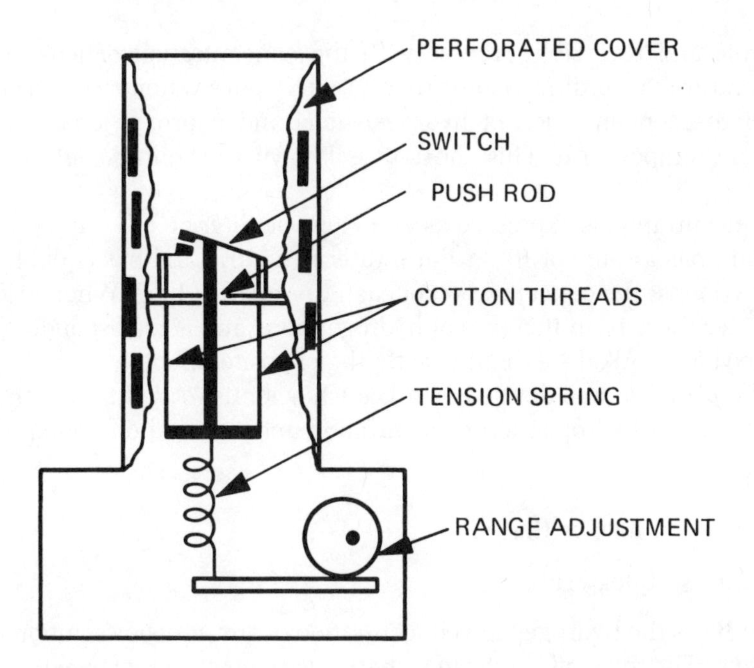

Figure 12.2
Humidistat

transducer. As the thread moistens, it contracts and stretches the restraining spring until at a point of preset relative humidity the force rod is moved sufficiently to operate a snap action switch. The operating point of the switch is adjusted by adjusting the restraining spring tension.

It is possible to provide this humidistat with a three-position switch. This may be set off in a preset neutral zone, switch one way when the humidity is high, and the other when the humidity is low. With this arrangement both humidifying and dehumidifying equipment may be operated from one humidistat.

pH MEASUREMENT

In many chemical and water purification processes it is necessary to know the acidity or alkalinity of aqueous solutions. These properties are produced by the presence in the solution of free (dissociated) hydrogen atoms carrying a positive charge and described as *hydrogen ions*. The more acidic the solution, the greater the hydrogen ion concentration.

If pure water is considered, all but a small fraction exists in its molecular form. The fraction ionizes according to the expression:

$$H_2O \leftrightharpoons H^+ + OH^-$$

This equation indicates that the water has split into two parts, a hydrogen ion carrying a positive charge and a *hydroxyl ion* carrying a negative charge. It is a dynamic and reversible process with the concentration of ions depending on temperature. According to ionic theory, where pure water is concerned, the product of the concentration of hydrogen ions and hydroxyl ions is a constant at constant temperature. This constant is known as the *dissociation constant for water.*

Concentration is expressed as gram ions per liter at 25°C, and the dissociation constant has a value of 10^{-14}. Pure water is neutral and has equal hydrogen and hydroxyl ions; the concentration of each therefore is 10^{-7}. When acid is added to the water there is an increase of hydrogen ions and a corresponding decrease in hydroxyl ions. Alkalis exhibit exactly the opposite effect.

The *pH scale* was devised by a Danish scientist called Sorenson to provide a linear scale of hydrogen ion concentration and is based on an expression:

$$(H^+) = 10^{-pH}$$

from which it follows that:

$$pH = -10 \log (H^+)$$

where H^+ is the hydrogen ion concentration of any aqueous solution in gram ions per liter. The *p* in *pH* indicates that it is a power or exponent value. If the complete range of hydrogen ion concentration from 1 to 10^{-14} is considered, then the corresponding range of *pH* will be 0 to 14, with pure water equal to 7. On this scale, solutions with pH values lower than 7 will be acid and above 7 will be alkaline, each unit step representing a tenfold change in hydrogen-ion concentration.

pH Electrodes

Measurement of *pH* relies on the buildup of a positive charge on an electrode by the free hydrogen ions. The difference in potential between this electrode and a reference electrode is a measurement of *pH*. The measurement electrode consists of a thin-walled bulb made of special conductive glass forming the end of a container with walls of nonconductive glass. The glass bulb contains a *buffer solution* connected to the instrument lead by an inner electrode (Figure 12.3).

If the measurement electrode is inserted into a solution with a hydrogen ion content that is lower than the buffer, hydrogen ions will diffuse from the buffer into the external solution until their concentrations are equal. This will lower the potential on the inner electrode when compared with a stable reference source. Similarly, if hydrogen ion concentration of the solution is greater than that of the buffer, hydrogen ions will migrate into the buffer until equilibrium has been restored. The potential of the electrode in this instance will have been raised.

Reference sources are modified measurement electrodes whose junctions are isolated from the solution by a *salt bridge.* One such bridge uses a barrier layer of potassium chloride solution which, although providing electrical contact with the solution being tested, creates no potentials due to ionization. The salt bridge

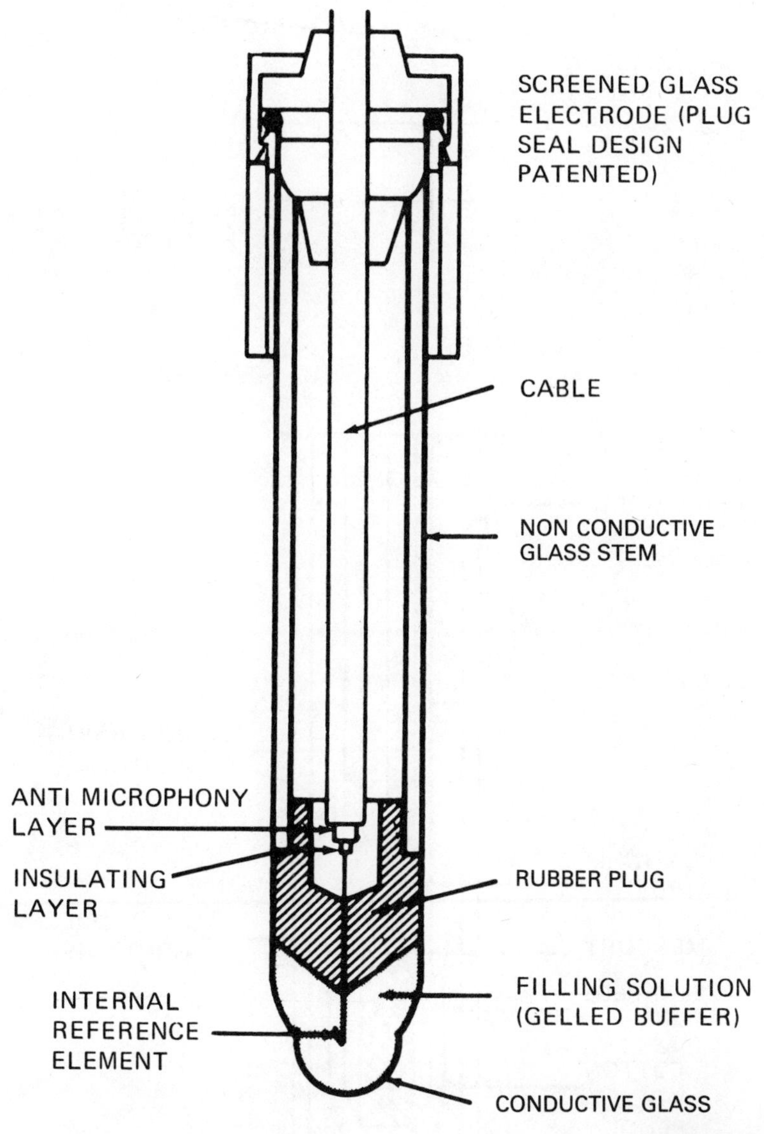

Figure 12.3
pH Glass Electrode.
(Electronic Instruments Ltd.)

solution is contained in an outer vessel, and contact is maintained with the test solution by allowing it to percolate slowly through a porous diaphragm. An inner vessel contains a platinum electrode wire that passes into a mercury pool to provide electrical contact with the measuring system. This measuring system consists of an electrode paste of mercurous and potassium chloride, providing a stable source of mercurous ions, which in turn produces a stable reference potential (Figure 12.4). The measurement of *pH* is temperature dependent, and although a manual correction is possible it is usual to make the correction electrically within the measuring instrument.

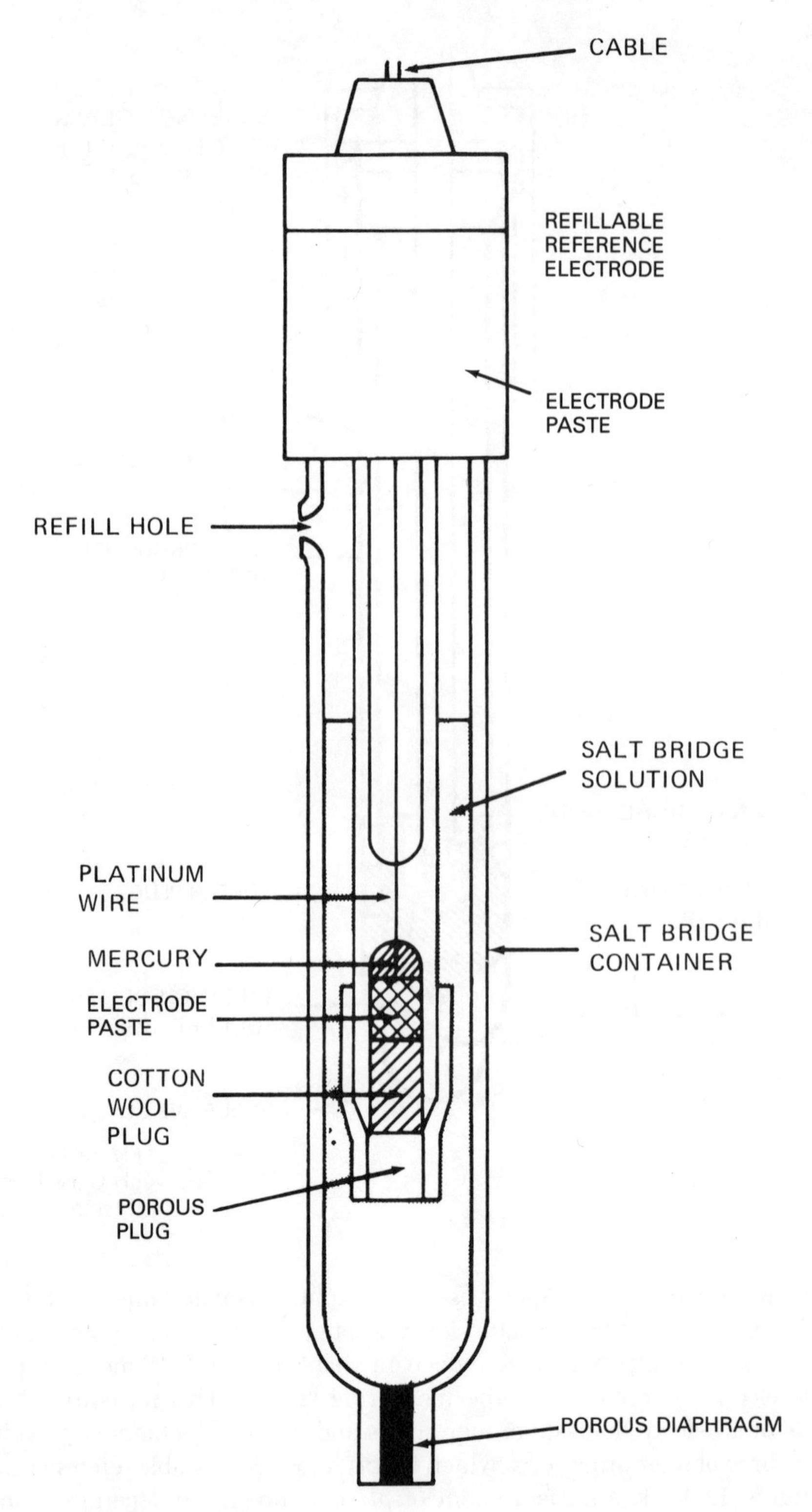

Figure 12.4
pH Reference Junction.
(Electronic Instruments Ltd.)

DENSITY MEASUREMENT

Several methods of measuring density are in use, one of which is the *continuous weighing densitometer* illustrated in Figure 12.5. This system employs a U-tube mounted horizontally and connected to a pipe system by flexible connectors. The U-tube is free to hinge vertically and this movement is reacted by a force balance motor. The U-tube represent a fixed volume through which the fluid whose density is being measured passes. Any change in density of this fluid is reflected by a change in weight of the volume and causes the tube to move vertically against the force applied by the motor. This movement is detected by a displacement sensor (often an LVDT) and is used to off-balance a servo amplifier. Depending on the direction of movement of the tube, the servo amplifier adjusts the current of the force balance motor to return the tube to the horizontal. Measurement of the motor current provides a measurement proportional to the density of the fluid.

In addition to measuring densities of fluids, the device is also used for measuring the densities of difficult substances such as slurries.

SPECIFIC GRAVITY MEASUREMENT

The simplest way of measuring the specific gravity of a fluid, and by far the best known, is the *hydrometer*. A number of adaptations of this principle are used, the simplest consisting of a ferrous tube fitted to the stem of the hydrometer and used as the core of a differential transformer.

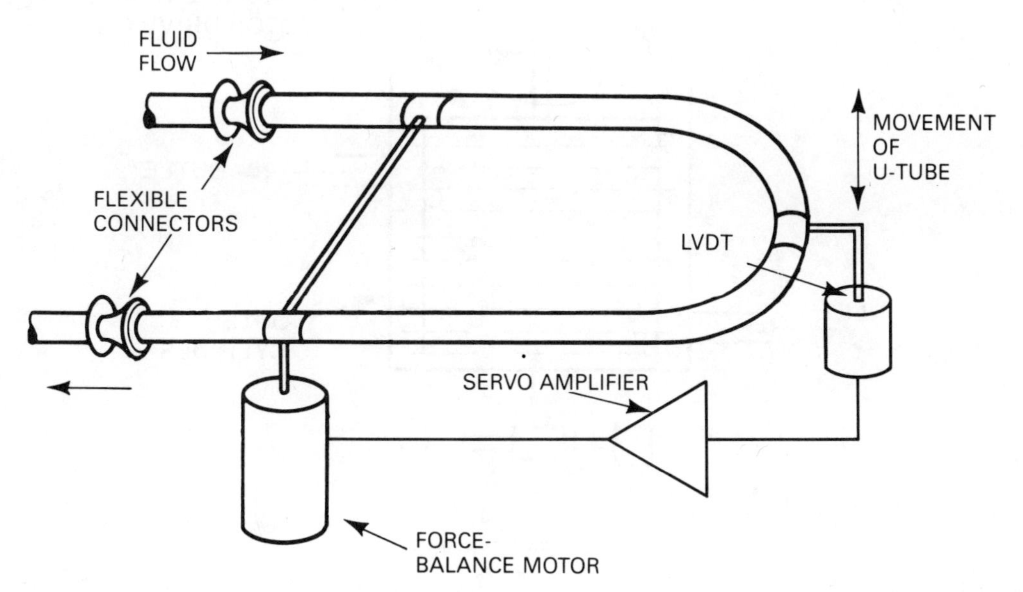

Figure 12.5
Continuous Densitometer

VISCOSITY MEASUREMENT

Drag Viscometers

A number of commonly used *viscometers* require visual observation and are of little use in a continuous process. They may not be operated automatically, nor do they easily adapt to electrical outputs. Of the traditional viscometers, the rotating cylinder type is probably of most importance (Figure 12.6).

Two concentric cylinders are filled with a fluid whose viscosity is to be measured. The inner cylinder is rotated by an electric motor; the outer is suspended on leaf springs resisting torsional movement. The viscosity of the fluid creates "drag" between the two cylinders which is balanced by the torsional couple. Measurement of this couple gives a measurement of viscosity and may be measured by using strain-gaged leaf springs or alternatively a servo force balance system.

The basic design of this viscometer presents certain problems. If the system is to be used in line, fluid must be allowed to enter and leave the device without interfering with the coupling moment. Inlet and outlet pipes cause a restraint and one solution is to mount both cylinders in a sampling chamber. (See Figure 12.7.)

This design uses a solidly mounted outer cylinder and an inner cylinder rotated by an electric motor. Either the motor current is measured to provide a measurement of viscosity, or the motor body may be mounted on bearings and the turning motion resisted by a spring. In this latter instance the angular movement of the motor body provides a measurement of viscosity.

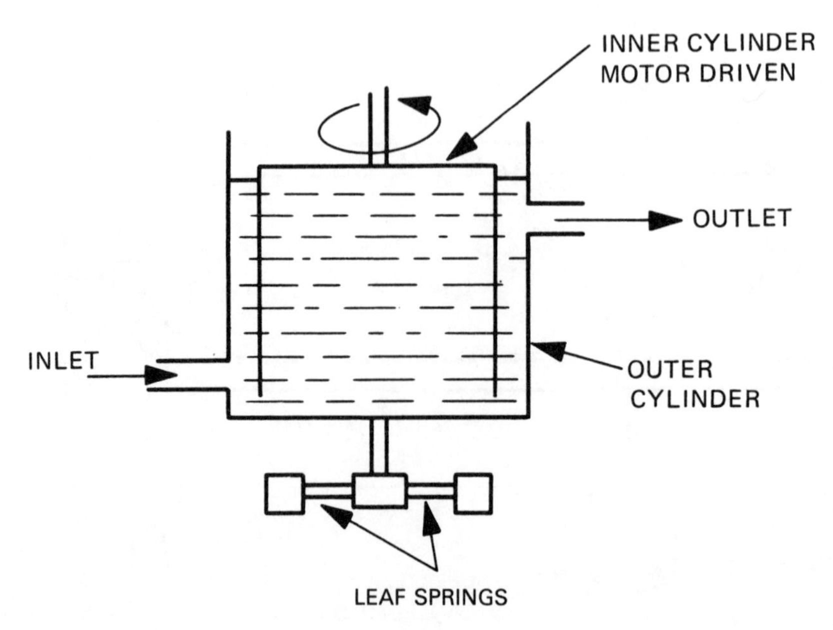

Figure 12.6
Rotating-Cylinder Drag Viscometer

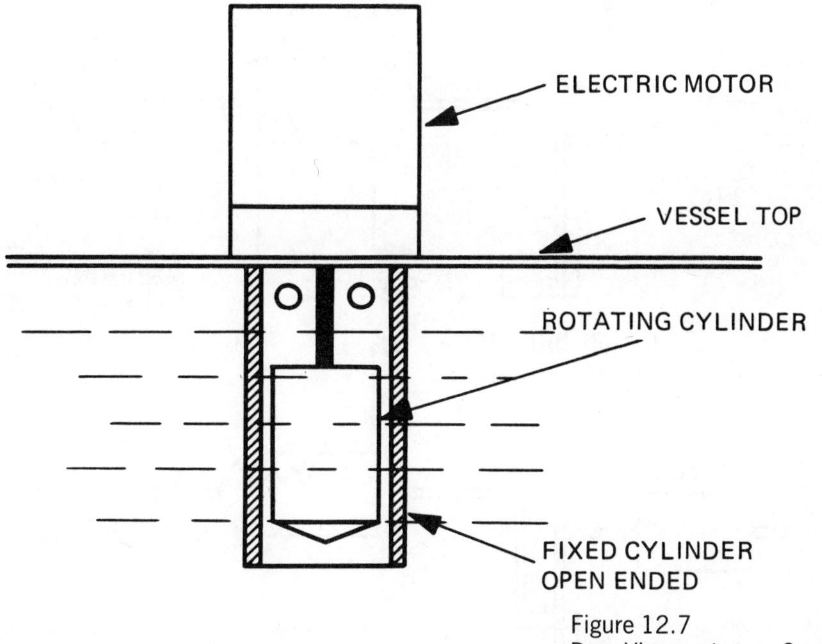

Figure 12.7
Drag Viscometer—a Sampling-Chamber Mounting

CARBON DIOXIDE MEASUREMENT

Measuring the amount of carbon dioxide in a flue gas is important in determining the efficiency of combustion. The maximum efficiency occurs when the flue gas contains between 10 and 14% carbon dioxide. One method of measuring carbon dioxide content uses a measurement of the *thermal conductivity* of the gas (Figure 12.8).

Two thermal conductivity chambers are used in this equipment, both identical in thermal mass and containing two platinum heater wires connected as adjacent arms of a Wheatstone bridge. Flue gas and air are metered through the chambers at equal flow rates, both having been previously saturated with water vapor and reduced to the same temperature in bubblers. With identical gases flowing through each chamber, the bridge remains in balance, but passage of a flue gas containing carbon dioxide removes heat from the sample chamber and creates a new equilibrium temperature depending on thermal conductivity of the gas. Oxygen and nitrogen have little effect on thermal conductivity, and the major source of difference is due to carbon dioxide. Consequently, the off balance of the bridge is due to the difference in temperature.

COMPONENT GAS ANALYSIS

The *gas chromatograph* (Figure 12.9) provides a useful method of analyzing component gases from a mixture. It is based on the principle that if a mixture of

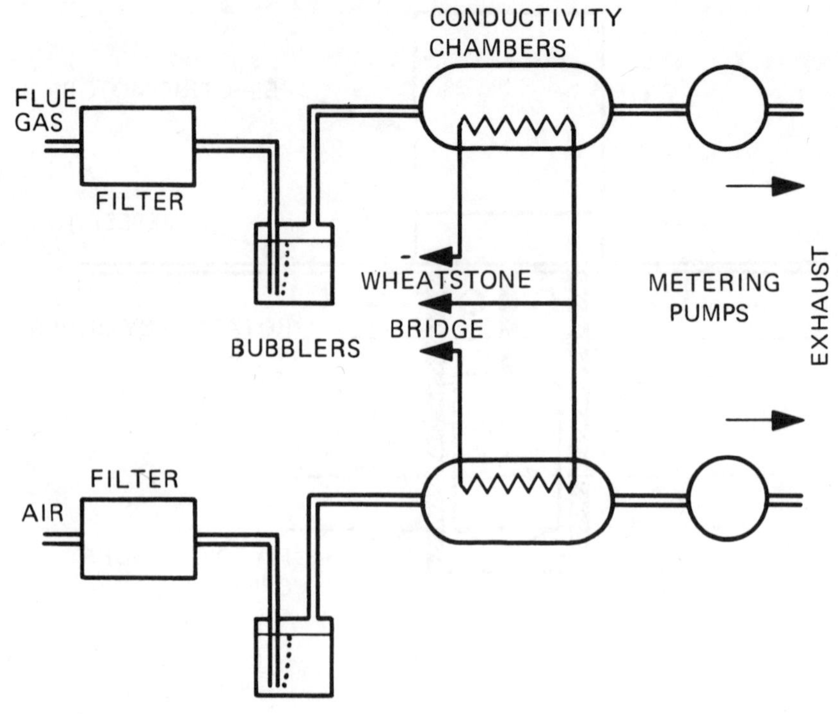

Figure 12.8
CO_2 Thermal Conductivity
Measurement Chambers

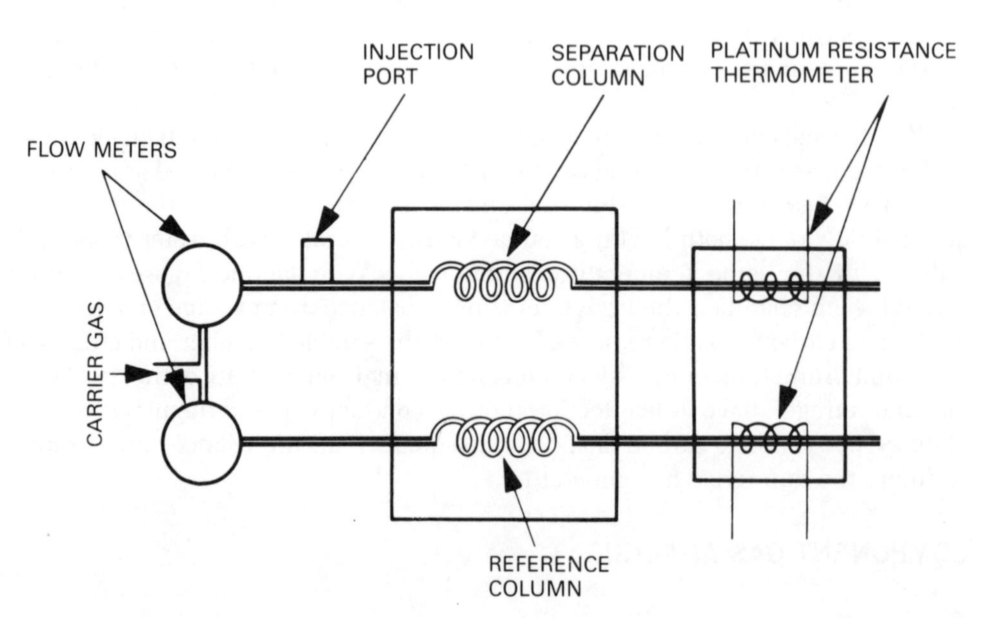

Figure 12.9
The Gas Chromatograph

gases is drawn through a material that resists its progress, lighter gases will pass through the material more quickly than heavier ones. The process by which separation occurs is known as *adsorption*, and for any particular gas and any particular material, the length of time a component gas takes to pass is a constant.

In the gas chromatograph a long column packed with *adsorption* material is installed in an oven and maintained at a constant temperature. As the gas passes through the column it separates into its components which emerge from the column into a temperature-measuring zone where the gas stream temperature is recorded against time. The length of time between the gas entering and leaving the column is an indication of its composition, and the total heat it contains is a measurement of the quantity of constituent present.

It is normal for a gas chromatograph to contain two columns, a reference column and a sampling column, both maintained at an accurately controlled temperature. The columns are connected to a common supply of an inert gas known as a *carrier gas* whose flow rate is metered and adjusted for equal flow. The temperature of the gases issuing from the columns is sensed by a detector, which may be a resistance thermometer, or an ionization sensor, or a thermal absorption sensor.

With the simpler types of gas chromatograph a minute quantity of sample is injected into an injection port and a mark is made on a chart recorder, indicating the injection time. This sample mixes with the carrier gas, and as it is carried into the column it vaporizes. The gas continues along the length of the column, the heavier constituents absorbing to the packing material longer than the lighter ones. By the time the lightest component has reached the detector, the remaining constituents are strung out behind in an increasing order of density. Emergence of the components is indicated on a chart record by a series of peaks. The time from injection to emergence identifies a particular constituent, and the area under the peak indicates its proportion of the total.

Analysis of chromatograph charts is a tedious process and is made difficult by peaks running into one another where a second constituent gas begins to emerge before the previous one is fully flushed. Drift of the baseline can also occur, which necessitates correction to all the peaks. In industries where a large number of samples are being analyzed daily, automatic methods of gas chromatography have been developed. With these systems, metered samples of gases are automatically injected into the instrument, and the sensor output is monitored by a computer. The computer analysis is usually more accurate than a manual one and is available within seconds of the emergence of the final component. Analyzed results are usually in the form of a printout, identifying each component by name, and indicating its proportion of the sample.

CHAPTER 13

VIBRATION MEASUREMENT

SIMPLE HARMONIC MOTION

One of the simplest examples of a vibration system is that of a mass suspended from the lower end of a spring, the upper end of which is attached to a rigid support. If the mass is pulled down below its rest level and then released, the system will oscillate in a vertical plane, the amplitude gradually diminishing until the mass comes to rest. The number of oscillations the system makes per second is described as its *natural frequency*, and the rate of decay of the amplitude is known as the *damping* in the system. If the effects of damping are ignored, it is possible to develop a simple mathematical formula describing the motion of the system.

Consider a spring mass system shown in Figure 13.1 with a mass m attached to a spring whose stiffness is k. If the mass is pulled down a distance x from its rest position and then released, restoring force acting on the mass is $-kx$. The resultant motion of the mass is governed by Newton's second law of motion. This states that the force acting on a body in motion is equal to the product of its mass and acceleration.

It follows, therefore, that at any instant:

$$m \frac{d^2x}{dt^2} = -kx$$

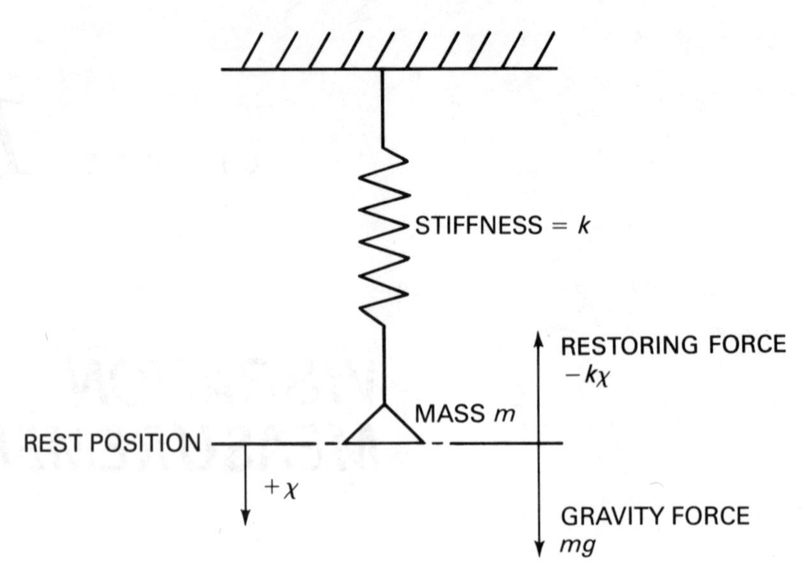

Figure 13.1
Simple Spring Mass System

or

$$\frac{d^2x}{dt^2} + \frac{kx}{m} = 0 \qquad (1)$$

This is a second-order differential equation, the general solution to which may be written as:

$$x = A \sin \omega_n t + B \cos \omega_n t$$

or

$$x = A \sin (\omega_n t + \phi) \qquad (2)$$

where

$$\omega_n = \sqrt{\frac{k}{m}} \qquad (2a)$$

The mass, therefore, will experience a cyclic motion with a natural frequency of ω_n radians/second, and since $\omega_n = 2\pi f$ the natural frequency in hertz is $\omega_n/2\pi$.

A and B are constants of integration determined by substituting specific conditions of the motion.

The motion described in this equation is known as *simple harmonic motion.* This is formally defined as the motion of the projection of a vector on its vertical axis, when the vector is rotating at a constant angular velocity.

Consider a vector of length A that rotates about a center at an angular velocity ω. After a time t it will have rotated through an angle ωt from the horizontal.

The projection of the vector on the vertical axis at this time will be s and is given by the expression

$$s = A \sin \omega t \tag{3}$$

(See Figure 13.2.)

The velocity v of the projection is given by ds/dt. Therefore,

$$v = A\omega \cos \omega t \tag{4}$$

and the acceleration given a is dv/dt. Therefore,

$$a = -A\,\omega^2 \sin \omega t \tag{5}$$

Substituting for s from (3)

$$a = -\omega^2 s \tag{6}$$

A graphical representation of the phase of these relationships is shown in Figure 13.3.

DAMPED HARMONIC MOTION

In the simple spring mass system in Figure 13.1 an assumption was made that there are no forces acting on the system resisting its motion. In fact, this is never the case, and the equation of motion

$$m\frac{d^2x}{dt^2} = -kx$$

must be modified.

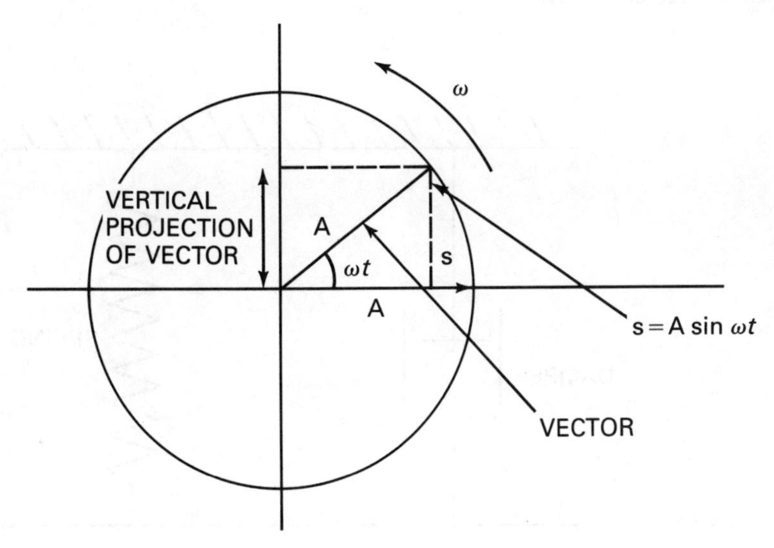

Figure 13.2
Simple Harmonic Motion Vector Representation

Suppose a spring mass system is represented as shown in Figure 13.4, where the motion is resisted by a damping device. In reality, the damper exists only in concept, its effect being produced by air resistance acting against the spring, and to a lesser extent by molecular friction in the spring material. This damping force is proportional to the velocity and can be written mathematically as $-cv$ or as $-c\ dx/dt$

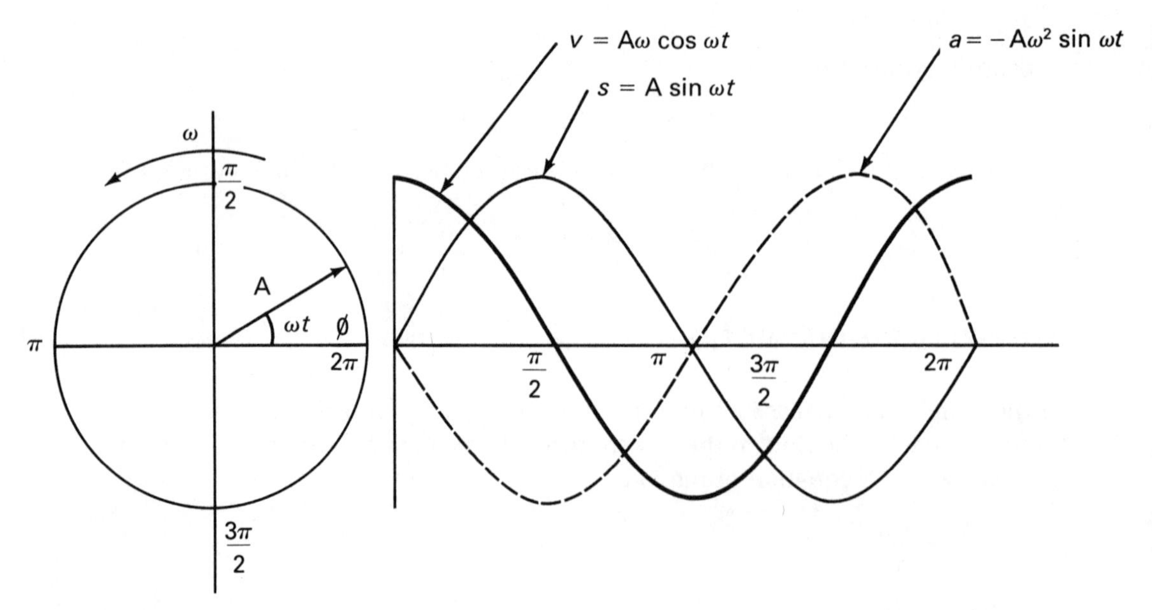

Figure 13.3
Relation Between Displacement Velocity and Acceleration Vectors

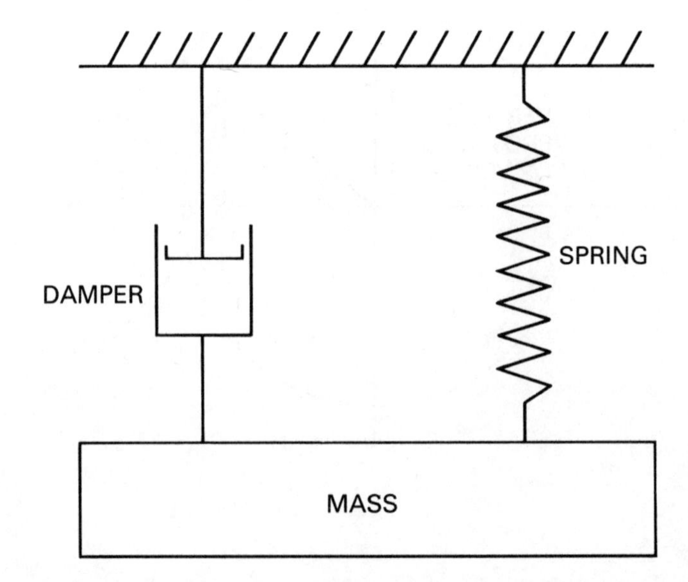

Figure 13.4
Damped Harmonic Motion

where

c = a constant known as the *damping coefficient*.

If equation (1) is rewritten to include this coefficient:

$$m \frac{d^2x}{dt^2} = -kx - c\frac{dx}{dt}$$

or

$$\frac{d^2x}{dt^2} + \frac{c}{m}\frac{dx}{dt} + \frac{k}{m} = 0 \tag{7}$$

There are three possible solutions to this equation. Where the damping is such that the mass describes a number of oscillations, each one smaller than the previous, until it comes to rest, the system is said to *subcritical*, or *underdamped*. In this condition the solution to equation (7) is:

$$x = Ae^{-ct/2m} \sin(\omega' t + \phi) \tag{8}$$

When a system is damped there is no real cycle of motion as the amplitude of vibration is progressively decreasing. The maxima and minima of the displacement curves, however, each recur after equal intervals of time equal to a value $2\pi/\omega'$.

ω' is described as the natural frequency of the damped system and has a value of

$$\sqrt{\omega_n^2 - \frac{c^2}{4m^2}} \tag{8a}$$

ω_n is the natural frequency of an undamped system and from (2a) has a value of

$$\sqrt{\frac{k}{m}}$$

Substituting in (8a):

$$\omega' = \sqrt{\frac{k}{m}\left(1 - \frac{c^2}{4mk}\right)} \tag{8b}$$

A graphical representation of the oscillation of a system with damped harmonic motion is illustrated in Figure 13.5.

If the amplitude of any two successive maxima of the decaying oscillation is x_1 and x_2, the relationship between x_1 and x_2 can be shown to be:

$$-\log_e\left(\frac{x_2}{x_1}\right) = \frac{\pi c}{m\omega'} \tag{8c}$$

where

$$-\log_e\left(\frac{x_2}{x_1}\right)$$

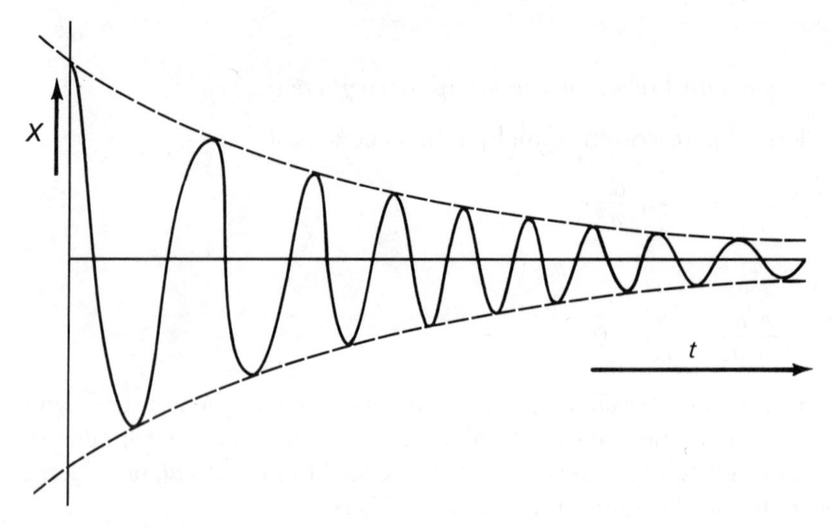

Figure 13.5
Graphical Representation of Damped Harmonic Motion

is known as the logarithmic decrement. This formula provides a useful method of calculating the damping coefficient from the record of a decaying oscillation.

Two further solutions to the equation are possible (see Figure 13.6). The first is where damping is adjusted so that the mass moves from its extended position to its static equilibrium position in minimum time without actually oscillating. This is described as a condition of *critical* damping and the solution to the equation becomes:

$$x = (A + Bt)\, e^{-ct/2m} \tag{9}$$

Critical damping occurs when

$$c = c_c = 2m\omega_n \tag{9a}$$

where c is the damping coefficient of the system, c_c is the coefficient of critical damping, and ω_n is the natural frequency of the undamped system.

From equation (2)

$$\omega_n = \sqrt{\frac{k}{m}}$$

and substituting in (9a):

$$\omega_n = \frac{c_C}{2m} \quad \text{and} \quad c_C = 2\sqrt{mk} \tag{9b}$$

The final condition, that of overdamping, has the following solution to the general equation:

$$x = e^{-ct/2m}(Ae^{\omega'' t} + Be^{-\omega'' t}) \tag{10}$$

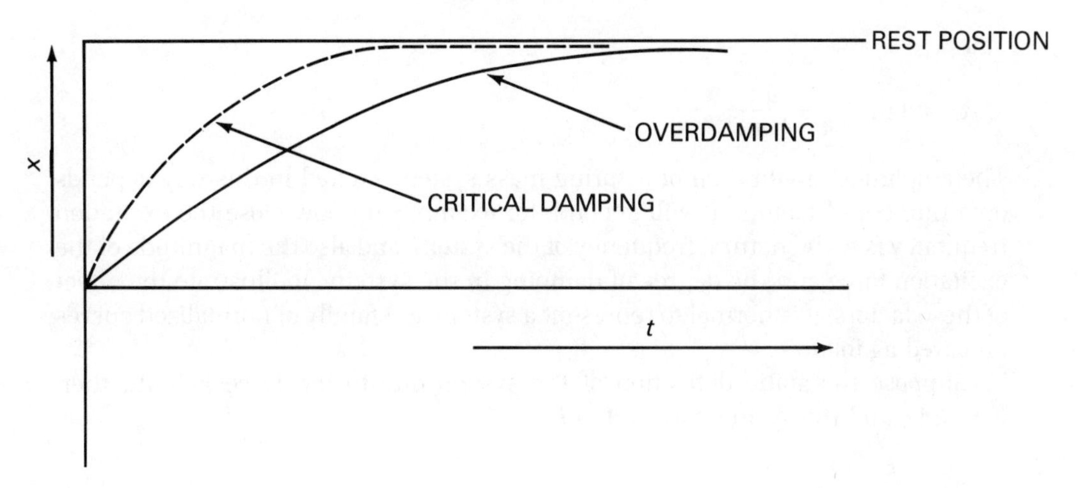

Figure 13.6
Critical and Overdamping

where

$$\omega' = \sqrt{\left(\frac{c}{2m}\right)^2 - \frac{k}{m}} \tag{10a}$$

FORCED VIBRATION OF A SIMPLE SYSTEM

The previous examples have assumed that the mass on the spring has been pulled down a fixed amount and then released. The more usual situation is one in which oscillation is maintained by a periodic disturbing force.

With the simple damped system shown in Figure 13.4 it is possible to sustain a motion by introducing a periodic disturbing force $F \sin \omega t$, where ω is some frequency not necessarily the natural frequency of the system.

The equation of motion for this system now becomes:

$$m \frac{d^2x}{dt^2} + c \frac{dx}{dt} + kx = F \sin \omega t \tag{11}$$

If the system is allowed time to settle into a state of steady oscillation, the solution to equation (11) is:

$$x = A \sin (\omega t - \phi) \tag{12}$$

where

$$A = \frac{F}{\sqrt{(k - m\omega^2)^2 + c^2 \omega^2}} \tag{12a}$$

and

$$\phi = \tan^{-1} \frac{c\omega}{k - m\omega^2}$$

The amplitude of vibration of a spring mass system excited in this way depends on a number of factors. It will depend, for example, on how close the excitation frequency is to the natural frequency of the system, and also the magnitude of the excitation force and the degree of damping in the system. To illustrate the effect of these factors, it is normal to represent a system as a family of normalized curves prepared as follows:

Suppose the static deflection of the system due to the force F is A_1, then $F = kA_1$, and the spring constant is k.

$$A_1 = \frac{F}{k} \tag{13}$$

Suppose also the undamped natural frequency of the system is ω_n:

$$\omega_n = \sqrt{\frac{k}{m}} \tag{14}$$

Dividing (12a) by (13) provides a dynamic-to-static-amplitude ratio. Therefore,

$$\frac{A}{A_1} = \frac{F}{\sqrt{(k - m\omega^2)^2 + c^2\omega^2}} \cdot \frac{k}{F}$$

$$= \frac{1}{\sqrt{\left(1 - \frac{m}{k}\,\omega^2\right)^2 + \frac{c^2\omega^2}{k^2}}}$$

But from (14) $\omega_n = k/m$, and the equation may be rewritten as

$$\frac{A}{A_1} = \frac{1}{\sqrt{\left(1 - \frac{\omega^2}{\omega_n^2}\right)^2 + \left(\frac{c\omega}{m\omega_n^2}\right)^2}} \tag{15}$$

and

$$\phi = \tan^{-1} \frac{c}{m\omega_n^2} \cdot \frac{\omega}{\left(1 - \frac{\omega^2}{\omega_n^2}\right)} \tag{16}$$

Considering equation (15), if $\omega = \omega_n$, the equation becomes

$$\frac{A}{A_1} = \frac{1}{\sqrt{\left(\frac{c}{m\omega}\right)^2}} = \frac{m\omega}{c} \tag{17}$$

This is described as a *resonant condition*, and if the damping coefficient c is small, the corresponding value of A/A_1 becomes very large. At this point the value of tan ϕ approaches infinity, that is, ϕ approaches 90°.

Resonance is a very important phenomenon in engineering, and much work is involved both in detecting it in machinery and calculating its value, with the object of avoiding the condition. At resonance, not only is the vibration amplitude at maximum, but the resulting stresses are also at a maximum. These stresses may manifest themselves either as discomfort to an operator because of excessive movement or noise, or, in the extreme, they may cause failure of the stressed component.

The expression (15) is more convenient if the value of c is normalized by expressing it in terms of critical damping.

Suppose

$$\gamma = c/c_c \tag{17a}$$

where

γ = the damping ratio

and

c_c = the value of the damping coefficient for critical damping.

Expression (15) now becomes

$$\frac{A}{A_1} = \frac{1}{\sqrt{\left(1 - \frac{\omega^2}{\omega_n^2}\right)^2 + \left(\frac{\gamma c_c \omega}{m\omega_n^2}\right)^2}} \tag{18}$$

From Equation 9a, at critical damping:

$$\frac{c_c}{2m} = \omega_n$$

or

$$c_c = 2m\omega_n \tag{18a}$$

Substituting for c_c in equation (18)

$$\frac{A}{A_1} = \frac{1}{\sqrt{\left(1 - \frac{\omega^2}{\omega_n^2}\right)^2 + \left(\frac{2\gamma\omega}{\omega_n}\right)^2}} \tag{19}$$

Similarly, from equation (16), ϕ becomes

$$\tan^{-1} \frac{\gamma\, c_c\, \omega}{m\omega_n^2 \left(1 - \frac{\omega^2}{\omega_n^2}\right)}$$

$$= \tan^{-1} \frac{2\gamma \sqrt{mk\omega}}{m\omega_n^2 \left(1 - \frac{\omega^2}{\omega_n^2}\right)}$$

$$= \tan^{-1} \frac{2\gamma\omega}{\omega_n \left(1 - \frac{\omega^2}{\omega_n^2}\right)}$$

By substituting values for A/A_1 and ω/ω_n in equation 19 it is possible to plot a family of nondimensional graphs for a series of damping ratios that illustrate the effect of damping on the relative response. A sample of such curves is shown in Figure 13.7.

Most sensors used for vibration measurement directly referenced to ground are damped harmonic systems. Their theory, which is a development of basic damped simple harmonic motion theory, is discussed in the following section.

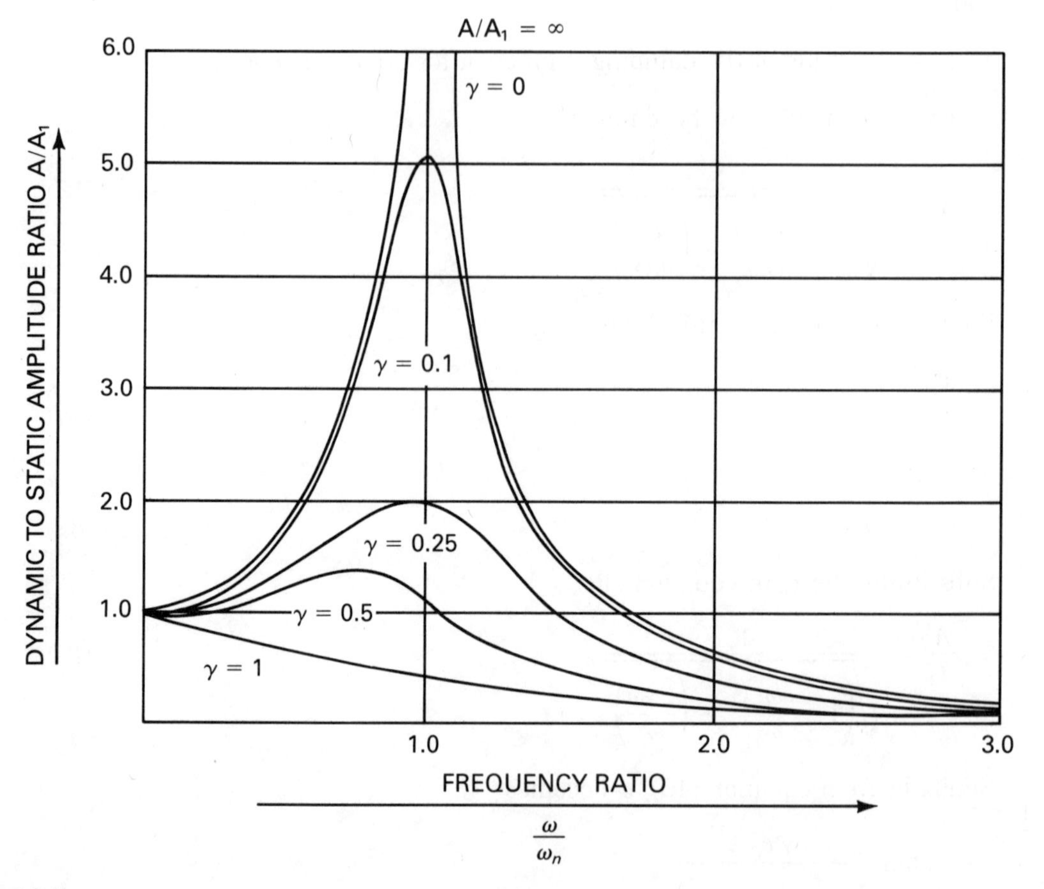

Figure 13.7
Effect of Damping on Forced SHM System

VIBRATION MEASUREMENT INSTRUMENT THEORY

Vibration of a structure may be measured by attaching the body of a sensor, such as a differential transformer, to a stationary datum, and attaching the armature to the moving structure. Unfortunately, it is not always possible to obtain a fixed datum; thus other types of sensors must be employed that use some other datum as a reference. Most sensors that fit into this category contain a spring mass system suspended inside the case. The amplitude of vibration is determined by measuring the difference in movement between the case and the mass. For very-low-frequency systems potentiometers are sometimes used to obtain an electrical output from these sensors. For higher frequencies it is more common to use some type of proximity measurement.

Consider a sensor illustrated in Figure 13.8. Suppose the case moves through a distance x relative to a fixed datum, and suppose the movement of the mass relative to the case is y. Let the damping of the system be c and the spring rating be k. The spring force acting on the mass is:

$$k = (x - y)$$

The damping force acting on the mass is:

$$c\left(\frac{dx}{dt} - \frac{dy}{dt}\right)$$

the equation of motion for the system is:

$$m\frac{d^2y}{dx^2} = c\left(\frac{dx}{dt} - \frac{dy}{dt}\right) + k\,(x - y) \tag{21}$$

Rearranging:

$$\frac{d^2y}{dt^2} + \frac{c}{m}\frac{dy}{dt} + \frac{k}{m}y = \frac{c}{m}\frac{dx}{dt} + \frac{k}{m}x \tag{22}$$

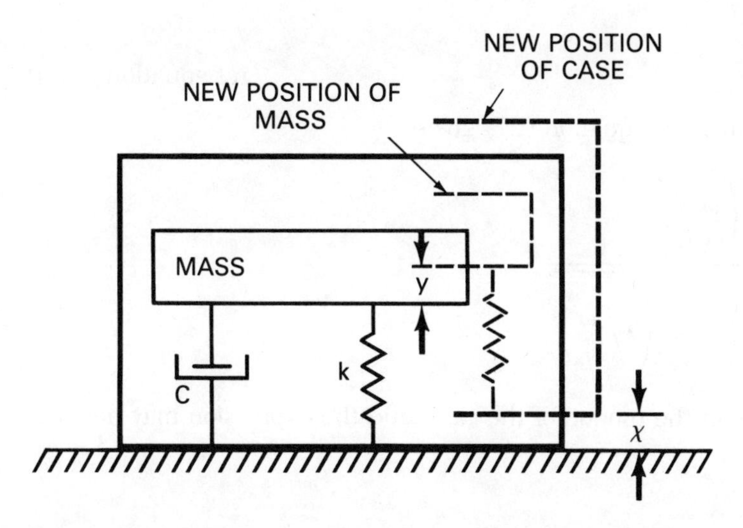

Figure 13.8
Basic Mechanics of a Vibration Measuring Sensor

If the case moves with a motion $x = A \cos \omega t$, equation (22) may be rewritten:

$$\frac{d^2y}{dt^2} + \frac{c}{m}\frac{dy}{dt} + \frac{k}{m} y = A\left(\frac{k}{m} \cos \omega t - \frac{c}{m} \omega \sin \omega t\right) \tag{23}$$

The solution to equation (23) is:

$$\mathcal{R} = (C \cos \omega' t + D \sin \omega' t)e^{-c/2m} + \frac{mA\omega^2 \cos (\omega t - \phi)}{\sqrt{(k - m\omega^2)^2 + c^2\omega^2}} \tag{24}$$

where $\mathcal{R}$ is the relative movement of the case to the mass $(x - y)$, C and D are constants of integration, and ω' is the natural frequency of the damped system

$$= \sqrt{\frac{k}{m}\left(1 - \frac{c^2}{4mk}\right)} \tag{25}$$

$$\phi = \tan^{-1} \frac{c\omega}{k - m\omega} \tag{26}$$

The solution of (24) is composed of two terms. The first term involves an exponential function, and is transient. After an initial settling period the relative motion will become that of the second term. The solution therefore resolves into its basic form:

$$\mathcal{R} = B \cos (\omega t - \phi)$$

where

$$B = \frac{m\,\omega^2 A}{\sqrt{(k - m\omega^2)^2 + c^2\omega^2}} \tag{27}$$

The amplitude of relative motion B may be normalized to an amplitude ratio by eliminating m and k, and reducing c to a damping ratio:

$$k/m = \omega_n^2 \qquad \text{ref equation (2a)}$$

$$\gamma = c/c_c \qquad \text{ref equation (17a)}$$

and

$$c_c = 2m\omega_n \qquad \text{ref equation (18a)}$$

Substituting these values in equation (27) gives

$$B = \frac{A\left(\dfrac{\omega}{\omega_n}\right)^2}{\sqrt{\left(1 - \left(\dfrac{\omega}{\omega_n}\right)^2\right)^2 + \left(2\gamma\dfrac{\omega}{\omega_n}\right)^2}} \tag{28}$$

A is the amplitude of the motion of the case, and the expression may now be normalized to

$$\frac{B}{A} = \frac{\left(\frac{\omega}{\omega_n}\right)^2}{\sqrt{\left(1 - \left(\frac{\omega}{\omega_n}\right)^2\right)^2 + \left(2\gamma\frac{\omega}{\omega_n}\right)^2}}$$

and

$$\phi = \tan^{-1}\frac{2\gamma\frac{\omega}{\omega_n}}{1 - \left(\frac{\omega}{\omega_n}\right)^2}$$

The ratio B/A may be used to plot the family of curves illustrated in Figure 13.9 demonstrating the relationship between the amplitudes of relative displacement of the mass to the displacement of the case over a range of damping and frequency ratios.

If now the motion is considered in terms of acceleration of the sensor, rather than its displacement, a modified expression may be obtained. The displacement of the sensor case is:

$$x = A\cos\omega t$$

and this produces an acceleration:

$$\frac{d^2x}{dt^2} = -\omega^2 A \qquad (6)$$

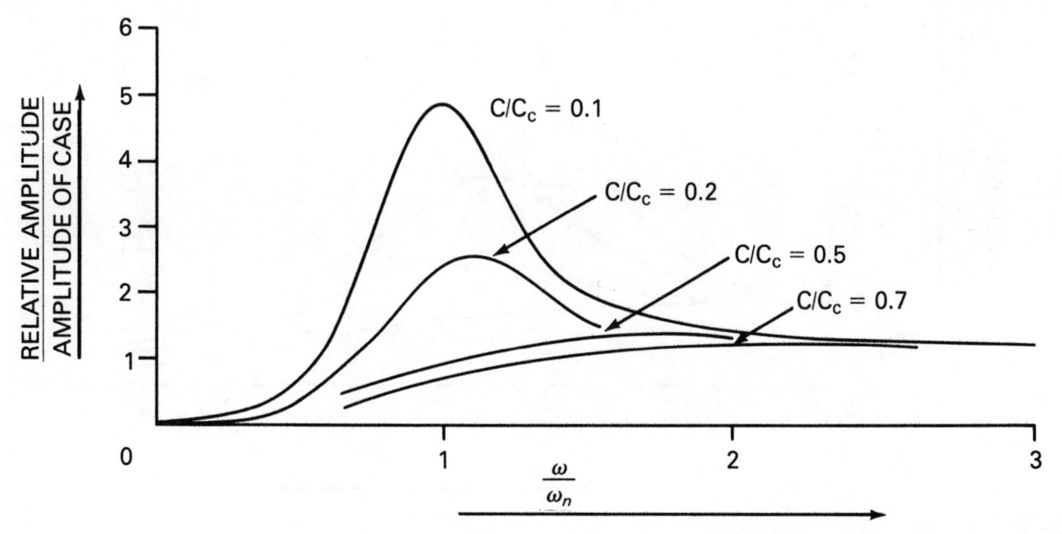

Figure 13.9
Effect of Damping on the Case to Mass Movement of Vibration Sensor Relative to Displacement of Case

Suppose this is written as

$$\ddot{A} = -\omega^2 A$$

where $\ddot{A}$ is the acceleration of the case. Then

$$A = -\ddot{A}/\omega^2$$

Substituting in equation (29)

$$\frac{B}{\ddot{A}} = -\frac{1}{\omega_n^2\sqrt{\left(1-\left(\frac{\omega}{\omega_n}\right)^2\right)^2+\left(2\gamma\frac{\omega}{\omega_n}\right)^2}} \tag{31}$$

The ratio $B/\ddot{A}$ may be used to plot a second family of curves illustrated in Figure 13.10 demonstrating the relationship of the relative displacement of the mass to the acceleration of the sensor case over a range of damping and frequency ratios.

Considering first Figure 13.9, the interesting portion of the curves lies above the natural frequency of the system. In this region the amplitude ratio approaches unity, implying that the mass become virtually stationary. This is often described as a seismic condition. At some frequency, therefore, the movement of the mass becomes sufficiently small for it to be used as a datum from which the movement of the case may be measured. The lowest frequency where this occurs depends on the damping of the system. With a damping ratio of 0.7 the instru-

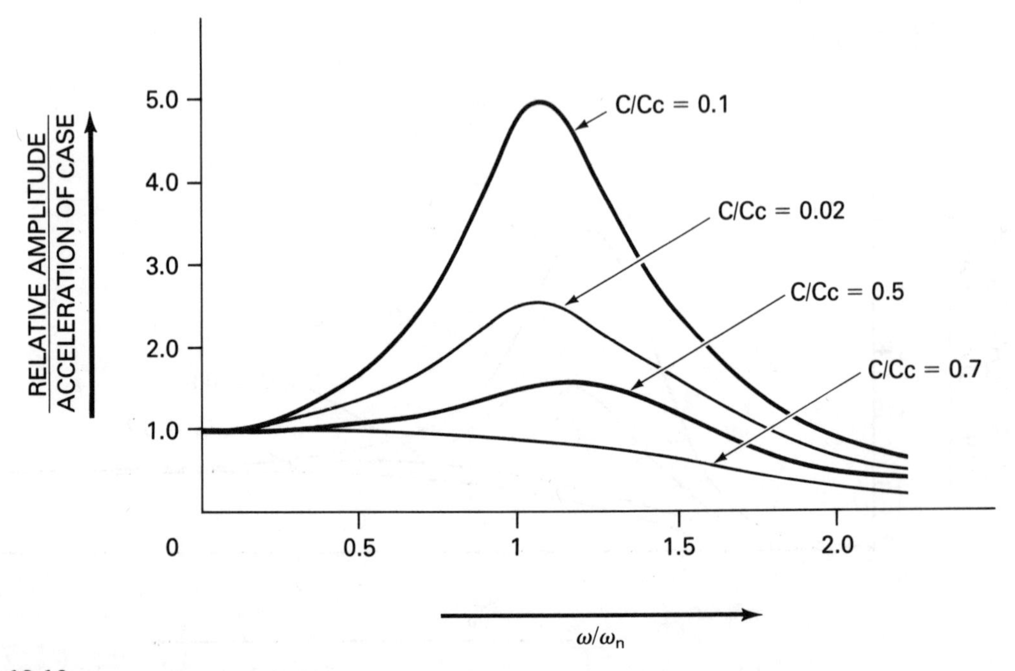

Figure 13.10
Effect of Damping on the Case to Mass Movement Relative to the Acceleration of the Case

ment may be used to measure the amplitude of vibration of a structure down to a frequency whose ratio $\omega/\omega_n = 2$ without introducing errors greater than 2.5%.

If now Figure 13.10 is examined, it can be seen that a similar situation exists below the natural frequency of the system where the relative amplitude/acceleration ratio is unity. In effect the relative movement of the mass to the case is such that it provides a measurement of the acceleration of the case. The useful range of such an instrument again depends on its damping. With a damping ratio of 0.7 it may be used up to a frequency ratio $\omega/\omega_n = 0.5$ without introducing errors greater than 2.5%

VIBRATION-MEASURING TRANSDUCERS

The previous section illustrates two ways in which spring mass systems may be used as transducers. The first type operates at frequencies above their natural frequency and are described as *seismic transducers*. The second operate below their natural frequency and are described as *accelerometers*. Seismic sensors produce signals proportional to the amplitude of a vibrating structure; accelerometers produce signals proportional to the acceleration experienced by the structure. The essential physical difference between the two types of transducer is one of size and weight. If a seismic device is to respond to low frequencies, the resonant frequency must also be low. In general, low resonant frequencies are associated with soft springs and comparatively large masses. Devices with high resonant frequencies are made with small masses and small, stiff springs.

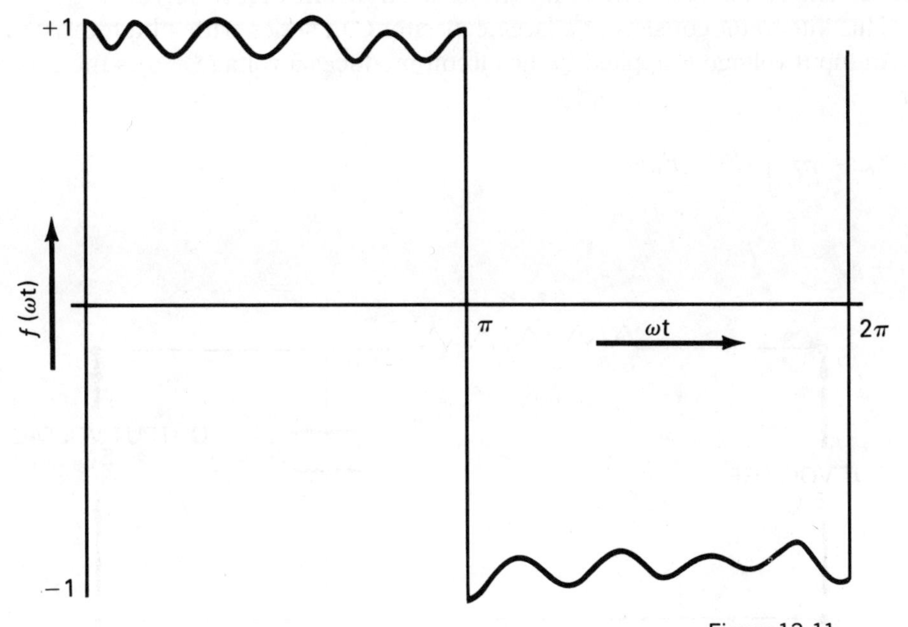

Figure 13.11
Graphic Illustration of Complex Harmonic Motion

If large structures are involved, the addition of comparatively large masses does not influence the characteristics of the structure significantly, and the size of the sensor is only of secondary importance. There are, however, many instances in which the addition of mass seriously affects the structural characteristics, and then the smallest sensor possible must be used. In these situations accelerometers have a decided advantage.

The principal disadvantage of the accelerometer is the nature of the signal. Acceleration is proportional to ω^2. Consequently, high-frequency accelerations of very small amplitude can produce large signals. If a structure suffers a vibration producing a signal of one frequency, the accelerometer response to that signal may be easily interpreted. Unfortunately, this is seldom the case. It is more usual for a vibration to be made up of a number of disturbances that add into a complex waveform, where significant low-frequency signals may be masked by higher frequency signals which are of little interest.

A number of techniques have been evolved that to some extent overcome the ω^2 problem, the simplest being the careful selection of accelerometer. Above the resonant frequency of the accelerometer, the response falls off rapidly until the mass becomes seismic to the disturbing frequency. Any frequencies above the natural frequency, therefore, will produce signals proportional to their amplitude, which if of no interest will be small. This allows a clearer record of low-frequency components to be obtained. Another method reduces the accelerometer output to a displacement signal by a process of integration.

By definition, an acceleration signal is obtained by differentiating a harmonic displacement signal twice. It follows, therefore, that if an acceleration signal is twice integrated, a signal proportional to the original displacement will be obtained. The simplest electrical integrator is shown in Figure 13.12.

This integrator consists of a large capacitor C in series with a high resistance R. An input voltage E applied to the circuit produces a voltage across the capacitor:

$$E_0 = \frac{1}{RC}\int (E - E_0)dt$$

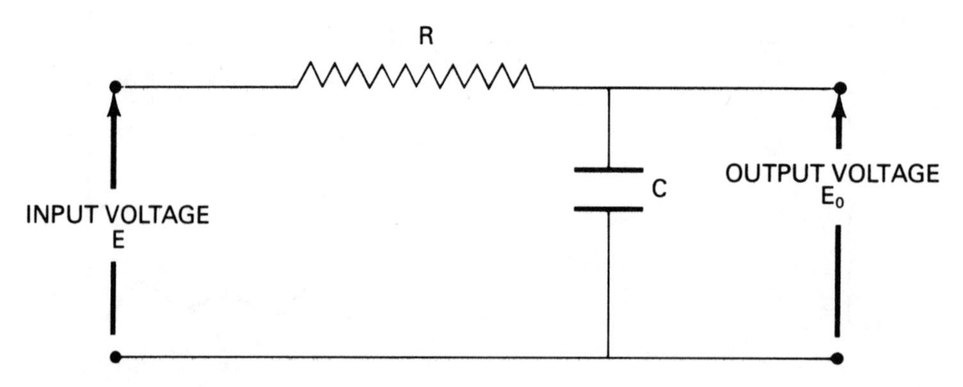

Figure 13.12
Simple Integration Circuit

When E_0 is small compared with E, the expression becomes:

$$E_0 \simeq \frac{1}{RC}\int E\,dt$$

This implies that R must be very large compared with the reactance $1/2\pi fC$ of the capacitor at the lowest frequency to be integrated.

Many commercial integrators are now available that can operate over wide frequency ranges with good accuracy. The cost of these devices has now been reduced to the point where they have made the simple circuit obsolete. The principle of an integrator based on an operational amplifier is illustrated in Chapter 17.

Velocity Sensors

Many velocity sensor designs are based on the interpretation of Faraday's law, which states that when a conductor moves in a magnetic field a voltage is generated across the conductor proportional to the product Bvl, where B is the flux density of the magnetic field, l is the length of a conductor moving at right angles to the field, and v is the velocity of the conductor.

The conductor in a velocity sensor is a coil mounted on springs and free to move between the poles of a magnet. The sensor operates above the natural frequency of the coil on its suspension and is, therefore, a seismic sensor with a stationary coil and the magnet experiencing the movement of the case. Velocity sensors are usually larger than accelerometers designed for similar frequency ranges, and less robust. To convert their output to a measurement of displacement, however, requires only one stage of integration whereas an accelerometer requires two. A typical velocity sensor design is illustrated in Figure 13.13.

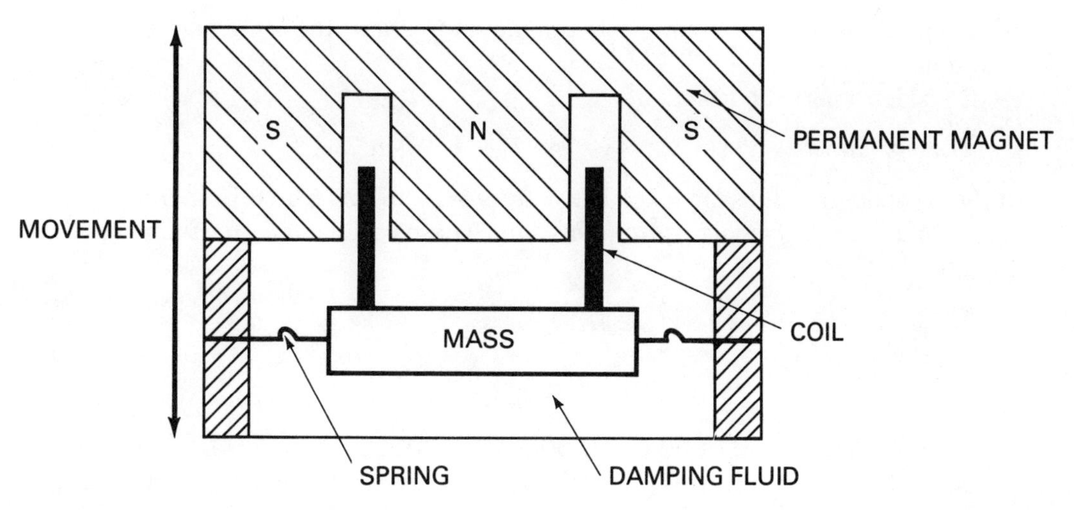

Figure 13.13
Velocity Sensor

MAKING VIBRATION MEASUREMENTS

Vibration measurements fall into two basic types. Measurements may be made of a structure that is already operational because the levels of vibration for some reason are unacceptable. Alternatively, measurements can be made on a structure to satisfy the designer that predictions of natural frequency and the degree of damping are correct. These latter tests are described as *resonance tests*, and the structure is forced into its various modes of vibration by the use of some kind of vibration generator.

The vibration measurements described in the following sections are intended to provide an indication of the range of vibration measurements and the instruments appropriate to those measurements.

COMPLEX MOTION

So far the discussion has been confined to the effect of sinusoidal disturbing forces on simple systems. In the practical case, very few vibration systems are simple, and the motion of a spring mass system is often required to respond to a number of disturbing forces, all contributing different amounts of energy, at different frequencies.

Such motion is described as *complex*, and though the motion may be periodic, it is unlikely to be harmonic. Fortunately for those involved in vibration analysis, a comparatively simple treatment of complex motion was discovered by Fourier, who proved that any periodic function could be represented as the sum of a series of harmonic functions. This may be expressed mathematically as

$$x = A \sin(\omega t + \alpha) + B \sin(2\omega t + \beta) + C \sin(3\omega t + \gamma) + \cdots$$

Although each term is a simple harmonic expression, the sum may be very different.

Consider a periodic function

$$f(\omega t) = \sin \omega t + \tfrac{1}{3} \sin 3\omega t + \tfrac{1}{5} \sin 5\omega t + \tfrac{1}{7} \sin 7\omega t + \cdots$$

If the component values are listed between $\omega t = 0$ and 2.1π, in 0.1π increments, the function can be calculated from this sum. These calculations are summarized in Table 13.1.

Table 13.1

ωt	$\sin \omega t$	$\frac{1}{3} \sin 3\omega t$	$\frac{1}{5} \sin 5\omega t$	$\frac{1}{7} \sin 7\omega t$	$\frac{1}{9} \sin 9\omega t$	sum
0.0	0.0000	0.0000	0.0000	0.0000	0.0000	0.0000
0.1π	0.3090	0.2697	0.2000	0.1156	0.0343	0.9286
0.2π	0.5878	0.3170	0.0000	−0.1359	−0.0653	0.7036
0.3π	0.8090	0.1030	−0.200	0.0441	0.0899	0.8460

Table 13.1—continued

ωt	$\sin \omega t$	$\frac{1}{3} \sin 3\omega t$	$\frac{1}{5} \sin 5\omega t$	$\frac{1}{7} \sin 7\omega t$	$\frac{1}{9} \sin 9\omega t$	sum
0.4π	0.9511	−0.1966	0.0000	0.0840	−0.1057	0.7328
0.5π	1.0000	−0.3333	0.2000	−0.1492	0.1111	0.8349
0.6π	0.9511	−0.1966	0.0000	0.0840	−0.1057	0.7328
0.7π	0.8090	0.1030	−0.2000	0.0441	0.0899	0.8460
0.8π	0.5878	0.3170	0.0000	−0.1359	−0.0653	0.7036
0.9π	0.3090	0.2697	0.2000	0.1156	0.0343	0.9286
1π	0.0000	0.0000	0.0000	0.0000	0.0000	0.0000
1.1π	0.3090	−0.2697	−0.2000	−0.1156	−0.0343	−0.9286
1.2π	−0.5878	−0.3170	0.0000	0.1359	0.0653	−0.7036
1.3π	−0.8090	−0.1030	0.2000	−0.0441	−0.0899	−0.8460
1.4π	−0.9511	0.1966	0.0000	−0.0840	0.1057	−0.7328
1.5π	−1.0000	0.3333	−0.2000	0.1429	−0.1111	−0.8349
1.6π	−0.9511	0.1966	0.0000	−0.0840	0.1057	−0.7328
1.7π	−0.8090	−0.1030	0.2000	−0.0441	−0.0899	−0.8460
1.8π	−0.5878	−0.3170	0.0000	0.1359	0.0653	−0.7036
1.9π	−0.3090	−0.2697	−0.2000	−0.1156	−0.0343	−0.9286
2.0π	0.0000	0.0000	0.0000	0.0000	0.000	0.0000
2.1π	0.3090	0.2697	0.2000	0.1156	0.0343	0.9286

The graph illustrated in Figure 13.11 shows the effect of adding harmonic waveforms of a specific series. In this instance, the addition approaches a square waveform and would have been more nearly so had higher harmonics been included. Other series produce different shapes, many far removed from the shape of the harmonics that produce them.

SEISMOLOGY

Of all the sources of vibration, the earthquake is probably the most spectacular and destructive. The mechanics of earthquakes are just beginning to be understood, but attempts to detect them have been made for almost 2000 years. Faults in the earth's surface are one source of earthquakes. In places the two sides of the fault move relative to one another, building up large strains in other parts of the fault that are stationary. Eventually, a stage is reached where the stresses are so great that the ground shears, producing violent shock waves which may be felt over a radius of several hundred miles. The point at which an earthquake originates is known as its *epicenter*.

In the eighteenth century the first serious attempt to record earthquakes used instruments based on the simple pendulum. A heavy pendulum bob was suspended by a long, fine wire from a support attached to a building. Any movements of the ground caused the building and the ground under the bob to move, while the inertia of the pendulum tended to maintain its position. Records were made on a smoked glass by a stylus attached to the bob. In effect a pendulum can be regarded as an accelerometer if operated below its natural fequency and as a displacement sensor if operated above. The natural frequency of the instrument is equal to $1/\tau$, where τ is the period of oscillation of the bob.

The limitation of the simple pendulum is that of providing a low natural frequency while still retaining a reasonable length. Typical frequencies generated by earthquakes are about four oscillations per minute.

The period of a simple pendulum is given by the formula

$$\tau = 2\pi \sqrt{\frac{l}{g}}$$

where l is the length of the pendulum and g is the gravity constant.

If the pendulum bob is to be seismic to a disturbance of four oscillations per minute, the natural frequency of the pendulum must be two oscillations per minute or greater (see page 229). If this value is substituted in the formula, it can be seen that a length of 600 feet (200 meters) is required. For this reason some of the early pendulums were suspended in old mine shafts in attempts to make them seismic. If, however, the pendulum is used as an accelerometer, it may have a natural frequency of about eight oscillations per minute or higher. This is equivalent to a length of about 40 feet (13 meters) and is obviously a more practical proposition.

An improvement on the simple pendulum was achieved by using the horizontal pendulum, Figure 13.14a.

The horizontal pendulum employs a mass that swings on a horizontal boom. The boom is constrained at the vertical support by a cone and cup bearing, and it is held horizontal by a wire suspension. A feature of the horizontal pendulum is that it will only respond to horizontal movement at right angles to the boom and will therefore measure only the component of a ground movement in that direction. To measure the true amplitude and direction of a movement, two components must be measured at right angles and added vectorially. All pendulums are susceptible to extraneous vibration, and therefore it is necessary to introduce artificial damping. Both viscous and electromagnetic damping are used.

Early seismographs employed a stylus that drew lines on smoked paper to record movements of the earth. Later these were replaced by optical recording systems which allowed substantial signal amplification without the frictional problems of the stylus. More recently, electrical recording techniques have been introduced using variable reluctance transducers to produce signals that can be amplified to drive galvanometers.

The seismographs described so far measure horizontal movements of the ground. In addition, vertical movements are also experienced and these must be measured by a different technique.

Figure 13.14b illustrates the principle of the spring mass seismograph for detecting vertical disturbances. This instrument employs a spring mass system with a low natural frequency, supported on a substantial frame. The indicating arm is attached to the mass by a pivot point P, and to the frame by a hinge H.

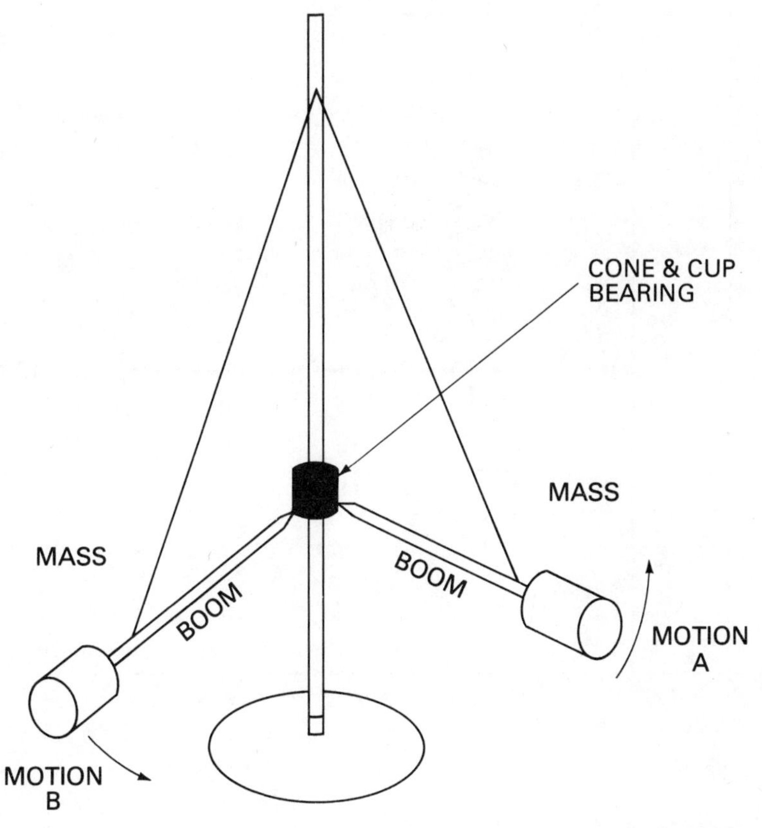

Figure 13.14a
Horizontal Pendulum Seismograph

Vertical movement of the ground moves the hinge point while the pivot movement is modified by the spring/mass system. This causes a corresponding differential movement of the arm, which is amplified by using a long arm.

Earthquake shock waves are characteristic. Figure 13.15 illustrates a typical record from a seismograph registering a distant earthquake. The first indication of disturbance occurs when a primary disturbance *P* is registered, indicating the arrival of a train of vertical movements of small amplitude. After some minutes a secondary disturbance *S* is experienced, which marks the arrival of horizontal disturbances. Both types of disturbance originate simultaneously, but they travel through the earth at different velocities.

Some time after the arrival of the *P* and *S* waves, large surface waves are recorded. These waves travel comparatively slowly and are responsible for the large amplitudes shown on the record.

A number of scales have been devised for the measurement of earthquake intensity, but the one most commonly used is the *Richter scale*. This is a quantitative scale based on the maximum amplitude recorded by a standard seismograph at a given distance from an epicenter.

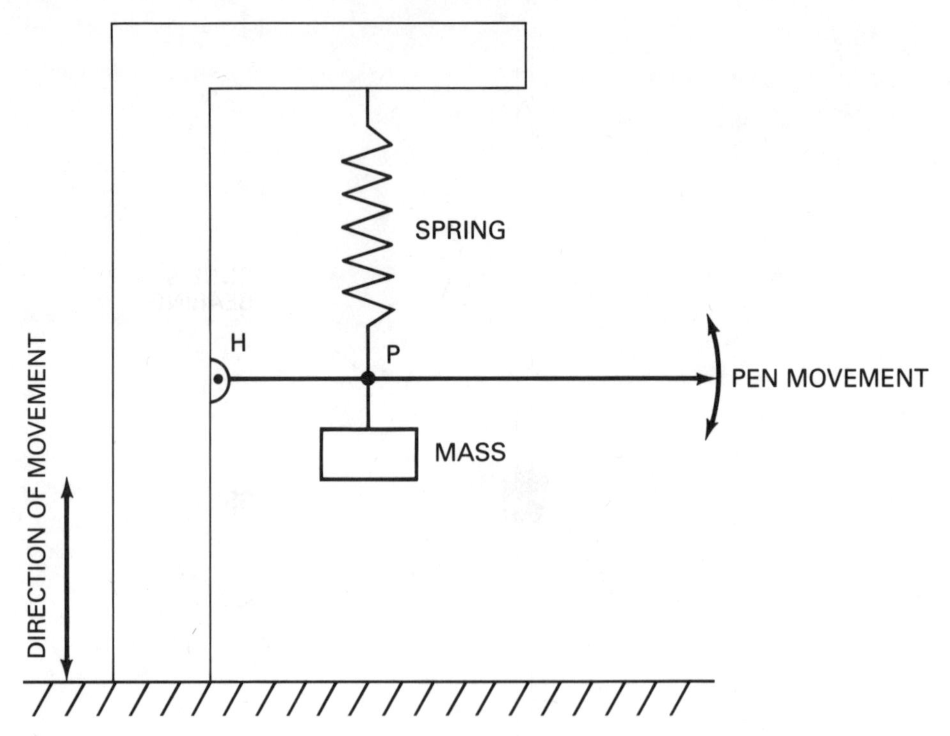

Figure 13.14b
Spring/Mass Seismograph

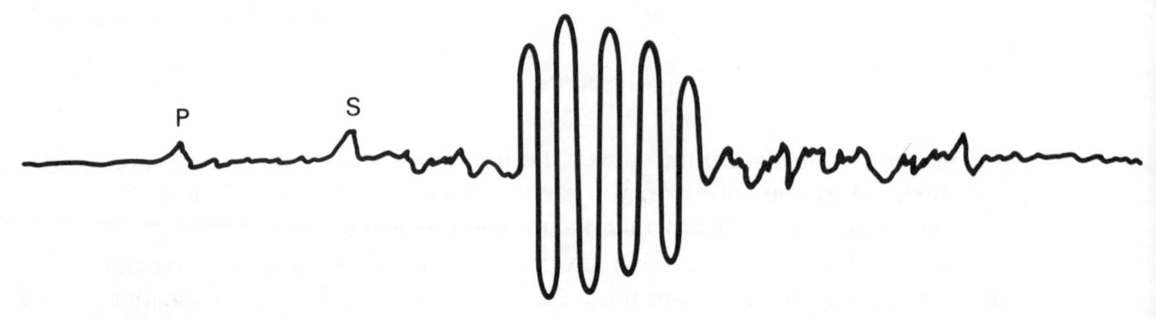

Figure 13.15
Typical Seismograph Record of Earthquake Shock

Seismic Geological Surveys

Investigations of the velocity of shock waves through different terrestrial materials were begun in the mid-nineteenth century. Charges of gunpowder (shots) were exploded below the surface, and the disturbances were measured with seismographs. From these investigations developed the techniques that are used by geologists to determine the nature of subterranean structure.

This method of survey relies on the fact that there is appreciable change in the velocity of shock waves as they pass from one kind of rock to another. The velocity through sand, for example, is about 3000 feet (1000 meters) per second. Through a massive rock such as granite it may be as high as 20,000 feet (6000 meters) per second.

When a shock wave is generated by a geologist's shot it travels through low-density surface layers until it meets a rock stratum. It may then behave in three ways: It may be reflected back to the surface; or if the angle of incidence is less than a certain critical value, it may be refracted through the high-density medium. At a certain critical angle the wave is transmitted along the boundary of the high-density medium, a phenomenon that may be described as boundary refraction. The mechanics of boundary refraction are somewhat complex. The shock wave on reaching the rock stratum is transmitted along its surface at a velocity characteristic of the high-density medium and in the process generates a progressive wave in the low-density medium that returns to the surface at a velocity determined by the surface layers. The time a boundary wave takes to return to the surface at any point, therefore, depends on the distance it travels through the high-velocity medium and the depth of the medium below the surface. Reflected waves are fairly straightforward in behavior. They travel down to the rock stratum and back at constant velocity.

Geologists measure reflected and refracted waves with sensors called *geophones*, and they record their results on multichannel recorders that incorporate accurate timing facilities. The three types of shock waves of principal concern must be resolved by an analyst interpreting the results of a survey. Clearly, it is impossible to separate the different signals with a single geophone, and it is necessary to employ groups of sensors sited at different distances from the disturbance. A simplified example is illustrated in Figure 13.16.

Where sensors are placed close to the shot, reflected and refracted waves arrive at the same time. At greater distances refracted waves reach the sensors earlier than the reflected. Surface waves progress at the velocity of sound in the surface medium. The difference in the arrival times of each of these waves is used to estimate the depth and the composition of the lower strata. Clearly this is a simplification of the real situation and the unpredictable nature of the earth's upper layers can make record analysis extremely difficult.

Geophones

Seismographs are unsuitable for field work in terms of both physical size and fragility. To overcome the problems involved in using seismographs, the geophone was developed. Geophones are better known to the vibration engineer as velocity sensors and have already been discussed under that title.

A number of variations of the basic principle illustrated in Figure 13.12 are in use, but essentially all contain coils free to move between the poles of a permanent magnet. The coil is usually part of a low natural-frequency spring-mass system, which can be locked to prevent damage when not in use.

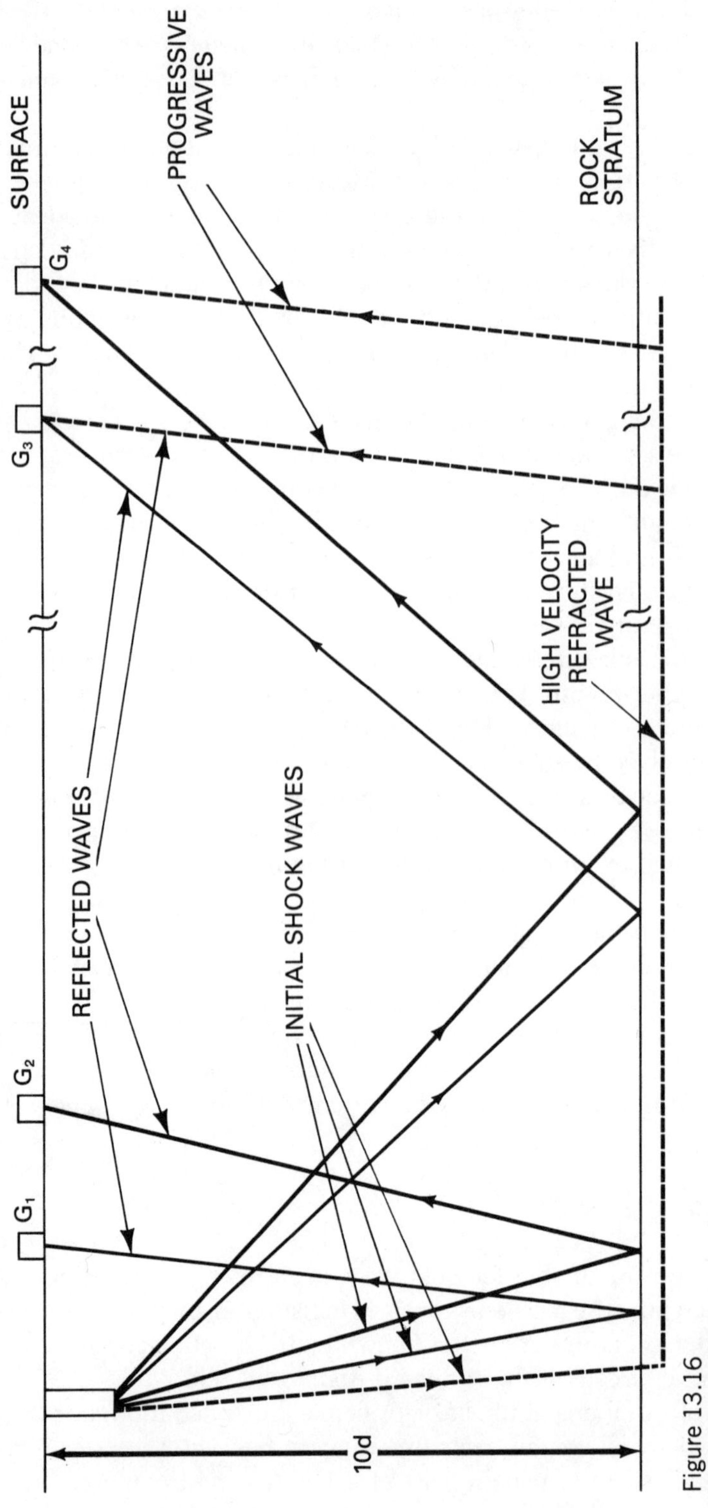

Figure 13.16
Propagation of Shock Waves from Survey Shot

ACCELEROMETER MEASUREMENTS

Structural vibrations between about 5 Hz and 10 kHz are often most conveniently measured by accelerometers. In general, accelerometers are robust devices capable of suffering considerable abuse without damage or loss of calibration. They also tend to be considerably smaller and lighter than alternative sensors, and therefore have less effect on the vibration characteristics of the structure on which they are mounted. If the effects of structural loading are critical, and if no stationary datum is available for reference purposes, they are often the only practical type of sensor that can be used.

STRAIN-GAGED ACCELEROMETERS

A simple accelerometer design is one based on the strain-gaged cantilever illustrated in Figure 13.17.

Strain-gaged cantilevers have been discussed at some length in Chapters 4 and 8, so it is only necessary to summarize the relevant formulas.

Consider a cantilever of length l from the center of mass to its support. Let

d = the distance of the free end of the cantilever to the center of the strain-gaged elements,

ν = the surface stress at the fixed end,

y = the distance of the neutral axis from the surface,

l = the length of the cantilever from the center of mass to the support,

t = the thickness of the cantilever,

w = its width,

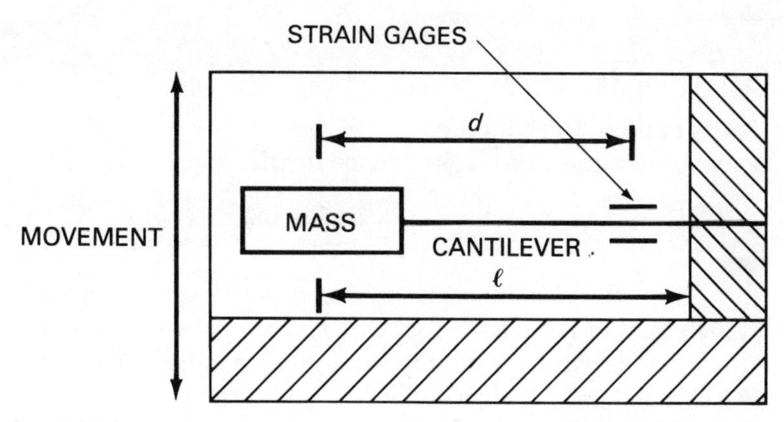

Figure 13.17
Strain Gaged Cantilever Accelerometer

m = the mass at the end of the cantilever,

ϵ = Young's modulus for the material.

The basic formula for calculating the strain in the cantilever is:

$$\frac{M}{I} = \frac{\nu}{y} \qquad (1)$$

where M is the bending moment in the beam and I is the second moment of area round the neutral axis:

$$M = md \qquad (2)$$

$$I = wt^3/12 \qquad (3)$$

$$y = t/2 \qquad (4)$$

Substituting in (1):

$$\frac{12md}{wt^3} = \frac{2\nu}{t}$$

or

$$\nu = \frac{6md}{wt^2} \qquad (5)$$

In terms of strain

$$\frac{\delta l}{l} = \frac{6md}{wt^2\epsilon} \qquad (6)$$

The deflection of a cantilever at the center of mass is given by the expression

$$\frac{ml^3}{3\epsilon I} \qquad (7)$$

The natural frequency of a cantilever with a concentrated load at the end numerically approximates:

$$\frac{1}{2\pi}\sqrt{\frac{3.04\ \epsilon I}{l^3(m + 0.224m_c)}} \qquad (8)$$

where m_c is the mass of the cantilever.

If m_c is small, then the expression may be further approximated to:

$$F \simeq \frac{1}{2\pi}\sqrt{\frac{3\epsilon I}{ml^3}}\ \text{Hz}$$

From this point, the design becomes one of successive approximation. A typical working strain of 0.1% is reasonable, but the remaining parameters will depend on the natural frequency required and the size of instrument that can be tolerated. For best accuracy the deflection should be small.

Vibration engineers usually measure vibration in terms of g, dividing the absolute value of acceleration by the acceleration constant due to gravity; that is,

32.3 ft/s^2 (9.81 m/s^2). In these terms the sensitivity of the instrument can be estimated directly. In the position illustrated in Figure 13.17 the force of gravity is acting on the mass. If the acceleration due to gravity is g, then the static deflection of the mass is equivalent to an acceleration of 1 g acting upon it. The strain induced in the cantilever as a consequence of the static deflection therefore represents an acceleration of 1 g.

Using the design parameters from equation (6) the strain experienced by a strain gage is

$$\frac{6md}{wt^2\epsilon} \text{ per } g \text{ of acceleration} \tag{9}$$

For this accelerometer to operate correctly it must be filled with a damping fluid. The viscosity of this fluid is usually chosen to make the damping of the cantilever with its mass of 0.7 of critical. Various types of oil are used, but their selection requires considerable care. Viscosity of an oil usually changes with temperature, and the choice, therefore, may not be made on its value of viscosity alone, but also on the expected changes over the working temperature range.

Commercial accelerometers have reached a high standard of design; strain gages, variable reluctance principles, and potentiometer transducers are all used to provide electrical output and are useful in appropriate circumstances. Strain-gaged cantilever accelerometers can be made quite small and will operate up to 1 kHz without difficulty. Some models, however, are susceptible to damage from shock loading. In applying them, care must be taken to ensure that impact loads do not exceed the specification. Variable reluctance devices are very robust and will withstand considerable abuse, but their response is limited by the carrier excitation frequency. A good many of these devices are designed to operate from 2000-Hz carriers and have frequency responses up to about 250 Hz. Potentiometers are intended for low-frequency applications and may have frequency responses flat to about 10 Hz. They are particularly useful for the measurement of large amplitude vibrations with frequencies of a few hertz. Potentiometers, however, are likely to be damaged by high frequencies and impact loading.

A variable reluctance accelerometer is illustrated in Figure 13.18. This device is a relatively standard design with two coils wired in opposition to produce an ac bridge with no output when the mass is in the central position. As the mass moves toward either pole, the balance of the bridge is disturbed, producing an output. This type of accelerometer is usually designed so that in conditions of gross mechanical overload the mass comes in contact with one of the pole pieces so as to prevent damage to the spring.

CALIBRATION OF ACCELEROMETERS

It may be necessary to calibrate accelerometers over their frequency range when the integrity of the spring mass system or their damping is suspect. In these situations it is necessary to mount the accelerometer on a vibrator table and make

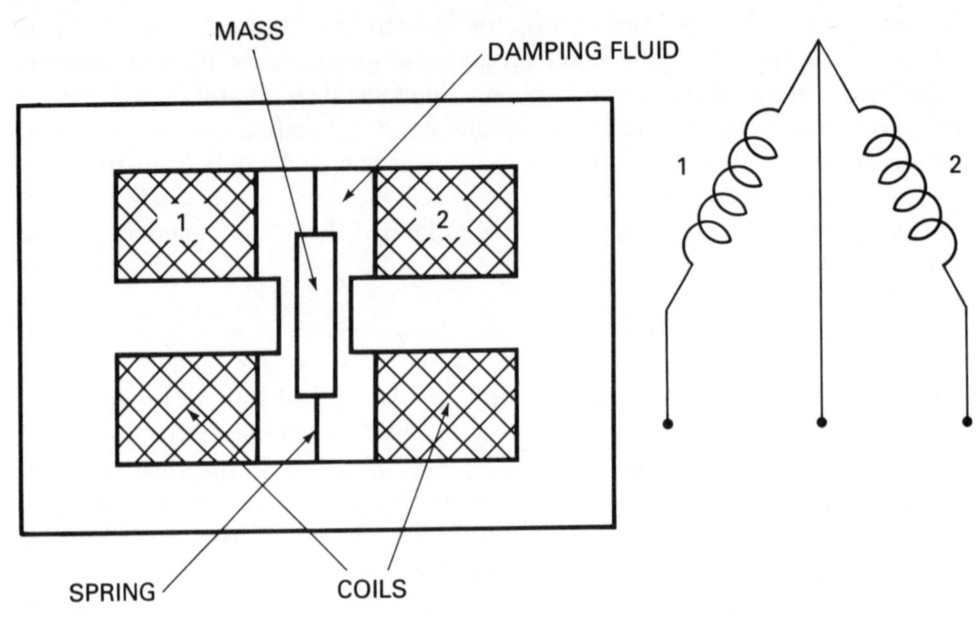

Figure 13.18
Variable Reluctance Accelerometer

measurements over the frequency range. This type of calibration will be discussed in Chapter 15.

If there is no reason to suspect malfunction of the sensor, only the simplest of calibrations is necessary. With the accelerometer standing as sketched in Figure 13.18, the electronic equipment can be balanced to zero. In this position gravity is acting at right angles to the spring mass system. Suppose now the accelerometer is turned on to one end, the output signal will be equivalent to the application of an acceleration of 1 g. If now the operation is repeated by turning the accelerometer onto its opposite end, the output will be equivalent to an acceleration of 1 g in the opposite direction. This calibration will be accurate to the specified limits of the sensor over its frequency range.

While this calibration is being performed, it is possible to check the functioning of the sensor to some degree. A different measurement in one direction than the other may indicate a faulty spring mass system. If the sensor is tapped gently against a hard surface when on its end, a decaying oscillatory output will indicate underdamping, possibly because of a loss of damping fluid.

PIEZOELECTRIC ACCELEROMETERS

Another type of accelerometer employs *piezoelectric* techniques and is used extensively when the addition of mass to a structure is an important factor or when high-frequency response is required. A piezoelectric material is one that generates an electrical output when subjected to mechanical stress. Some natural crystals such as Rochelle salt and quartz exhibit piezoelectric phenomena, but the crystals used for accelerometer construction are manufactured.

Certain polycrystalline ceramics such as barium titanate, lead titanate, and zirconate when polarized by a high voltage suffer a crystal realignment which is retained after the external field is removed. These polarized crystals are now piezoelectric and will remain so unless the crystal is heated to a temperature above its *Curie point* where the crystal changes its structure.

Piezoelectric crystals may be built into accelerometers in a number of ways, the most common being the compression and shear types. Figure 13.19 illustrates the simplest of the compression designs. In this design a piezoelectric crystal is sandwiched between a brass or steel mass and an electrically conductive base, the joints made with an epoxy resin adhesive. The base and the mass are used as electrical connecting points for the output, and the crystal serves as a spring as well as a transducing element. This basic design suffers a number of shortcomings, including sensitivity to acoustically transmitted vibration and changes in sensitivity due to humidity affecting the surface leakage resistance. An improvement on this design that eliminates most of the disadvantages is illustrated in Figure 13.20.

In this design the crystal is insulated from the base, and the bottom face connected to an output terminal not shown in the Figure. The mass rests on the uninsulated top face of the crystal and connects to the case through an isolation spring compressed to a desired preload by adjustment of the top cap. This design produces high sensitivity with a high natural frequency. The cross-sensitivity is also low and it has high immunity to acoustic noise and humidity.

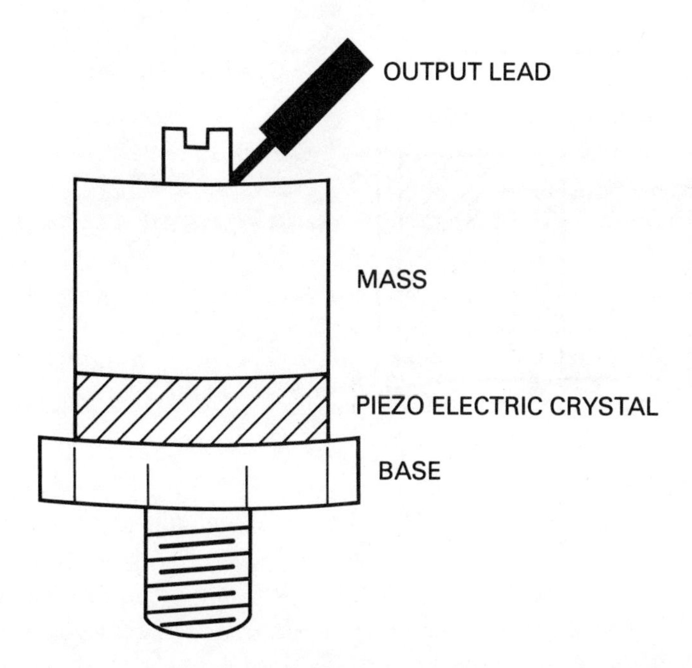

Figure 13.19
Basic Compression Piezoelectric Accelerometer

An alternative configuration uses the crystal in shear. A cylindrical crystal (Figure 13.21) is bonded to a central post, and a concentric central mass is bonded to the crystal. As with the isolated compression type, the spring mass system is protected from case effects and external noise. Accelerometers of this type may be reduced appreciably in size, and allow flexible mounting arrangements when manufactured with a hollow center post. The smallest unit of this range is less than 0.3 inches (8 mm) high and weighs about 0.018 oz (0.5 gram).

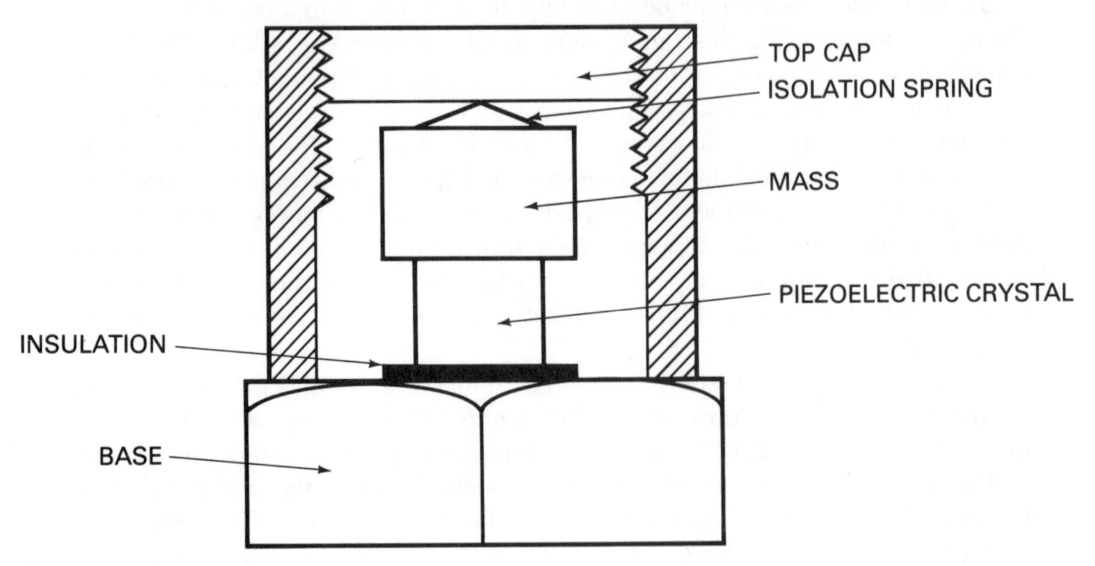

Figure 13.20
Piezoelectric Compression Accelerometer (Endevco Corp.)

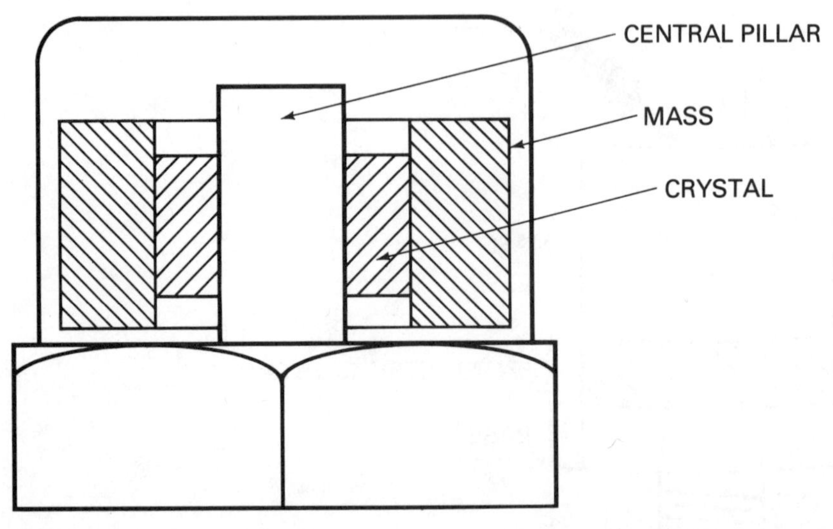

Figure 13.21
Piezoelectric Shear Type Accelerometer (Endevco Corp.)

Sensitivity

The sensitivity of a piezoelectric accelerometer may be expressed in two ways. When expressed in terms of voltage per unit of acceleration it is normal to quote sensitivity as rms voltage output when subjected to a sinusoidal acceleration whose maximum (peak) value is equivalent to 1 *g*. This value is quoted with a stated value of capacitance connected to the accelerometer. The rms voltage converts to peak voltage by multiplying the sensitivity factor by $\sqrt{2}$. An alternative method takes advantage of the fact that the accelerometer may be treated as a charged capacitor and quotes the sensitivity in terms of charge generated when subjected to a sinusoidal acceleration with a peak value of 1 *g*.

Application of Piezoelectric Accelerometers

The two methods of quoting sensitivity allow two methods of modeling the accelerometer in terms of *equivalent circuit*. Equivalent circuits are used by designers of electronic equipment to reduce circuits to basic networks so that they may be more easily understood, and their parameters calculated. Where voltage output of the accelerometer is being used to provide a measurement of acceleration, the equivalent circuit is illustrated in Figure 13.22a. The accelerometer is regarded as a voltage generator E with a capacitor C_a in series. The accelerometer is also connected across the input of a measuring circuit whose resistance is R, by a cable whose capacitance is C_e. When the charge of the accelerometer is being used as a measurement of acceleration, the accelerometer becomes a charge generator q coupled to a parallel capacitor C_a (Fig. 13.22b).

The circuit parameters are represented as follows:

E is the open-circuit output measured in millivolts/*g* of acceleration.

q is the charge output measured in picocoulombs/*g* of acceleration.

C_a is the accelerometer capacitance.

C_e is the external capacitance (primarily the cable capacitance).

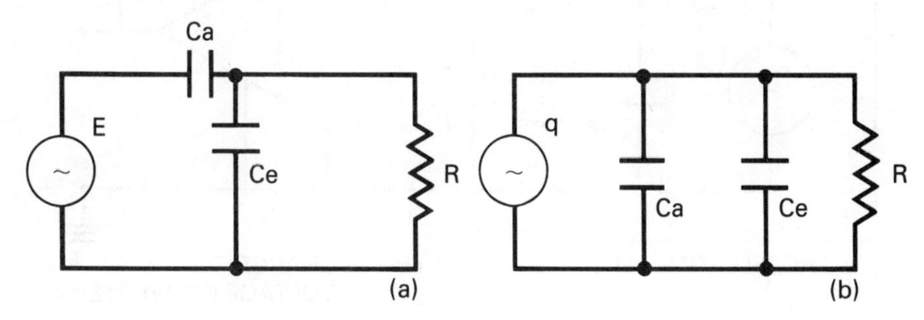

Figure 13.22
Piezoelectric Accelerometer
Equivalent Circuits

R is the resistance input of the measuring instrument (normally sufficiently high to be ignored).

E_R is the voltage appearing across R.

Considering just voltage sensitivity, the voltage appearing across R depends on the value of the external capacitance. In effect the accelerometer and cable capacitances act as a potential divider such that:

$$E_R = E\left(\frac{C_a}{C_a + C_e}\right)$$

System sensitivity, therefore, is a function of cable length and type. To obtain accurate measurements, it is necessary to calibrate the accelerometer and its cable as a system. This is obviously not a convenient procedure as it requires a new calibration each time the cable length is changed.

The alternative approach uses an instrument to measure the charge generated by the accelerometer when subjected to an acceleration. This is considered a far better technique than voltage measurement as it allows the effect of cable capacitance to be ignored.

The basic electrical relationship between current flowing in a circuit and the charge of a capacitor is given by:

$$i = \frac{dq}{dt}$$

It is possible, therefore, to determine the charge on a capacitor by integrating the current flowing in a circuit at any instant with respect to time. That is,

$$q = \int i dt$$

The charge to voltage converter illustrated in Figure 13.23 is essentially an

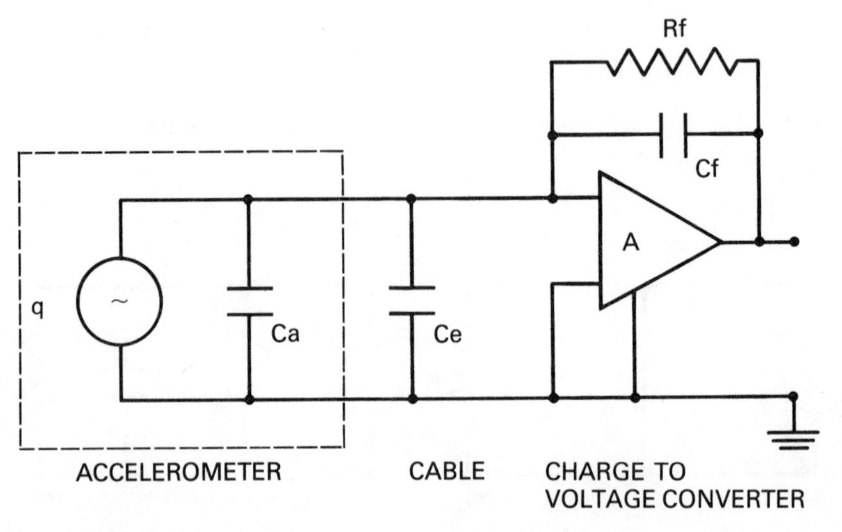

Figure 13.23
Charge/Voltage Converter and Accelerometer Circuit Schematic Diagram

integrator based on an *operational amplifier*. Operational amplifiers are further discussed in Chapter 16. Basically they are integrated circuit dc (direct-current) amplifiers with high input impedance, low output impedance, and very high gains. The characteristics of these amplifiers may be changed by connecting their output to one of their input terminals through some type of circuit. This is described as applying *feedback* to an amplifier. One example described as *negative feedback* requires that a resistance be connected between the output terminal and the amplifier negative input terminal. Depending on the value of this resistor, the connection reduces the gain of the amplifier, but improves its stability, and extends its *frequency response*. Frequency response is the frequency range over which the ratio of amplifier output to input is constant. Where dc amplifiers are concerned, this ratio extends down to and includes dc levels of input and output.

In the circuit illustrated, the amplifier has been turned into an integrator by the introduction of a resistor R_f and a capacitor C_f in the feedback circuit. The operation of these circuits is such that they respond to changes of current in the input circuit and produce an output proportional to the integral of the input. With the accelerometer connected, the input circuit adjusts itself so that no current flows. In this condition, the accelerometer and cable may be regarded as two charged capacitors in parallel. If now the accelerometer is subjected to an acceleration, there will be a corresponding change in charge that will cause a current flow in the circuit. Providing the voltage across the input does not change, the current will be entirely due to the change in charge and proportional to dq/dt. The output of the integrator, therefore, will be proportional to the change in charge and also acceleration. When integrators are used as charge converters, they are usually provided with subsequent stages of amplification, and in this form they are known as *charge amplifiers*.

In operation, the voltage across the input of a charge to voltage converter is maintained at constant and very low level, by suitable choice of feedback components. Under these conditions, the cable takes up a charge that is unaffected by, and has no influence on, the operation of the circuit. This implies that within normal operational limits, cable capacitance is of no concern, and that an accelerometer calibrated by this method may be subsequently used with other cables.

ANALYSIS OF VIBRATION RECORDS

It is common practice to use accelerometers to drive multichannel galvanometer recorders that reproduce the signals as fluctuating lines along the length of a moving strip of paper. Such recorders usually incorporate accurate timing facilities that draw lines across the paper at specified intervals.

If the trace shows a single frequency, the determination of amplitude and frequency of the signal is one of straightforward measurement. The more usual situation, however, is one in which the record is that of a complex waveform that must be reduced to its component frequencies if the vibration is to be fully understood.

It is possible to analyze records by Fourier techniques, but these involve tedious measurement and extensive mathematical operations. If Fourier analysis is required, it is better to replace the galvanometer recorder with an analog tape recorder and subsequently use electronic or computer analysis techniques.

In many instances the delays introduced between taking the record and obtaining analysis from Fourier techniques are unacceptable; for example, it may be desired to test a piece of moving machinery at a series of speeds up to its maximum. At each speed, an analysis is required to ensure no dangerous resonances are experienced or being built up. To record results on a tape recorder and wait for off-line analysis at each speed may extend the tests to an unacceptable time scale. Even in instances that warrant ultimate detailed analysis, a preliminary examination of the wave-form usually provides sufficient data to continue the test through its speed range, with a consequent reduction in overall testing time.

One such method of analysis, developed by R. G. Manley in his book *Waveform Analysis* (Chapman and Hall, London), is described as *the envelope method of waveform analysis*. It provides a quick method of determining the component frequencies in a complex waveform and their approximate amplitude. Manley reduces complex waveform to a series of *beat frequencies* from which he extracts the two highest frequency components. He then demonstrates that the waveform left after the removal of the higher frequencies may again be subjected to the same treatment until only the fundamental frequency is left.

The accuracy of the technique depends on the clarity and quality of the records, and on the skill and experience of the analyst. Even so, under favorable conditions it is possible to resolve a waveform of four frequency components.

Beat Frequencies

Consider two signals that have the form

$$A\sin(2\pi f_1 t) \qquad \text{and} \qquad B\sin(2\pi f_2 t)$$

Suppose f_1 is a higher frequency than f_2 and very close in value. If the two signals are summed over a number of cycles, a resultant signal illustrated in Figure 13.24 is produced, with the appearance of a sinusoid whose amplitude

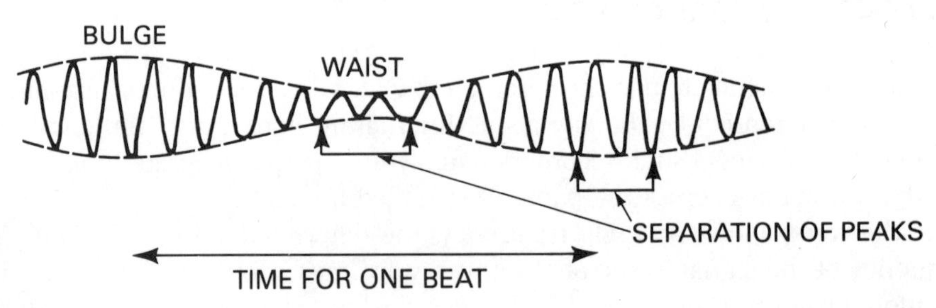

Figure 13.24
Beat Frequencies

varies sinusoidally with time. The frequency of the amplitude variation is described as the *beat frequency* and it can be shown mathematically that it equals $|f_1 - f_2|$. Where the peak-to-peak amplitude of the waveform reaches a maximum is known as the *bulge*, and the minimum as the *waist*. The peak-to-peak amplitude at the bulge can be shown to be $2(A + B)$ and at the waist $2(A - B)$.

The apparent frequency of the waveform when over the period of a beat frequency is that of the component with the larger amplitude usually described as the *major component*. The hidden frequency is that of the component with the smaller amplitude (*minor component*). Unfortunately it is not immediately apparent from the waveform whether the hidden frequency is of a higher or lower value than the visible one, and this must be deduced from a comparison between peak separation at the waist and the bulge. These follow the rule:

1. If the minor component has a lower frequency than the major, then the measurement between successive peaks at the bulge is greater than a corresponding measurement between peaks at the waist.
2. If the minor component has a higher frequency, the measurement between successive peaks at the bulge is less than at the waist.

This type of waveform is produced providing the ratio f_1/f_2 of the component frequencies is less than 2:1.

Waveforms with a High-Frequency Content

When two sine waves have a frequency ratio greater than 2:1, their sum produces a waveform such as that shown in Figure 13.25, and both components are apparent in the resultant waveform. This is true regardless of which is the minor component.

The measurement of the amplitude of these two components may be made directly, and their values are indicated at 2a and 2b. For envelope analysis purposes, the envelope enclosing the high frequency may be regarded as a special case of the beat frequency envelope in which the waist and the bulge are of equal amplitude.

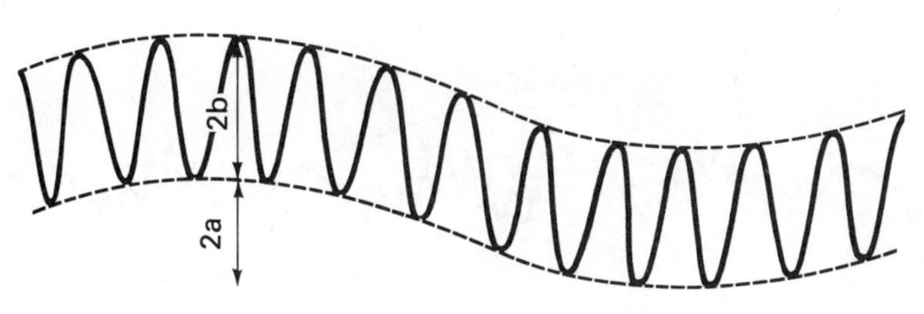

Figure 13.25
Waveform with High Frequency Imposed

Determination of the Fundamental Cycle

The first step in a waveform analysis is the determination of the fundamental cycle length of the waveform. This is usually easily recognized, but in certain instances it is possible to misinterpret the waveform. Figure 13.26 illustrates a waveform liable to misinterpretation. Examination of the waist at the points *A*, *B*, and *C* indicate a similarity, but only *A* and *C* have a characteristic shape. The fundamental cycle, therefore, is probably *A*, *C*, although it could actually be longer.

Although the interpretation of records based on such small differences may appear hazardous, it should be remembered that the analyst will be looking at a record considerably longer than one fundamental cycle, making repetitions more easily recognizable. This implies that no final conclusions should be drawn from the examination of any single cycle of a waveform. Much greater accuracy and confidence is achieved by analyzing several fundamental cycles of each record and comparing the results of each. A knowledge of the mechanics of the motion of the structure being analyzed will also provide clues to the validity of the analysis.

Peak Masking

Some waveforms at first sight may appear to be a fundamental carrying a single high-frequency component. Closer examination of the record often indicates that some of the high-frequency peaks are missing. The location of such missing peaks is important, for they indicate the presence of a second high-frequency component close in frequency and of nearly equal amplitude.

The location of missing peaks is not difficult, providing the analyst examines a length of record rather than a single fundamental cycle. In some places the peak will be missing completely; in others the remnant may still be recognizable.

Figure 13.27 shows two traces, the first where peaks are obviously missing, and the second where the existence of a peak is less obvious.

Figure 13.26
Waveform with Ambiguous Fundamental Period

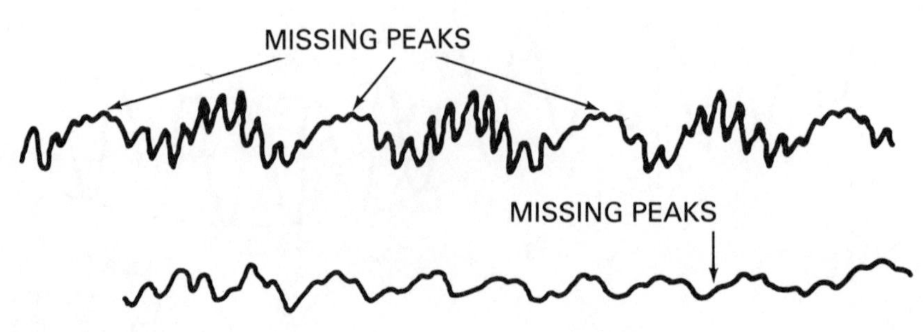

Figure 13.27
Waveforms with Peak Masking

APPLICATION OF ANALYSIS TECHNIQUE

The operation of the envelope analysis technique depends on the fact that when two frequencies with a ratio of less than 2:1 are present in a record they will form a beat frequency. This beat frequency may ride on a lower component frequency, or even on more than one component. Although the shape of the beat frequency envelope will be distorted, the method of analysis illustrated for a simple combination is still valid.

After the higher frequencies have been analyzed from the waveform, they may be eliminated completely by drawing an average curve through the envelope produced in the process of analysis of the higher frequencies. The lower order frequencies present in the resultant waveform are then reduced by a repeat operation.

The simplest way of performing the analysis uses transparent paper on which is drawn the envelope enclosing the highest frequencies. A second piece of paper placed over the envelope allows an average curve to be drawn, which may in turn be analyzed by a repeat operation.

The principles are probably most easily illustrated in a series of examples.

Analysis of a Simple Beat Frequency Waveform

Consider the waveform shown in Figure 13.28:

Peak-to-peak amplitude of bulge = 12 mm = $2(A + B)$

Peak-to-peak amplitude of waist = 3 mm = $2(A - B)$

Solving for A and B $A = 3.75$ mm

$B = 2.25$ mm

The visible frequency in the waveform = 12 Hz

Beat frequency = 1 Hz

The frequency of the minor component is 12 Hz ± 1 Hz.

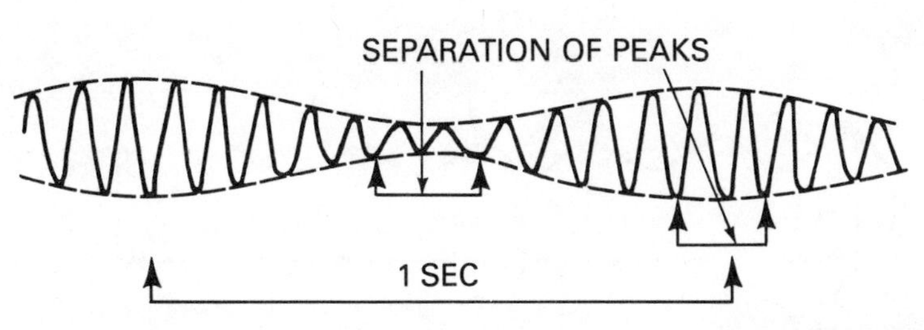

Figure 13.28
Analysis of Waveform with Simple Beat Frequency

Applying the separation rule: The separation of the peaks at the bulge is less than at the waist. The minor component, therefore, is the higher frequency. The second frequency is 13 Hz.

Analysis of a Three-Component Waveform

Consider the waveform in Figure 13.29a. The highest apparent frequency is 36 Hz.

Figure 13.29b illustrates the envelope produced by laying a piece of transparent paper over the original waveform and enclosing the high-frequency component. This envelope represents a distorted envelope of a beat frequency of 18 Hz. If the peak separations at the bulges and waists of these beats are examined on the original record, the width at the bulge is seen to be greater than the waist. The minor frequency is therefore the lower. The amplitude of the bulge is 6 mm. The amplitude of the waist is 1 mm. It follows, therefore, that

$$f_1 = 36 \text{ Hz}$$

$$f_2 = 18 \text{ Hz}$$

$$2(A + B) = 6 \text{ mm}$$

$$2(A - B) = 1 \text{ mm}$$

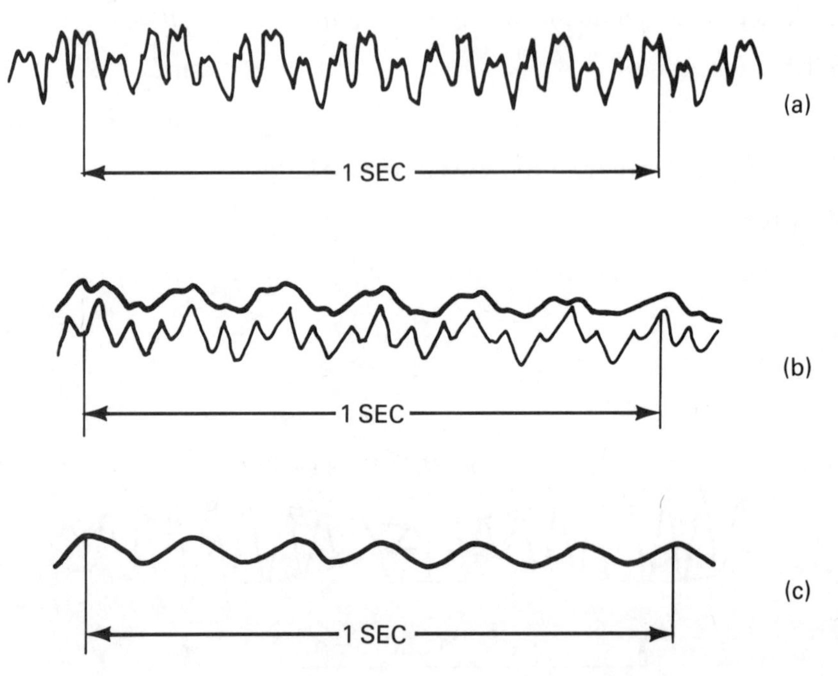

Figure 13.29
Analysis of Three-Component Waveform

Therefore,

$$A = 1.75 \text{ mm}$$

$$B = 1.25 \text{ mm}$$

The final stage of this analysis requires that a curve be traced midway between the two lines of the envelope. This produces Figure 13.29c, which is a fundamental frequency component of 6-Hz frequency and an amplitude of 3.0 mm.

The analysis of the waveform therefore is

f_1 = 36 Hz Amplitude = 1.75 mm

f_2 = 18 Hz Amplitude = 1.25 mm

f_3 = 6 Hz Amplitude = 3 mm

The second decimal place is academic and in no way reflects the accuracy of the analysis.

Analysis of a Four-Component Waveform

The final example is a waveform with four components, which represents the limits that this analysis can achieve with any degree of confidence.

The highest recognizable frequency present on the waveform in Figure 13.30 is 30 Hz. This example indicates very clearly the phenomenon of peak masking.

The first envelope illustrated in Figure 13.30b has a very small peak-to-peak amplitude across the waist. This implies that the two components of the waveform are of similar size.

From this envelope, the following measurements were made:

Amplitude at bulge = 5 mm
Amplitude at waist = 1 mm
Beat frequency = 2 Hz

Peak separation at the waist is less than at the bulge, therefore the minor frequency is lower than the major. From which:

Amplitude of major component = 1.5 mm Frequency = 30 Hz
Amplitude of minor component = 1.0 mm Frequency = 28 Hz

The waveform in Figure 13.30c is a curve drawn through middle of the envelope in Figure 13.30b. From Figure 13.30c the envelope in Figure 13.30d is produced. The apparent frequency of c is 8 Hz, and the envelope shows no sign of a beat frequency.

This represents an 8-Hz signal of a fundamental of 2 Hz, the amplitude of which may be obtained by direct measurement 13.30(e).

Summarizing the analysis of the waveform:

f_1 = 30 Hz Amplitude = 1.5 mm

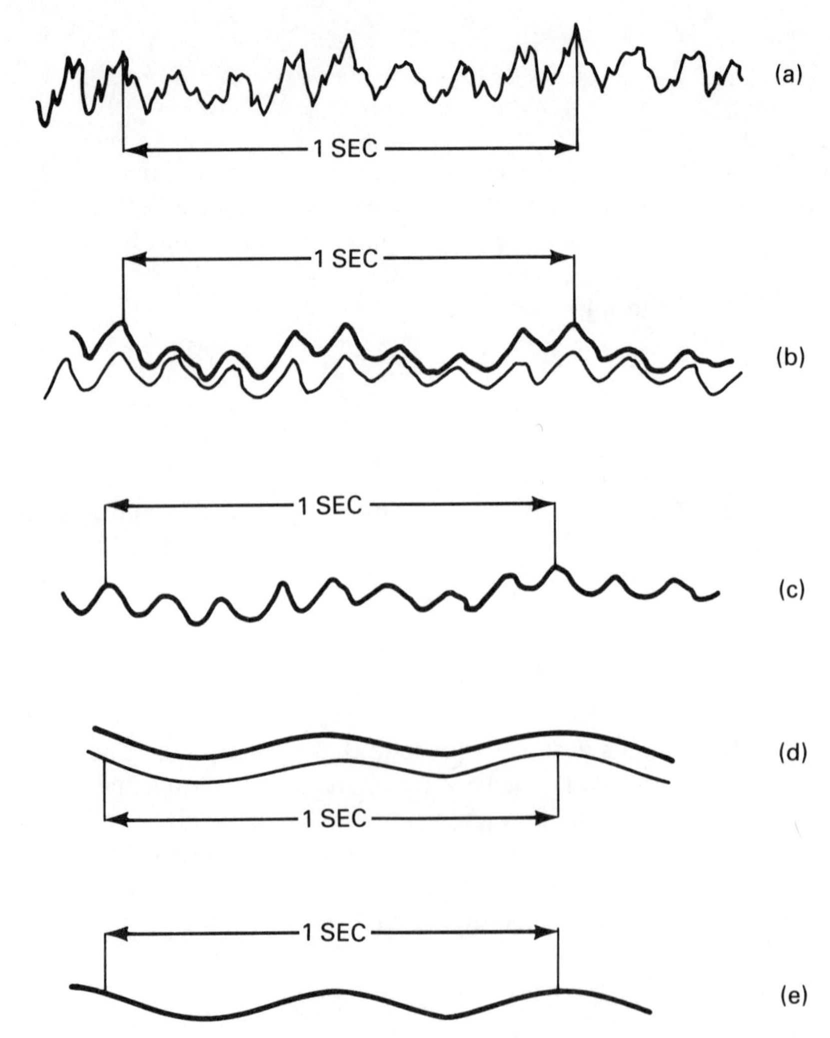

Figure 13.30
Analysis of Four-Component Waveform

f_2 = 28 Hz Amplitude = 1.0 mm

f_3 = 8 Hz Amplitude = 1.5 mm

f_4 = 2 Hz Amplitude = 1.5 mm

RESONANCE TESTING

Another type of vibration test is known as the *resonance test*. A structure is forced into vibration by a mechanical exciter, and the response to the excitation is measured to determine the characteristics of the structure.

Considering the simplest case, a vibrating cantilever at resonance may take up one of a number of dynamic shapes called *modes*. Figure 13.31 illustrates four resonant modes of vibration of a simple cantilever system.

Suppose the cantilever is excited by a vibration generator whose frequency may be varied over a wide range. As the frequency is increased from zero, at some point the amplitude of the vibration will increase as the cantilever reaches its first resonant mode; above that frequency the amplitude decreases and then increases until a second resonance is reached where the amplitude again increases. The shape the cantilever assumes at these two frequencies, however, is very different. At the lower frequency the maximum amplitude of vibration is measured at the

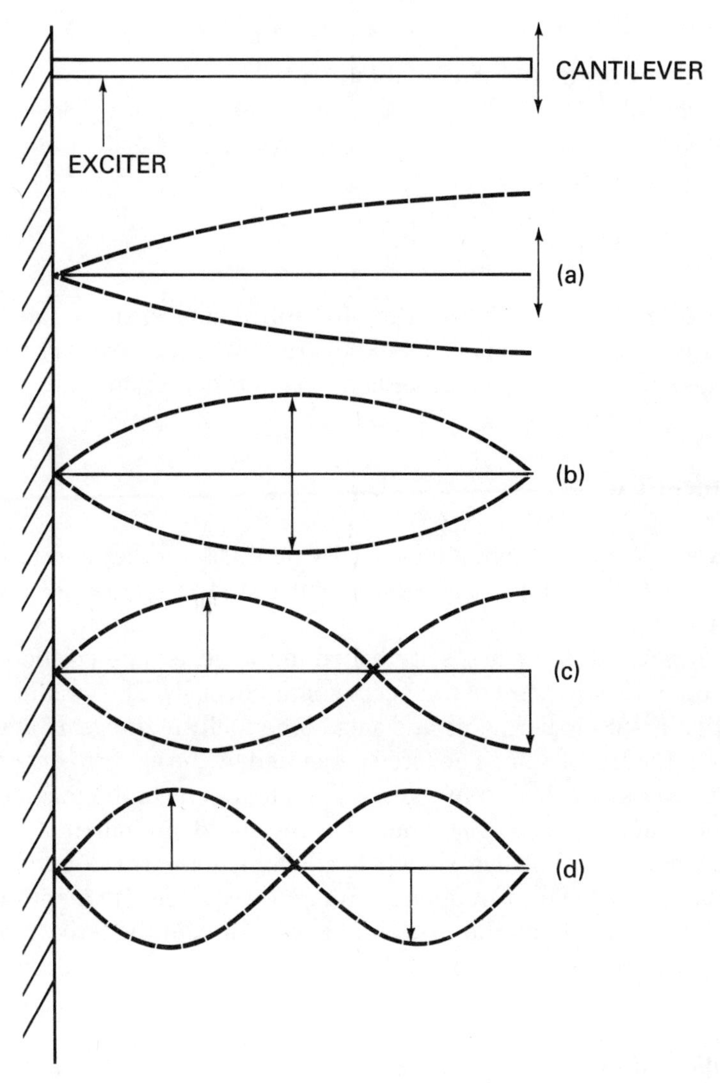

Figure 13.31
Resonance Modes of a Simple Cantilever

free end of the structure; at the higher frequency the end of the cantilever is stationary, and the maximum amplitude is recorded in the middle. Further resonances occur at higher frequencies, each with its own characteristic movement.

These resonant frequencies have strict relationships to one another, the higher ones being integral multiples of the fundamental.

Suppose the fundamental frequency of the cantilever is f:

The frequency of the fundamental resonant mode of the cantilever is f (Figure 13.31a)

The frequency of the fundamental second resonant mode will be $2f$ (Figure 13.31b)

The frequency of the fundamental third resonant mode will be $3f$ (Figure 13.31c)

The frequency of the fundamental fourth resonant mode will be $4f$ (Figure 13.31d)

EXCITERS

The choice of exciter depends on the structure being excited, the frequency range, and the forces required to excite the structure. Many types of exciter have been designed; the following represents a small cross section.

Mechanical Exciters

The mechanical exciter is sometimes used where large forces and low frequencies are required. The simplest of these is illustrated in Figure 13.32a.

This type of exciter employs two intermeshed gears driven by an electric motor. Two out-of-balance masses are bolted to the gears, and positioned so that they reach the meshing point of the gears simultaneously. This produces a resultant force that is simple harmonic and acts tangentially to the gears at the meshing point (see Figure 13.32b). The force generated by the exciter is $m\,\omega^2 R$, where m is the total mass attached to two gears, R is the radius of the masses from the center of the gears, and $\omega = 2\pi f$ where f is the speed of rotation of the gears in revolutions per second. The use of such an exciter is somewhat limited. The force generated is proportional to frequency squared, requiring frequent changes of mass in the radius at which the mass rotates to maintain the force at an acceptable level as the frequency is varied.

Hydraulic Exciters

For systems with low frequency range, hydraulic jacks are capable of providing very large sinusoidal forces. Jacks used for this purpose are driven by elec-

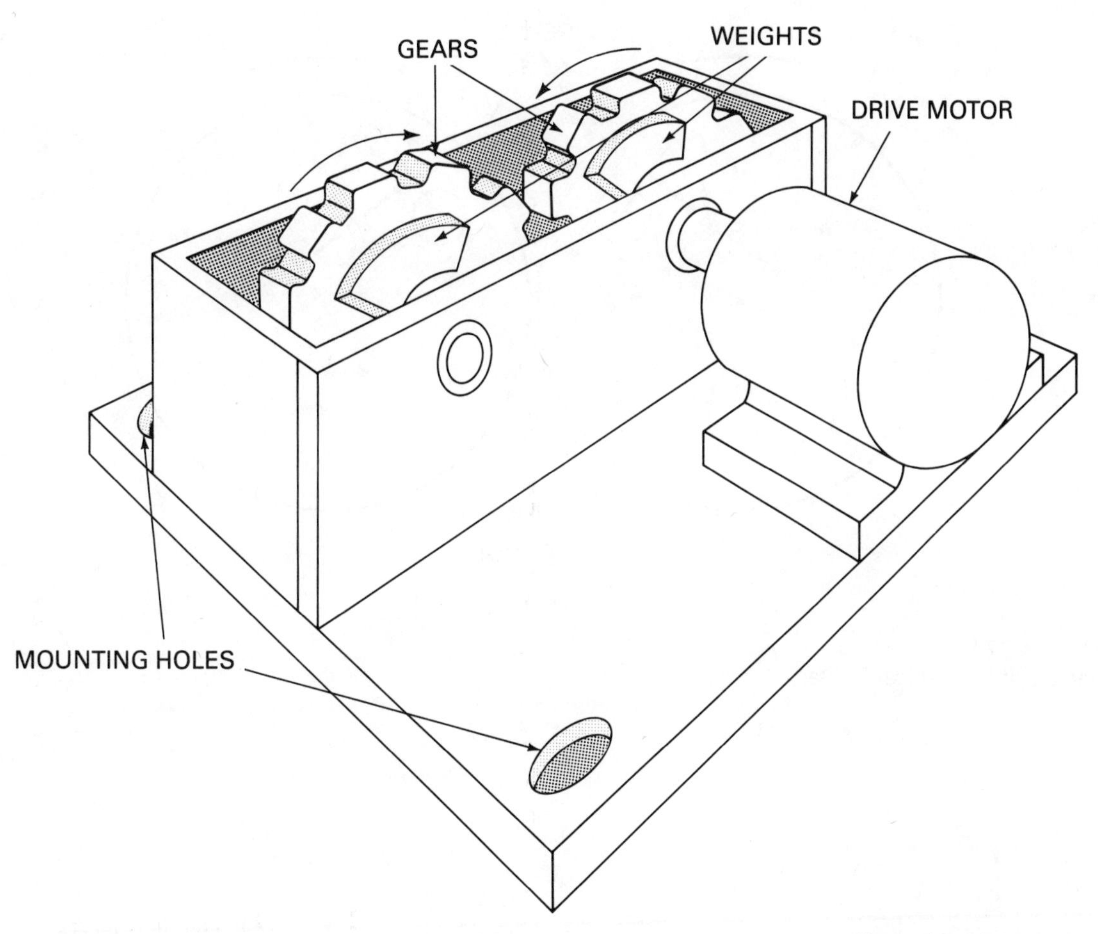

Figure 13.32a
Electromechanical Exciter

trohydraulic servo valves. The input of the servo valve is an amplified ac signal; the output is a varying hydraulic pressure that controls the main jack inlet valve.

Electromagnetic Exciters

By far the most versatile exciter is the electromagnetic exciter that operates in much the same way as an audio loudspeaker. A typical design is illustrated in Figure 13.33.

A coil wound on an armature is suspended between the holes of a magnet by springs called *spiders*. The coil moves vertically in the magnetic field driven by the influence of a second oscillating field produced by current flowing through the coil. This current, generated by a power amplifier driven by an oscillator, is easily and accuratly controlled, for both frequency and amplitude. High lateral stiffness is provided by placing the spiders at either end of the exciter and connecting them with a coupling shaft.

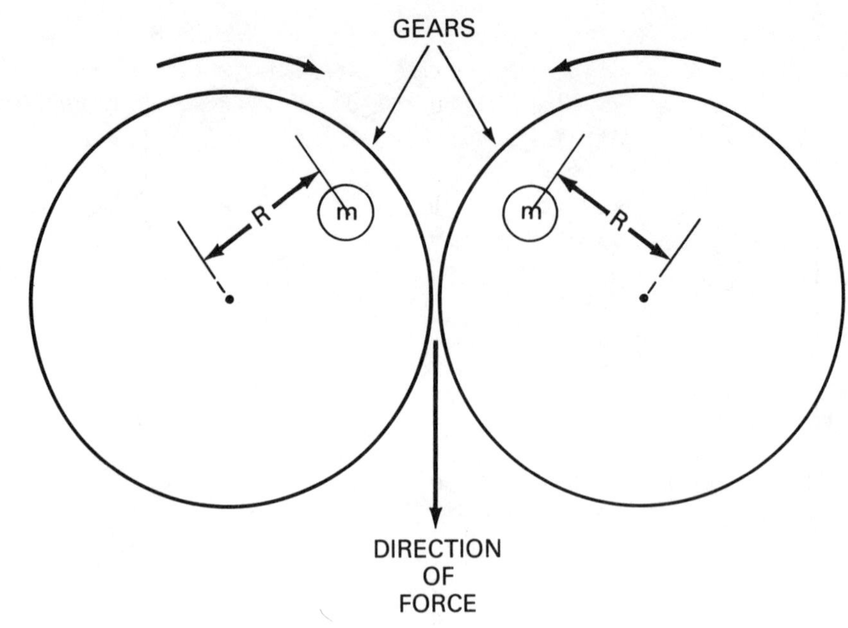

Figure 13.32b
Principle of Mechanical Exciter

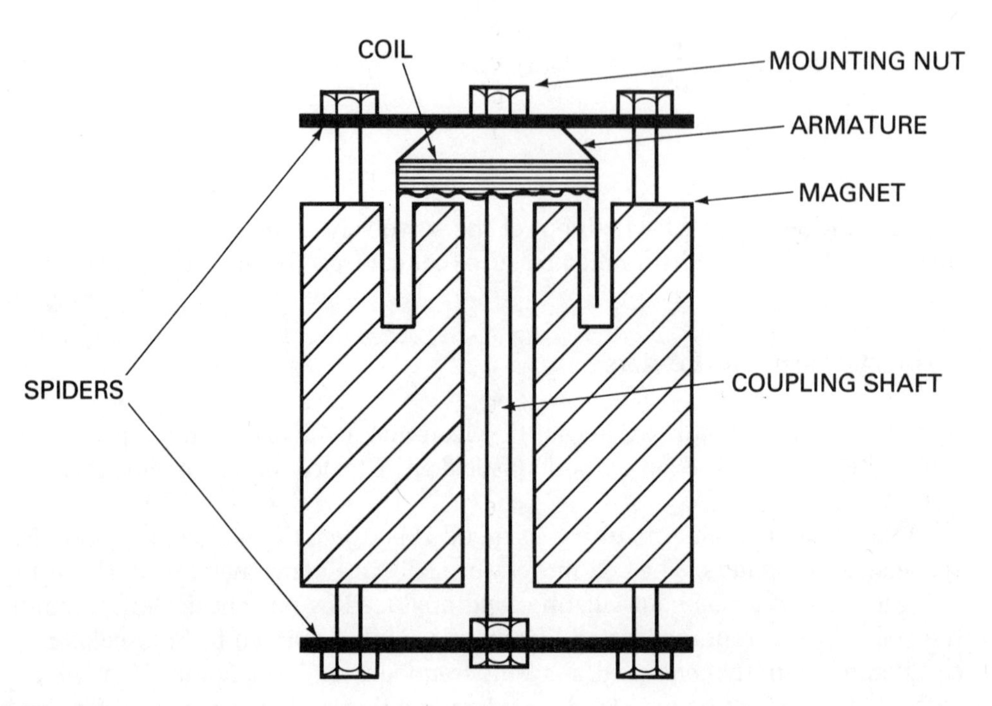

Figure 13.33
Electromagnetic Exciter

A number of variations of the basic design are common. Where larger forces are required, a second air gap and coil can be installed at the opposite end of the exciter. The coils are wired to allow tandem operation, effectively doubling the force output of the unit. Small exciters employ permanent magnets; larger ones employ coils energized from dc supplies. Larger models are also air cooled.

The air gap operates at high flux density, and to protect it from damage by ferrous particles, the whole assembly is enclosed in a case with rubber seals.

Electromagnetic exciters may be designed to operate at frequencies up to tens of kilohertz and deliver forces of several hundred pounds.

Acoustic Excitation

In some circumstances it is possible to excite specimens acoustically. Specially designed loudspeakers are used for this application, mounted in close proximity to the specimen.

Acoustic excitation is advantageous in that there is no mechanical connection between the specimen and the exciter and consequently there is no modification to the structural characteristics. The power levels that may be transmitted this way are small, however, and the sensor signal levels produced are also small. Such excitation therefore is used only in special circumstances.

Another form of acoustic excitation sometimes used for special applications is the noise generated by a jet engine, or rocket motor. Power levels obtained this way are much higher than with a loudspeaker, but the type of signal is very different. With a loudspeaker, excitation is usually a pure tone. A jet engine generates *random noise*, that is, noise containing a large number of frequencies of random amplitude. This type of noise causes all the modes of the specimen to be excited at the same time. Special types of analysis are required, and these are very different from the more common resonance test analysis.

Shock Excitation

In certain instances vibration engineers are reluctant to use continuous excitation of a structure. The measurement of the natural frequency of the tail of an aircraft in flight is one example. Even though it may be possible to mount and drive an exciter, there is always a risk of building up uncontrollable resonances that can become catastrophic before the exciter can be brought to rest. One alternative uses shock excitation to provide an oscillation of the tail from which a measurement of damping may be made.

A common way of inducing shocks in a specimen of this nature is to mount a slow burning explosive charge on either side of its surface. The force generated by each charge approximates a half sine wave. A second charge fired on the opposite surface after an appropriate time period completes an approximately sinusoidal force of one cycle to stimulate the natural frequency of the member. The response is recorded and the damping calculated from the rate of decay of the signal. This is illustrated in Figure 13.34.

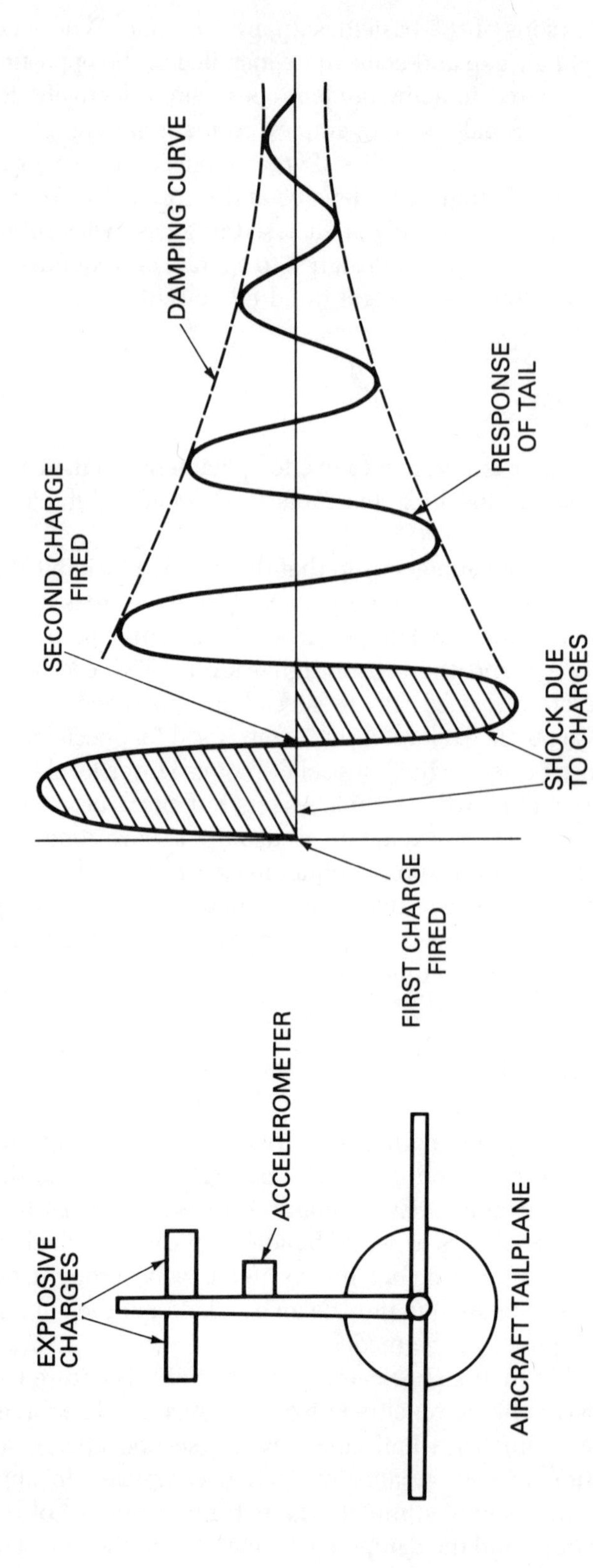

Figure 13.34
Shock-Induced Vibrations

GROUND-REFERENCED RESONANCE TESTS

A very large proportion of resonance tests are carried out on structures that are either fixed to the ground or mounted on supports close to the ground. In these cases, the body of sensors can be mounted to structures bolted to the floor, and their armatures attached to the structures being tested. The differential transformer is a useful device for this measurement.

Figure 13.35 shows a typical mounting for a differential transformer sensor. The body of the sensor is held in a substantial clamp stand that stands on the floor and allows vertical adjustment of the sensor. Because it is not always convenient to attach the armature of the sensor to the specimen, a shaped coil spring is often used to hold the armature against the specimen. Care must be taken when adopting this type of fixing that the spring natural frequency is well above the measurement frequency range and also that the added stiffness of the spring does not influence the specimen characteristics.

Such a sensor mounting is easily positioned and the sensor balance point is easily adjusted. Where it is necessary to move a restricted number of sensors around a structure to measure different modes, they are also easily repositioned.

Sensor Positioning

If a number of vibration modes are to be measured, the problem of positioning sensors arises. Considering Figure 13.31, a sensor at the end of the cantilever

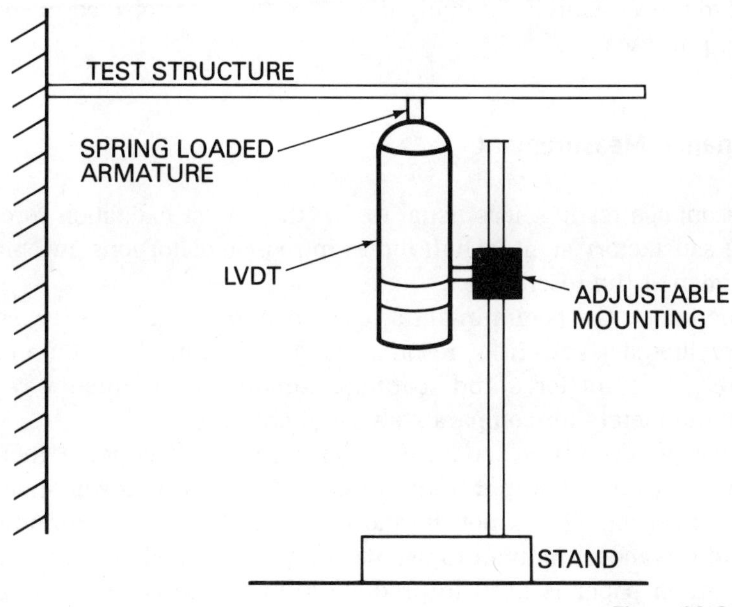

Figure 13.35
Ground Resonance Test Sensor Mounting

would provide the largest signal when the system is resonant in its first mode. It would not be clear, however, whether the system was in its first or third mode. To determine this, at least one additional sensor is required and some means provided to compare the relative direction of motion.

In the case of the simple cantilever, the position and number of sensors required to recognize the different modes are easy to estimate. In more complex structures this estimation can be very difficult and much cross-checking of results is necessary to ascertain the modes with certainty. In these complex situations, the stroboscope is a useful tool providing the vibration engineer with the means of examining resonant modes visually. This allows the engineer to select the sensor mounting points with more certainty.

Measuring Equipment

To determine a resonant mode it is necessary to measure the amplitude of vibration at several points on the specimen and also to determine the relative direction of movement of adjacent sensors. This is normally done by using one sensor as a reference and comparing the phase relationship of the others with it. It is usual to use a load cell for this purpose which is part of the vibrator attachment to the specimen. Figure 13.36 provides a schematic diagram of an installation which may be used to resonance test a simple cantilever structure.

An exciter mounted at a convenient point on the specimen is driven by a power amplifier and a variable-frequency oscillator. It is possible to use a number of types of displacement sensor but in this instance differential transformer sensors are illustrated as being mounted along the specimen. Their signals are returned to the measurement equipment, and they are selected in sequence for recording purposes.

Resonance Measurement

When resonance testing, it is usual to use the lowest excitation force that will provide a satisfactory signal. High forces introduce distortions and may damage the specimen at the resonance.

It is normal at the beginning of a resonance test to select a few sensors and make a preliminary search for resonances. This is done by maintaining an approximately constant force and scanning through the frequency range, while watching the meters for comparatively large changes.

One way of detecting resonances is to increase the frequency incrementally, maintaining a constant output from the exciter and noting the changes in amplitude from a sensor. This is not the most accurate way of determining resonant points and it is more common to use the Nyquist method.

Polar graph paper is used to produce a Nyquist diagram. A line is drawn through the origin of the paper, providing the reference from which the phase may be measured. At each increment of frequency, a measurement is made of a

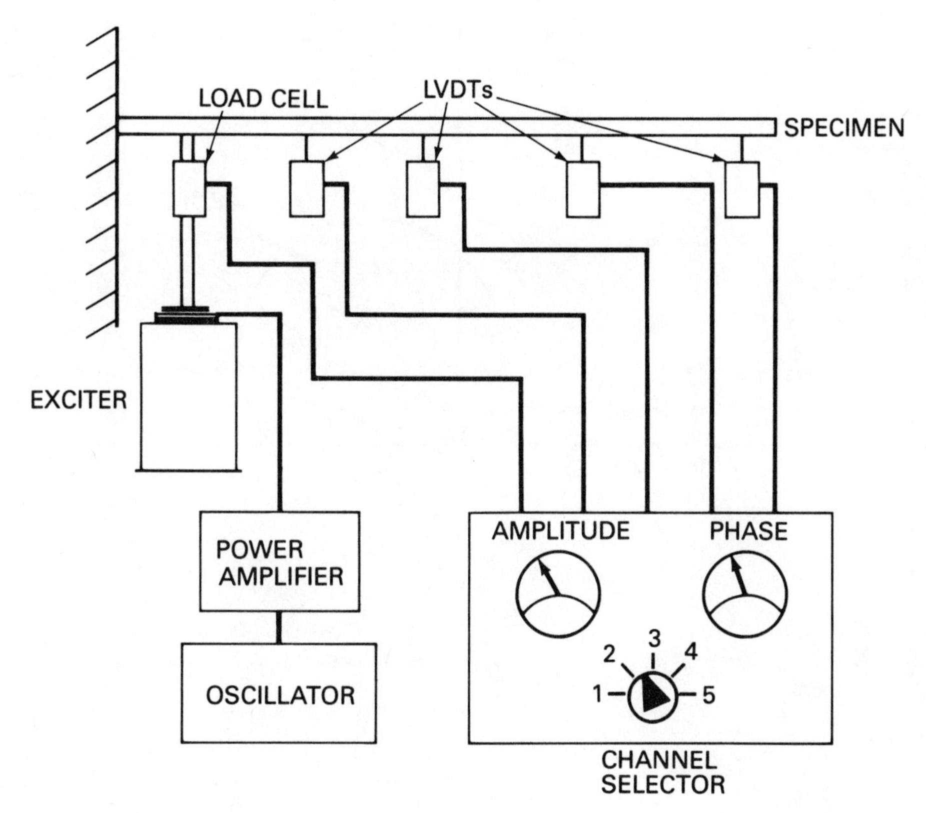

Figure 13.36
Resonance Testing Equipment

sensor output and its phase relationship to the exciter load cell. These are plotted on the diagram as an angle from the reference line and a distance from the origin. As the specimen approaches resonance, amplitudes become larger and phase changes become greater for equal increments of frequency until they both reach a maximum at resonance. The plot of these two parameters produces a curve that can be extrapolated back to its starting point to form a circle.

Figure 13.37 illustrates a typical Nyquist plot of a resonant mode from which it is possible to calculate the damping coefficient of a structure with low damping. This is illustrated in Figure 13.38.

Suppose the circle in Figure 13.38 represents a Nyquist plot of a resonance, and f_R represents the resonant frequency. If two frequencies are selected equidistant from f_R, both representing a small frequency displacement, two lines may be drawn joining each point to the resonant point. If two perpendiculars are drawn from these lines at f_H and f_L, they will intersect on the circumference of the circle.

Let $f_H - f_L = \delta f$.
Let a_1 be the length of a line joining the resonant point f_R to the intersection, and a_2 be the length of the line joining f_H to the intersection.

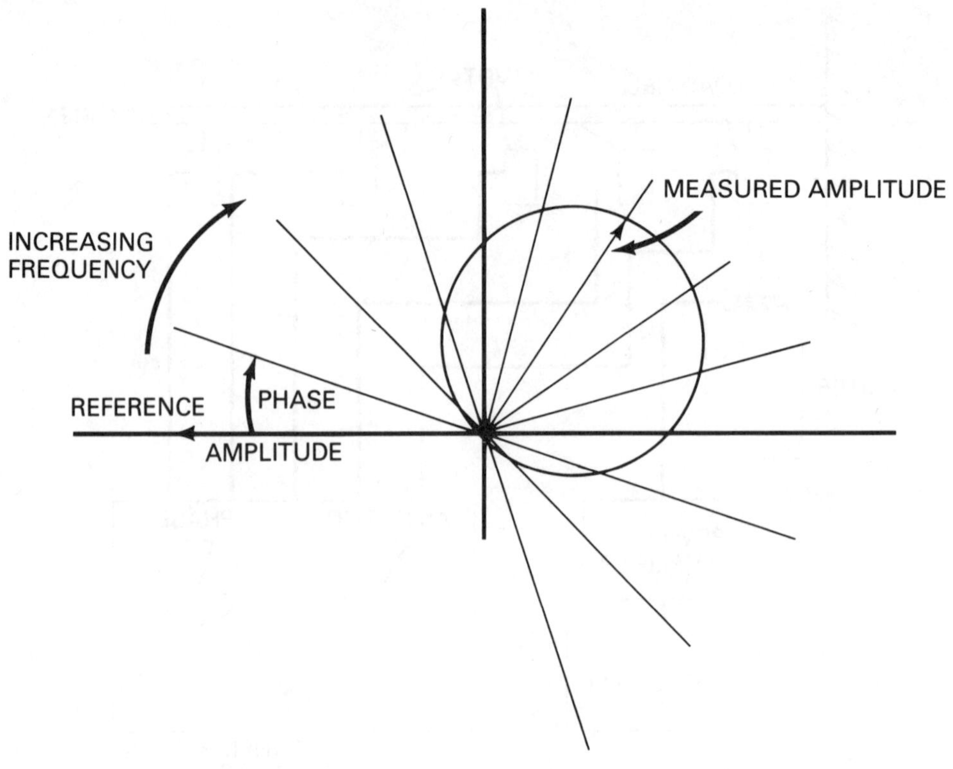

Figure 13.37
Nyquist Plot of a Resonant Mode

For a structure with low damping the damping coefficient is given by the formula:

$$c = \frac{\delta f / f_R}{\sqrt{\left(\frac{a_1}{a_2}\right)^2 - 1}} = \frac{\delta f}{f_R} \cot \phi$$

Mounting of Structures to Be Resonance Tested

A structure being resonance tested must be mounted on vibration-free supports, in a way which will not change the vibration characteristics of the structure. The normal method of mounting uses some type of seismic platform whose natural frequency is well below that of the lowest frequency being measured. One method of mounting uses jacks seated on inflated tires. Floating the specimen on inflated bags is another method, and pontoons floating in water troughs also provide good seismic characteristics. The introduction of seismic mountings can

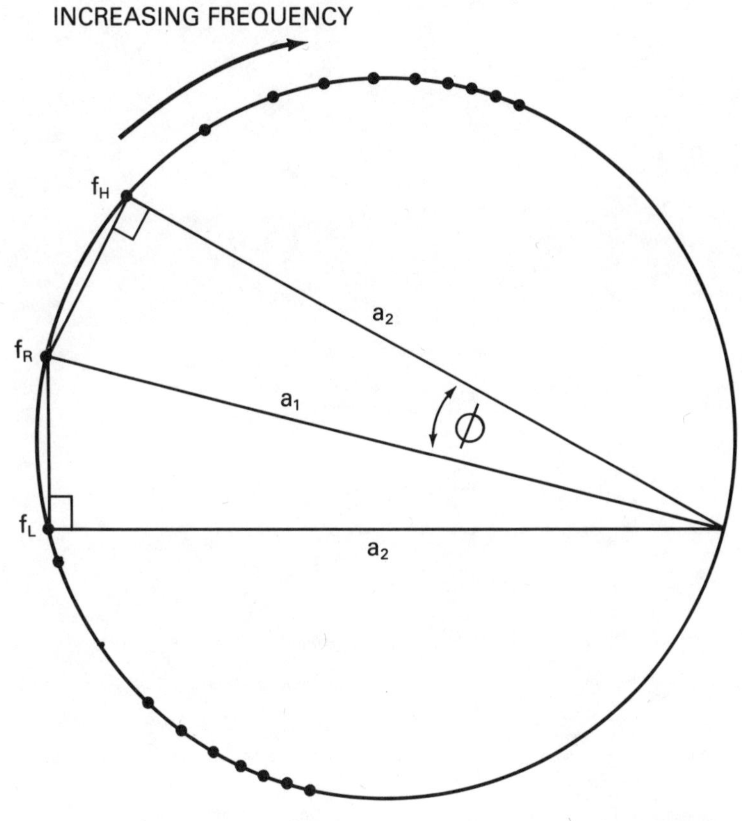

Figure 13.38
Determination of Damping from a Nyquist Plot

in themselves create a number of acute problems of attachment. Mounting structures for exciters and mountings for sensors also become more complex, with implied setting-up problems. The selection of mounting positions for a specimen can also be difficult. Supporting the weight of a structure without affecting its characteristics is a task that can only be considered on the merit of each case.

CHAPTER 14

RECORDING TECHNIQUES

ANALOG RECORDERS

Early analog recorders were for the most part "pen and ink" devices. The signal set a pen in motion across a clockwork-driven moving chart to produce an ink trace showing the variation of a signal. These recorders suffered several defects that were eventually overcome to produce the accurate and reliable instruments in current use.

Many of the early instruments had unreliable pens and small ink reservoirs that often produced lines of variable quality and intermittent continuity. Pens were also sources of high friction and inertia, and their limited length introduced inaccuracies and nonlinearity into records. Consequently such recorders could only be used where comparatively high drive power was available and accuracy was not a prime consideration. The introduction of the servo-driven pen significantly improved recording accuracy but suffered the limitation of restricted frequency response. This confined their application to relatively slowly moving parameters. If more rapid response was required, the servo was abandoned in favor of direct recording techniques with low inertia movements. Galvanometers were an obvious choice, using photographic methods of recording, but this could be done only by sacrificing the immediately visual record. Replacement of the traditional photographic paper with ultraviolet-sensitive paper produced recorders with semipermanent records that could be viewed within a few seconds of an event. This considerably improved the situation and in many instances ultraviolet recorders have replaced the photographic types. There are still occasions, however, when the developing time of ultraviolet-sensitive paper is unacceptable, so to meet these situations the ink jet recorder was created.

Galvanometers can operate up to frequencies in the kilohertz range, but at these frequencies high paper speeds are required to resolve the waveforms. Consequently, records of high frequencies become costly to produce and cumbersome to handle, so at these frequencies the tape recorder is used with oscilloscope display of the signals. Permanent records still employ techniques using sensitized paper, but these are simply *frames* of sections of the record that are of interest.

PEN RECORDERS

The earliest recorders were variations of the simple pen recorder illustrated in Figure 14.1

The mechanical movements of these recorders are similar to those of moving coil meters. Current passing through a coil pivoted about its axis produces a magnetic field that distorts the field of the permanent magnet. This distortion provides a force couple causing the coil to rotate against the spring, the direction of movement depending upon the direction of current. A pen on the end of an arm possibly 6 inches (15 cm) long is used to draw a line on a moving chart. As a consequence of the limited arm length the pen movement is an arc across the paper rather than a linear deflection at right angles to the paper. The resultant waveform is therefore distorted and nonlinear, similar to the one shown in Figure 14.1. In many instances the waveform distortion is of no consequence and the linearity problem is overcome by using chart paper with a nonlinear calibration scale.

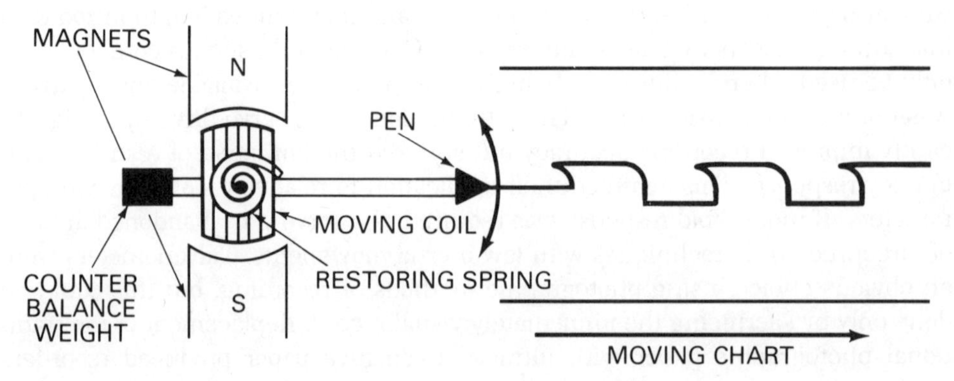

Figure 14.1
Principle of Early Pen Recorder

SERVO RECORDERS

To overcome the restrictions of the simple pen recorder, the servo recorder was developed. This type of recorder uses the principle of a *self-balancing* potentiometer to drive a pen. The potentiometer is used as the balancing potentiometer of a Wheatstone bridge, and when the bridge is off-balanced by a signal, balance is restored by a servo motor driving the potentiometer to a position of minimum bridge output. A simplified schematic diagram is shown in Figure 14.2.

The input circuit of the recorder is designed to accept different types of signal; the one illustrated is a simplification of the practical case. This circuit assumes a sensor that changes its resistance proportionally to some parameter and has been connected to form one arm of a bridge. Bridge excitation is provided by a stable dc supply integral to the recorder. In operation, the off-balance dc output of the bridge is converted into an alternating current by a device called a *synchronous converter*. This alternating current is either in phase or out of phase with the ac voltage energizing the drive coil of the converter. The converted signal is transformed and amplified prior to its connection to one winding of a two-phase servo motor, the other winding of which is capacitively coupled to the ac supply. Depending upon the phase relationship between the two windings, a rotating field is set up that drives the motor in one direction or the other. A mechanical drive connects the motor to the balancing potentiometer and rotates it in a direction that will restore the balance.

In its elementary form, the motor will *overshoot* the balance point and oscillate. This is overcome by introducing damping. In this instance, a tachometer is used to produce a phase-sensitive output with an amplitude proportional to the speed of the servo motor. The output of the tachometer is used to oppose the restoring signal produced by the synchronous converter and will have maximum effect when the servo is running at maximum speed. As the system approaches balance, the torque driving the motor falls, and also its speed. This in turn is reflected in the output of the tachometer, and provided circuit parameters are chosen correctly, the potentiometer is driven to its balance point without overshoot.

MULTIPLEXING RECORDER SIGNALS

Servo recorders by their nature are relatively expensive, and also relatively large. Where a number of slowly varying signals are to be recorded, considerable economies of space and cost are achieved by recording a number of signals on a single chart. The normal procedure is to switch or *multiplex* the signals in turn to the pen, displaying each signal long enough to draw a short line on the chart. The pens are sometimes left in contact with the chart to produce a stepped record of the signals, but more often the pen is raised during the switching and settling operation. Each trace then appears as a dotted line that, in the context of the rate of change of the signal, can be regarded as a continuous trace. Where signals can

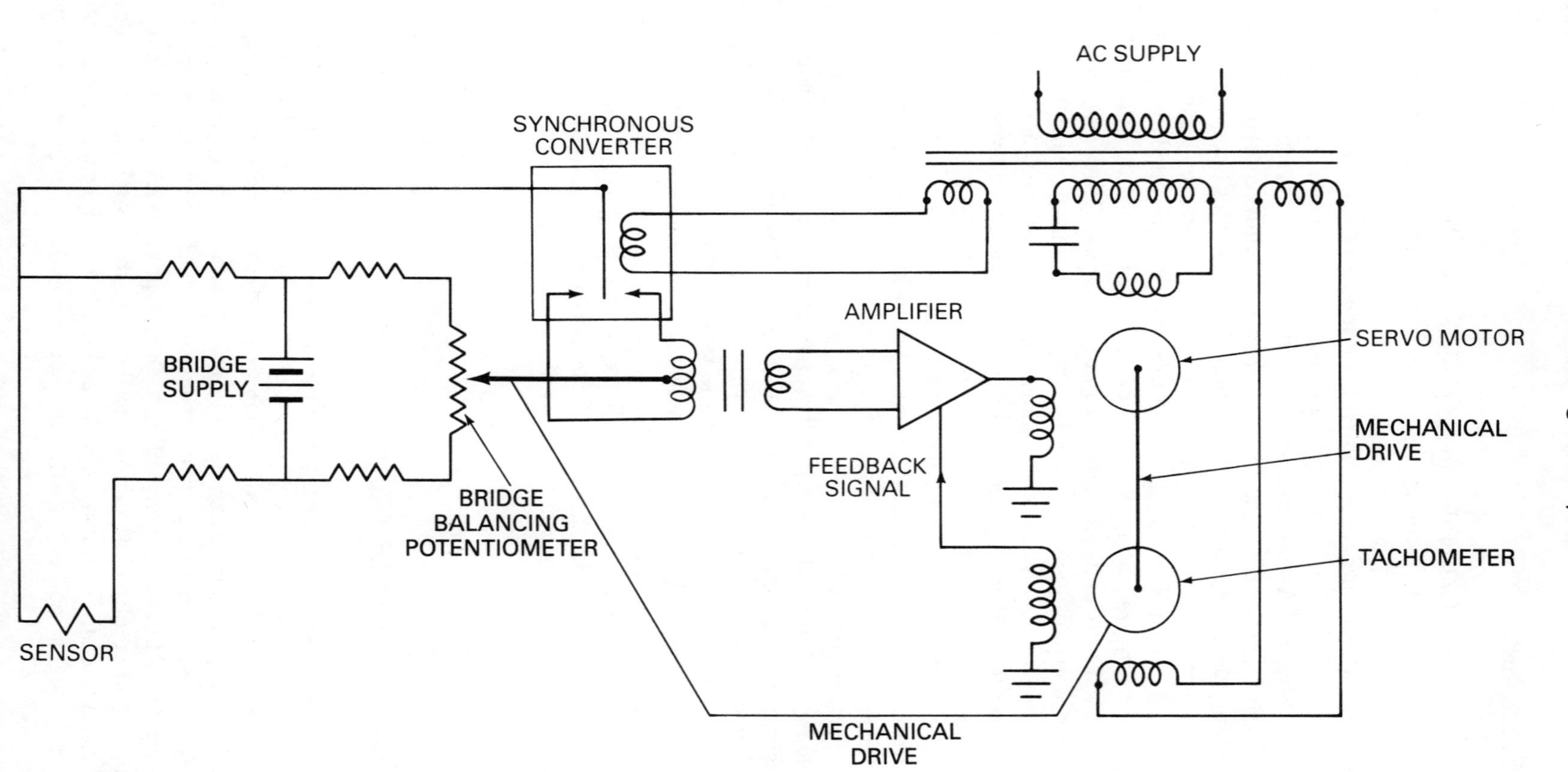

Figure 14.2
Principle of Servo Recorder

swing violently it can become difficult to follow individual traces. A number of methods have been devised to improve trace recognition, including multisymbol pen heads that print a unique symbol for each trace. Much can be done, however, during initial adjustment to ensure that traces do not interfere with each other excessively.

Many servo recorders are marketed with built-in multiplexing switches. But, if conversion of a single-channel recorder is required, one suitable type is that illustrated in Figure 14.3

This multiplexer used a sequencing switch described as a uniselector and widely used in telephone switching circuits. The actuator of the uniselector is a spring-loaded ratchet operated by a solenoid. In effect, the operation of the solenoid "winds up" the spring which, on reaching the limit of its travel, causes a pair of contacts to open and release the solenoid. The ratchet pull in returning to its rest position rotates the contacts one step. In the rest position, the actuator contacts are again closed and the sequence repeated. The advantage of this method of operation lies in the fact that contacts are changed by a constant spring force, and consequently movement of the switching contact is predictable and at relatively high speed. Wind-up time of the spring is of little consequence to the operation of the switch, and it is therefore possible to build time delays into this part of the sequence without affecting the switching characteristics. Figure 14.3 illustrates a switch with three sets (or banks) of contacts; two sets are used for switching sensors, and the third is used to provide a controlled switching action.

Considering the operation of the uniselector in Figure 14.3, closing the control switch S_1 causes a current to flow in the actuator circuit. At some point the capacitor charges to a value where the voltage across the solenoid is sufficient to produce a magnetic field strong enough to overcome the spring force. This causes the rachet to move to the limit of its travel, and simultaneously cause the actuator

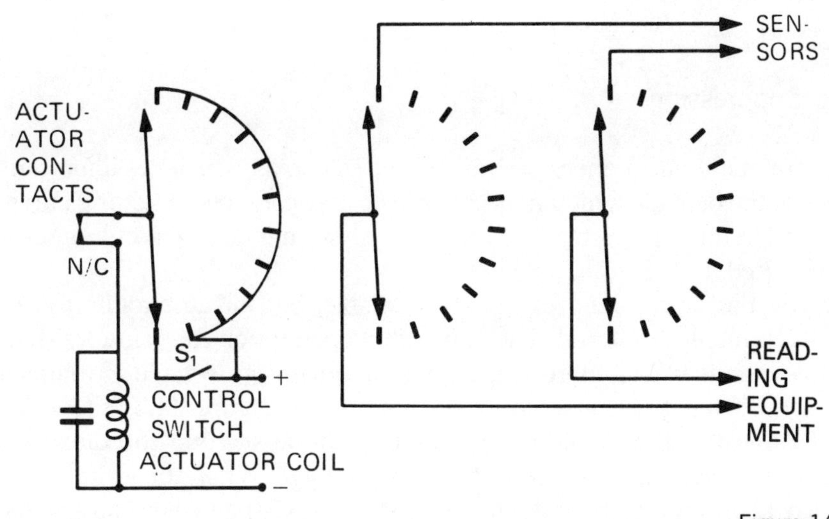

Figure 14.3
Uniselector Multiplexers

contacts to open. The capacitor now begins to discharge through the coil until the field can no longer retain the position of the armature against the spring, and then a change of contacts occurs. From the time S_1 is closed, the stepping action is automatic, and will continue until S_1 is opened, when it will return to the position illustrated in Figure 14.3.

Uniselectors can be built with a considerable number of banks of contacts; therefore, it is feasible to add another two sets to the three shown. The first additional set may be used simply to provide a pulsing circuit to lift the pen during stepping operations. The second set of contacts may be used to modify the operational characteristics of the switch.

Uniselectors are purchased with a fixed number of contacts that may not be equal to the number of sensors to be switched. Suppose a 12-position uniselector were used to switch eight sensors. To switch the uniselector through all its contacts at constant speed would produce an uneven trace through all its contacts at constant speed would produce an uneven trace due to the time spent scanning the unused contacts. It is common in these situations to use *homing contacts* to cause the switch to return to its starting position at high speed from position 8. This may be achieved by disabling the capacitor during the last four steps and allowing the uniselector to run freely. Figure 14.4 shows the modified circuit.

An examination of Figure 14.4 shows that in the first eight switching operations the capacitance is in circuit and will introduce delays in the operation. The final four positions disconnect the capacitor and the switch will therefore return to its first position at maximum speed. At this point the capacitor is again switched into circuit.

The problems with uniselector multiplexers are those of uncertain switch contact resistance and the comparatively high power required to switch the contacts. Contact resistance in some types of uniselectors can restrict their use to high-level sensors, but millivolt levels may be switched with those types having gold-plated contacts.

Arc Suppression

The power required to energize uniselector solenoids is usually important only because of the damage which may occur to control contacts switching currents in inductive circuits. To protect against this, some method of arc suppression is normally used.

Many types of arc suppressors are available, but for dc circuits the *reverse diode* is commonly employed. The suppressor operates on the principle that in an inductive circuit where there is a change of current, a voltage is induced that opposes the change. Considering the circuit in Figure 14.5, the current I normally flows in the direction indicated by the arrow and the diode will not conduct because of the polarity of the supply voltage. When the switch is opened the decay of current in the inductance produces a voltage whose polarity is such that it attempts to maintain the current. This current is discharged through the

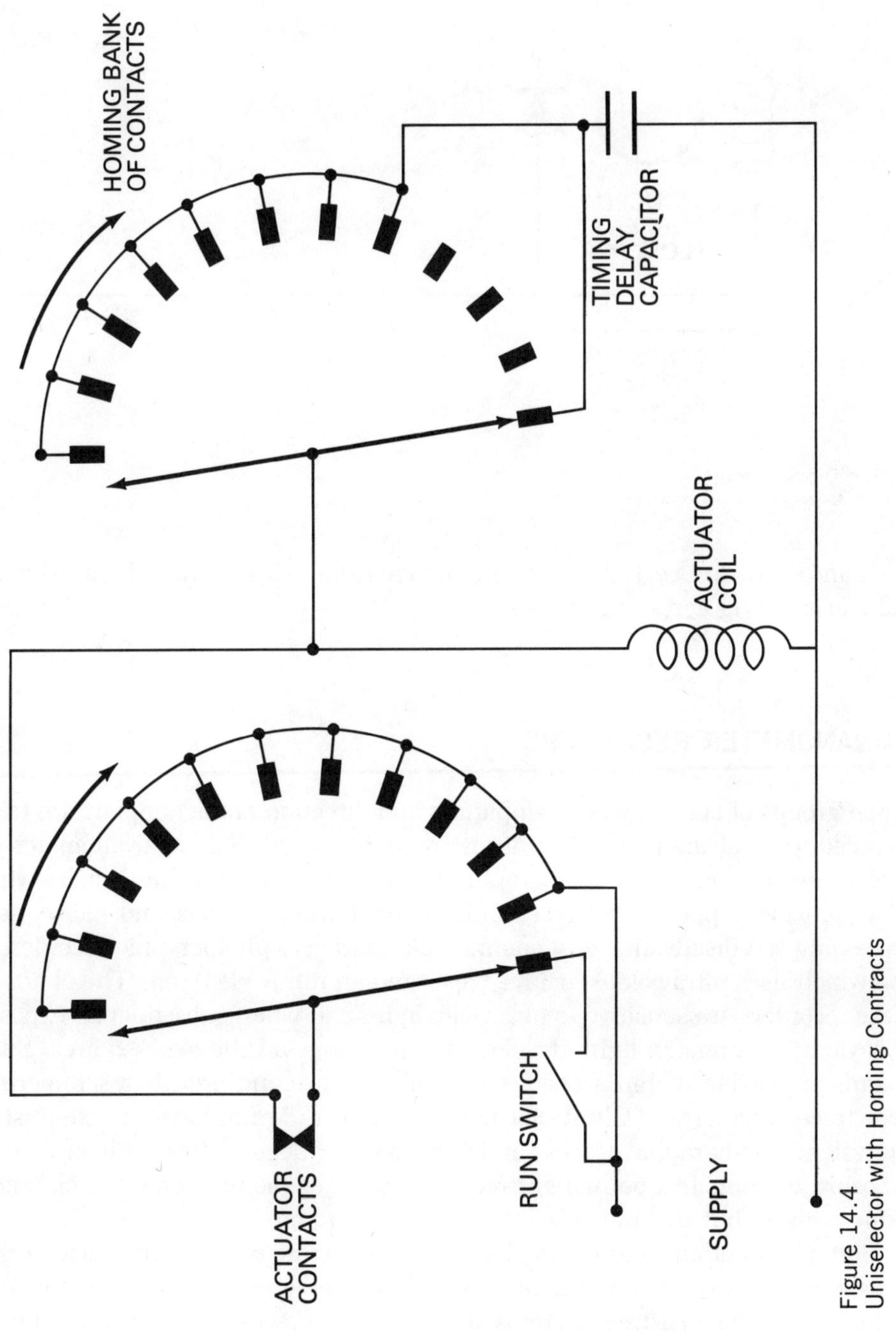

Figure 14.4
Uniselector with Homing Contracts

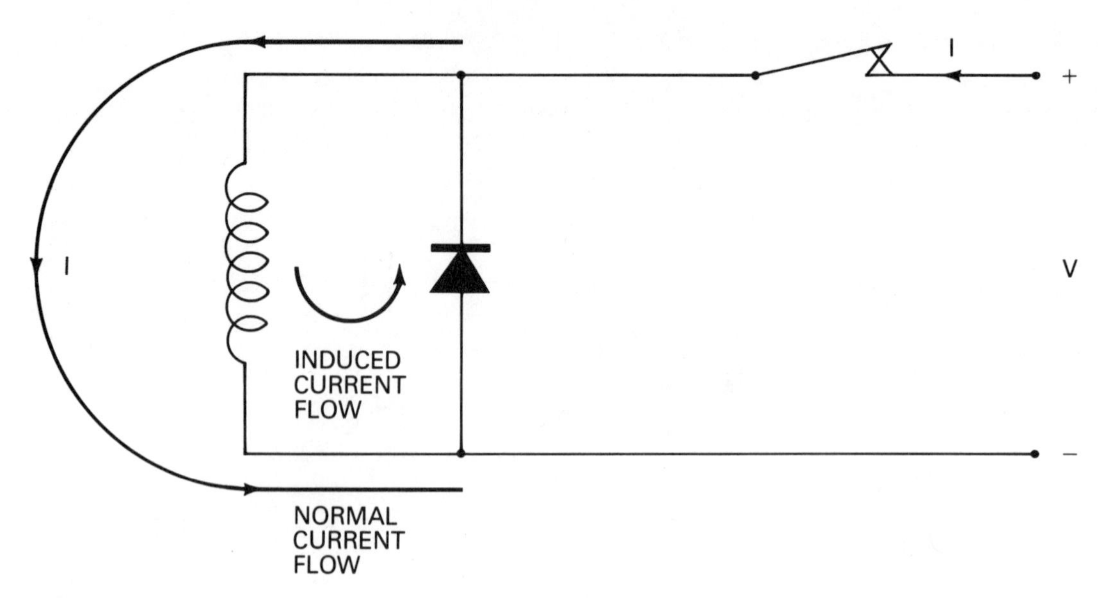

Figure 14.5
Arc Suppression Diode Circuit

diode and, in the process, prevents the voltage from rising to the point at which it will cause arcing across the switch.

GALVANOMETER RECORDERS

When groups of signals whose amplitude and direction change rapidly are to be recorded, the galvanometer recorder is frequently used. Some galvanometer recorders use conventional photographic techniques to record the signals; while these provide a permanent record, the cost of the materials and subsequent processing is a disadvantage. A common alternative to photographic recorders is one which uses ultraviolet-sensitive paper and an ultraviolet lamp. The charts of these recorders are sensitive to ultraviolet light, and when subsequently exposed to daylight or tungsten light, develop blue markings on the exposed area. Such records are visible within a few seconds of exposure and provide a semipermanent trace for analysis. Ultraviolet traces, however, do not have the contrast of equivalent photographic traces, and the contrast degenerates with prolonged exposure to light. If a permanent record is required, some method of chemical fixing is used, but this increases the cost of recording.

Both photographic and ultraviolet recorders operate on the same basic principles: A *magnet block* allows the mounting of a number of *mirror* galvanometers, and a *light path* consisting of lamps and an optical system causes beams of light to be reflected by the mirrors and focused onto the sensitive paper. It is also usual for both to have special features, such as accurate timing lines, event-marking facilities, and, often, trace identification devices.

Magnet Blocks

Galvanometer recorder magnet blocks have evolved to a high degree of precision. The block is a strong permanent magnet into which is machined a line of pole pieces that provide mounting for a number of mirror galvanometers. Individual pole pieces are often adjustable to provide accurate alignment of galvanometers. Magnet blocks may provide mountings for two or three galvanometers, or as many as 24. If more channels are necessary, more than one block may be fitted to the recorder. Some of the better units are temperature controlled to maintain constant galvanometer response characteristics over wide temperature operating conditions.

In addition to galvanometers it is common to mount two fixed mirrors that draw reference lines at either edge of the paper. Reference lines eliminate any errors introduced by lateral movement of the paper during recording. Often event markers are also fitted to the block with mirrors that can be deflected through a small angle by a switch on the recorder. These markers allow the operator to mark important events on the record.

Timing indication on recorders of these types is a very important feature. Some recorders use an event marker connected to a pulsing unit which marks the edge of the trace at equal intervals of time. Others use a tuning fork that vibrates at some chosen frequency (50 Hz is typical), to which a mirror is mounted. These systems are designed to draw lines across the entire width of the chart at 0.01-second intervals.

Galvanometers

Mirror galvanometers are very sensitive instruments with good linear characteristics over a wide frequency range. The galvanometer illustrated in Figure 14.6 is typical of the modern quality device used in many types of recorders. It is produced in a number of ranges that allow the choice of high sensitivity with low natural frequency, or decreasing sensitivity with increasing frequency ranges up to 8000 Hz.

In this type of galvanometer a small mirror (17) is mounted with a coil (18) on a torsional suspension (16 and 20), inside a hermetically sealed case. Current to the coil is carried through the upper and lower suspension wires to insulated contacts at the top and bottom of the case. A lens in the outer case focuses light on the mirror and transmits reflected light to the paper.

Galvanometers are spring mass systems in which the mirror and coil form the mass, and the suspension a torsional spring.

As with any spring-mass system the galvanometer will, if critically damped, faithfully follow excursions of a disturbing force up to a frequency of about half its natural frequency. This range may be extended somewhat if the tolerance of the measurement will allow a degree of underdamping. Usually a damping ratio of 0.64 of critical damping is used to provide a response within 5% of the dc sensitivity to about 60% of the natural frequency.

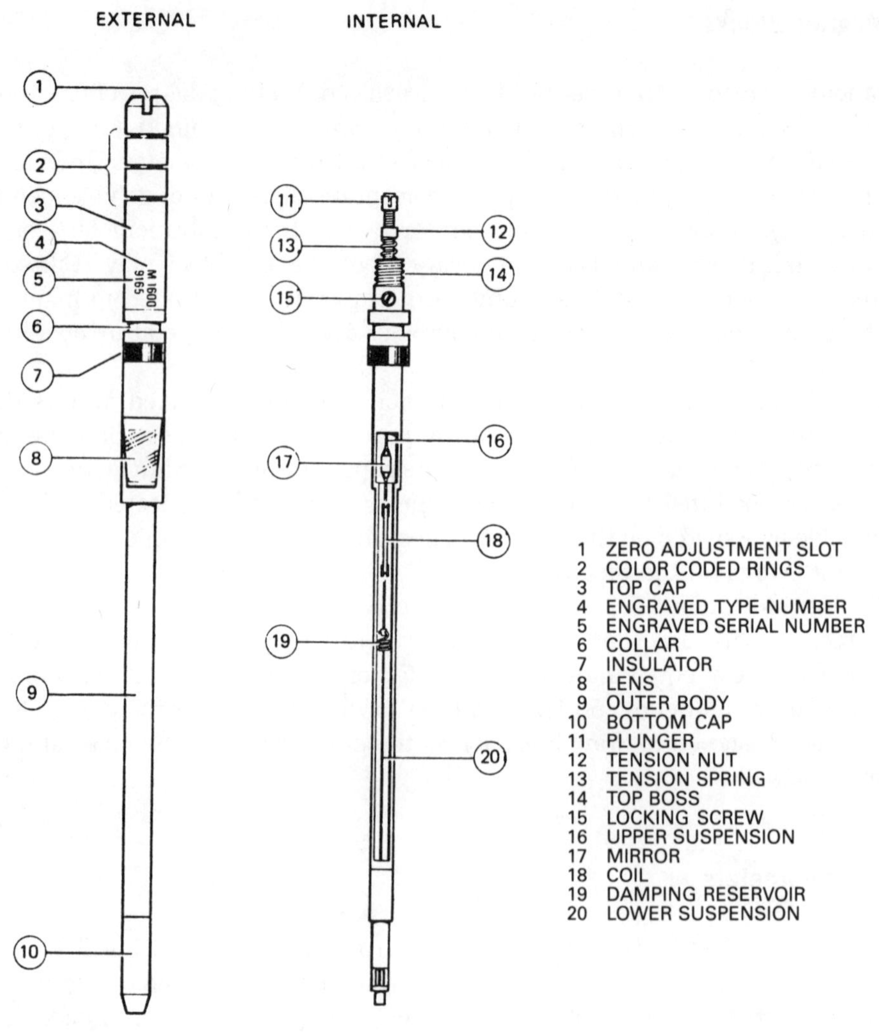

Figure 14.6
Mirror Galvanometer
(Micro Movements)

It is normal to use electromagnetic damping with the lower frequency galvanometers (i.e., those with a natural frequency below 1 kHz). Above this value viscous damping is necessary.

Galvanometer Recorder Optical Systems

The reflected light beam from the galvanometer mirror moves in an arc across the paper and consequently introduces nonlinearity into the recording. This nonlinearity, however, is quite small if the light path is long compared with the angle of rotation required to provide a usable trace.

The geometry of a light path is governed by the shape of the recorder and its internal components. Therefore, it follows that a recorder is unlikely to accommodate a straight light path from the source to the mirrors to the paper. When the light paths for timing markers and event markers are considered in addition to the basic galvanometer path, the complexity of the recorder optic can easily be appreciated. Figure 14.7 illustrates diagramatically the light path of one recorder.

INK JET RECORDERS

One method of obtaining an immediate display of a record, and also reducing the paper costs, is to combine the advantages of the pen recorder with those of the mirror galvanometer. The ink jet recorder makes this possible.

Figure 14.8 illustrates the principle of the ink jet recorder. The "pen" of this particular type of recorder is a fine glass capillary tube, one end of which is bent at right angles to provide the jet. The other end of the capillary tube is firmly fixed to a solid support. A coil and magnet similar to that of a galvanometer provides the angular movement using the capillary tube as a torsional spring. Ink is pumped from a reservoir through the capillary tube at high pressure, and the jet is directed at a moving chart to produce a trace. Jet length is more restricted than the galvanometer light path and nonlinearity is more apparent. In many applications, however, the instant display and the reduced cost of paper are considered as outweighing the sacrifice in accuracy.

MAGNETIC TAPE RECORDING

Magnetic tape recorders operate on the principle that a magnetic field of varying intensity can be imprinted on a moving magnetic tape passing over an electromagnetic *recording head*. The intensity of the magnetic field generated on the tape is a function of the current flowing through the energizing coil, and this in turn may be a representation of a sensor signal. A second head, described as a *playback head*, senses the magnetic variations when the tape is subsequently replayed and a voltage is induced in the playback coil which is a representation of the original signal.

Magnetic Hysteresis

Unfortunately, the physics of recording are not quite as straightforward as this explanation suggests. The ferromagnetic materials with which tapes are coated display magnetic hysteresis which causes distortion of the signal. This is illustrated in Figures 14.9a and 14.9b.

Consider an unmagnetized magnetic material that is to be magnetized incrementally by current flowing through a coil wound around the material. Suppose at each increment the strength of the magnetic field β is plotted against the magnetizing force H. As the current in the coil is progressively increased, it will produce a corresponding increase in β represented by the curve $0A$. At some value the material will saturate, and further increases in H will produce only negligible increases in β.

If now H is reduced to zero by reducing the current, the curve does not return to zero, but rather to the point B. This value represents a condition of permanent magnetization that can only be removed by a reversal of the magnetizing force. Increasing the current in the opposite direction produces the curve BC where a negative saturation is reached. A continuation of the process completes the curve CA and from this point the loop is cyclic.

Different magnetic alloys exhibit very different hysteresis properties and these are selected for appropriate applications. The coating on a recording tape, for example, would be selected for its ability to *retain* its magnetization while the magnetic circuit of a recording head would be designed from materials with very low retentivity.

Recording Heads

A recording head, illustrated in Figure 14.10, is basically a magnetic circuit with a small air gap in its *face*, forming the magnetic poles of the head. The recording gap is usually filled with some nonmagnetic material such as glass to prevent the ingress of magnetic dust, and the whole unit is usually encapsulated in epoxy resin.

The magnetic field produced in the recording gap varies with the signal current, and if a previously unmagnetized tape is used, the field strength induced in the tape above the recording gap will follow the curve $0A$ (Figure 14.9a). Moving the tape from over the gap has a similar effect to switching off the current at any instant. As a consequence the magnetic field will decay along some intermediate curve, cutting the vertical axis of the hysteresis curve at a point between 0 and B. Although this residual field is a representation of the current at the point of cutoff, it is unlikely that the relationship is linear. Figure 14.9b illustrates the type of distortion a sinusoidal signal suffers during transfer.

Biasing

Clearly, the distortion apparent in a simple signal is not acceptable for recording purposes; therefore, another method is required to reproduce pure signals. One method applies a dc bias to the signal and allows a limited signal incursion to be operated only on the more or less linear part of the hysteresis curve. This is not very satisfactory. Head sensitivity is poor, some distortion is still present in the

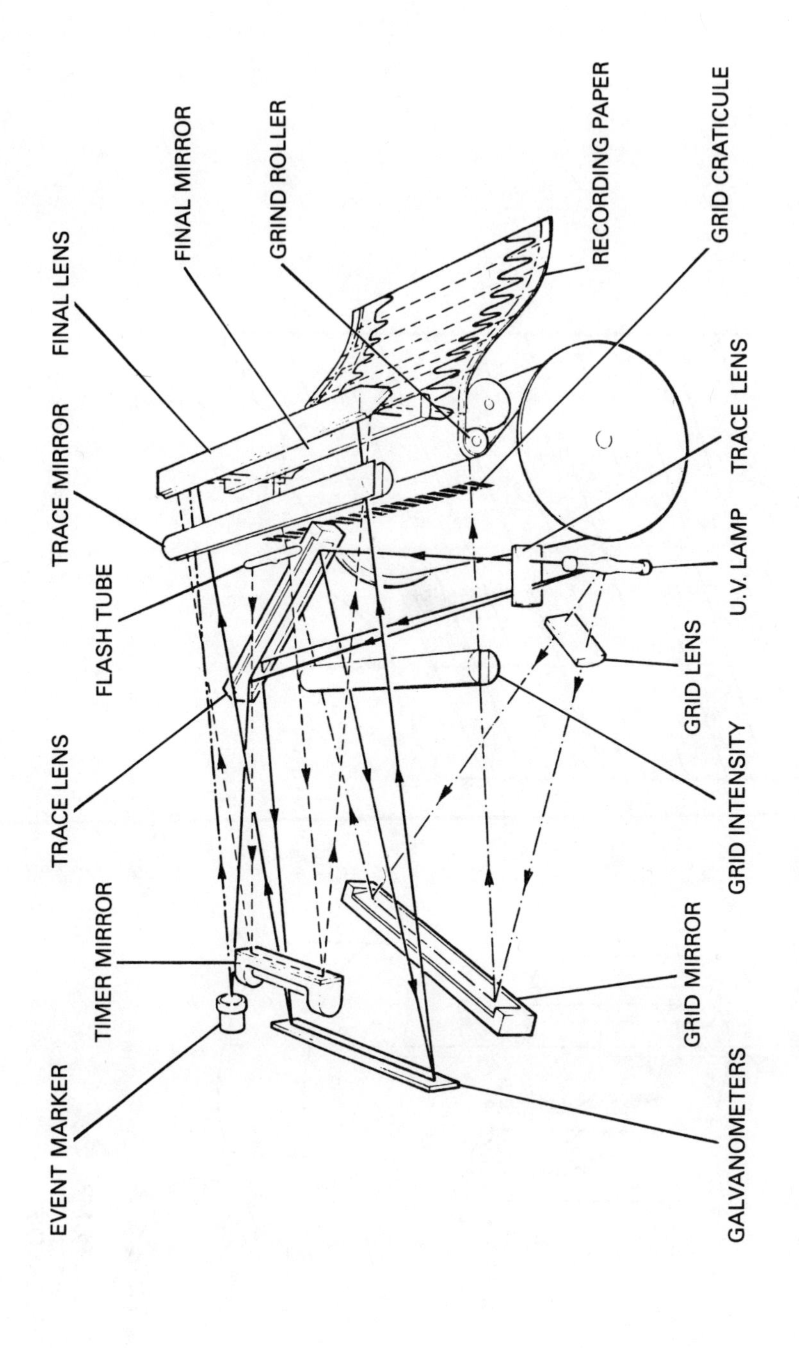

TIMING SYSTEM - - - - - GRID SYSTEM —·—·—
TRACE SYSTEM ——— EVENT SYSTEM —··—

Figure 14.7
Galvanometer Recorder Optical System
(SE Laboratories)

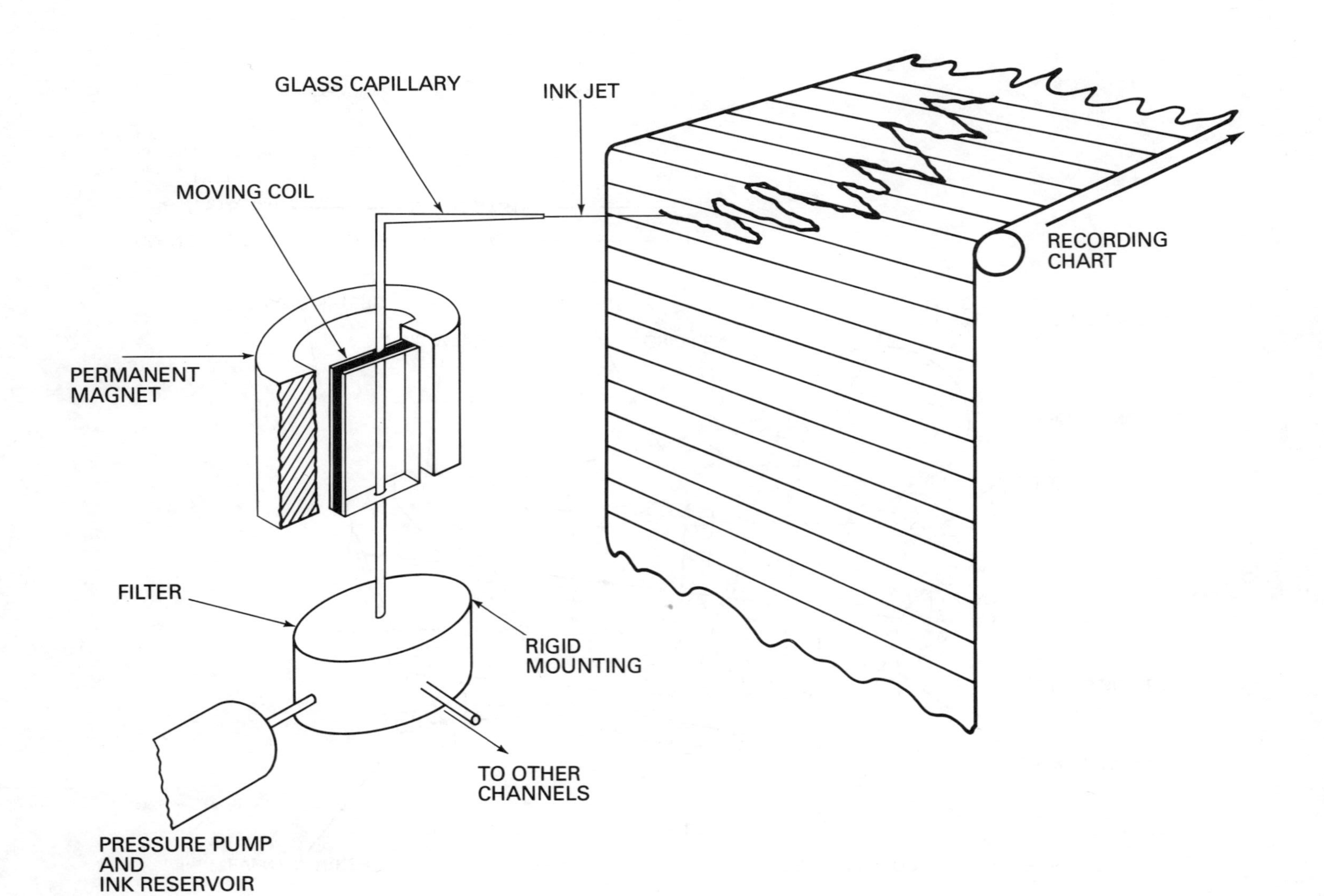

Figure 14.8
Ink Jet Recorder

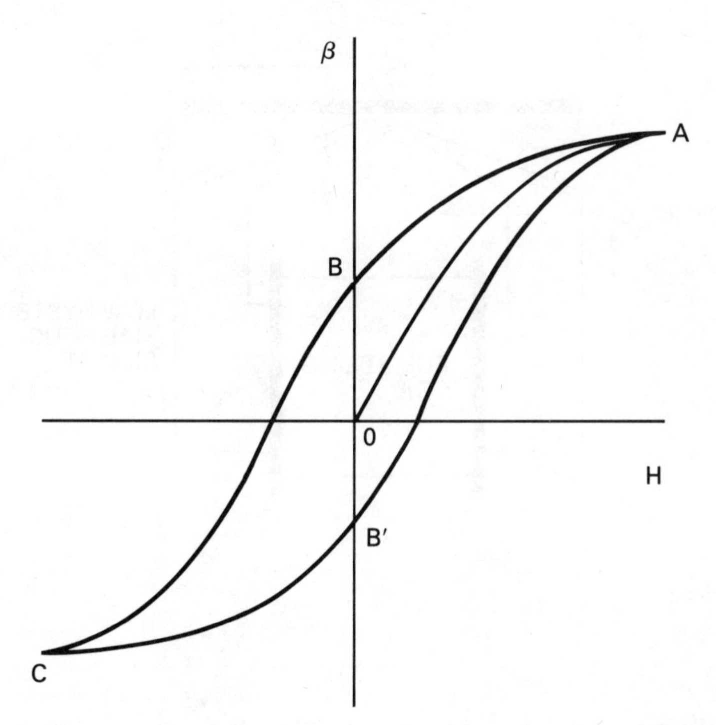

Figure 14.9a
Magnetic Hysteresis Curve

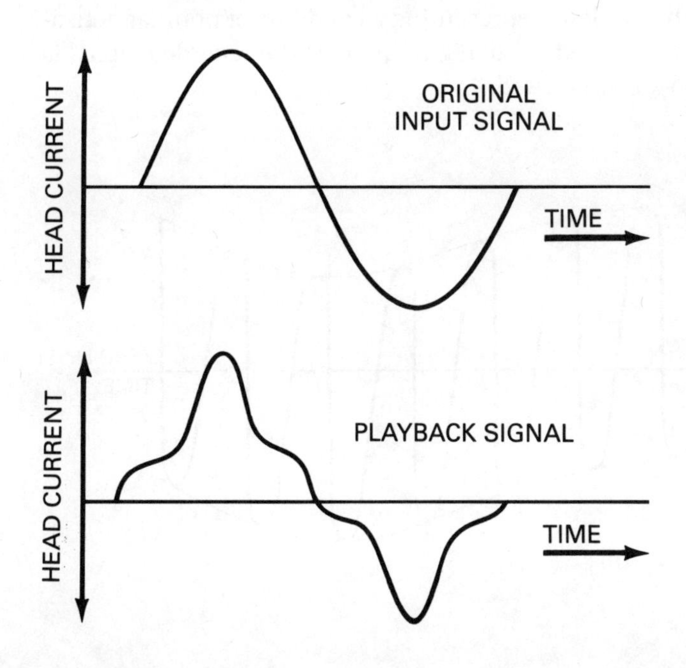

Figure 14.9b
Effect of Hysteresis on Recorded Sine Wave Signal

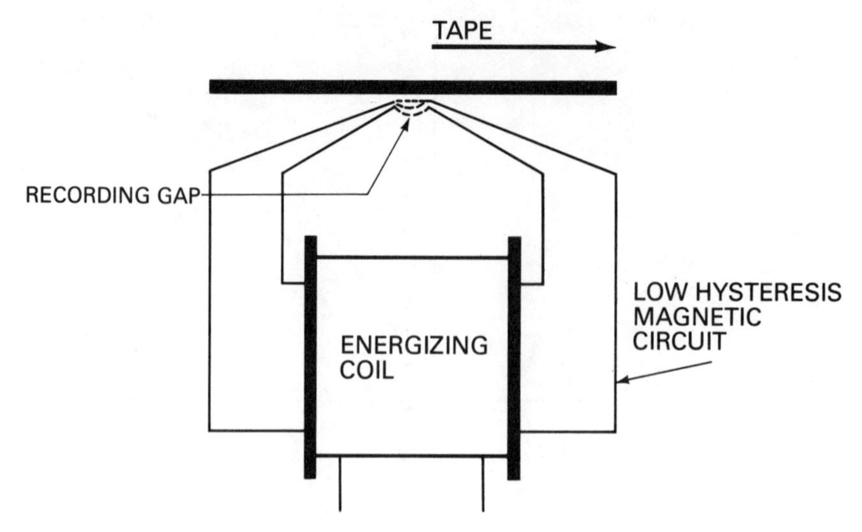

Figure 14.10
Magnetic Recording Head

signal, and signal-to-noise ratio is also poor. This is a method of recording, therefore, only used in poor-quality recorders.

A much improved method of recording uses an ac *bias frequency* well above that of the highest frequency to be recorded. A typical bias frequency may lie between 30 and 100 kHz. Suppose a magnetic tape is passed over a head that is energized by a high frequency of constant amplitude. As the tape leaves the head the magnetism induced at any point on the tape is determined by the instantaneous point on the hysteresis loop. Over a finite length the orientation of the magnetic particles will be random, representing a condition of nonmagnetization. If a prerecorded tape is exposed to such a frequency, the recorded signal is destroyed and is said to have been *erased.*

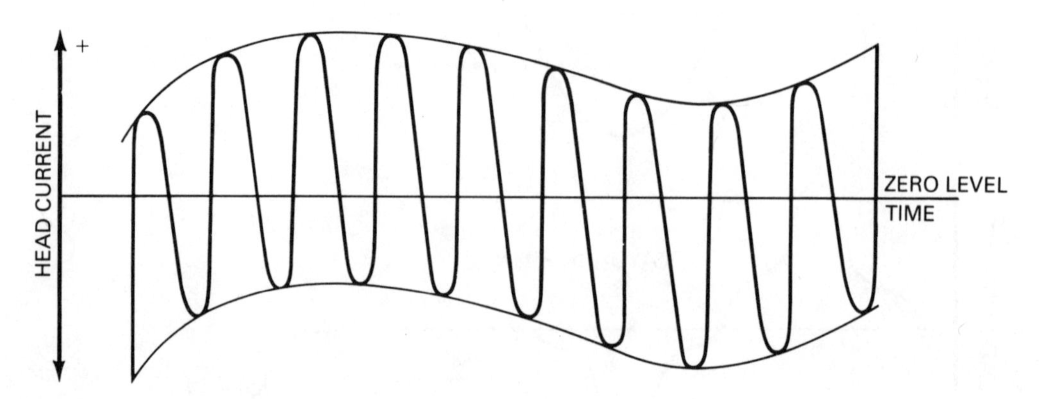

Figure 14.11
Modulated Signal for Direct Recording

It is possible to modify the ac bias by superimposing on it a signal frequency as shown in Figure 14.11. This technique is described as amplitude modulation and for recording purposes, the peak-to-peak amplitude of the signal is restricted to a value less than the peak amplitude of the bias. As the tape passes over a head energized by such a signal the magnetic particles suffer hysteresis cycling at the bias frequency whose instantaneous amplitude is biased in relation to a zero level. The intensity of this remanent field will be proportional to the instantaneous bias, and its direction will depend on the relative positive and negative peak amplitudes in relation to zero. Provided the frequency of the bias and its amplitude are chosen correctly, and provided the amplitude of the signal is limited, a remanent magnetic field will be recorded on the tape, faithfully reproducing the original signal. This method of recording is known as *direct* recording.

Playback

If during the recording process, tape is moved across a recording gap at constant speed, a sinusoidal signal will record a remanent magnetic field whose intensity and direction can be represented by the profile in Figure 14.12. The wavelength of this recorded signal is denoted by λ and has a value of ν/f where ν is the linear tape velocity and f is the signal frequency. To reproduce this signal the tape is moved across the gap of a playback head with the same speed as when recorded. Variations in the magnetic flux across the gap of a playback head cause corresponding variations in its core and these variations in turn induce a voltage in the playback coil. The playback head is very similar to the recording head and often one serves both purposes.

The voltage induced in a replay coil is proportional to the rate of change of the flux across the head gap. Therefore, if a tape traveling with constant velocity has impressed upon it a constant amplitude sinusoidal input, the induced voltage will double if the frequency is doubled. This relationship will hold until $\lambda/2$ equals l, the head gap length, at which point maximum signal is produced. At frequencies higher than this the signal falls and becomes zero when λ is equal to the gap length. Figure 14.13 illustrates the interrelation between head output, wavelength, and gap length.

The graph illustrates a typical playback head response to a sinusoidal input. Above the $l/\lambda = 1$ point a series of *lobes* are produced by higher frequencies that produce no output when the gap length to wavelength ratio is 2, 4, etc. These lobes are progressively attenuated by tape and head characteristics and in general are not used; the useful *bandwidth* of the recorder is considered that below the first cutoff frequency.

The frequency range over which a recorder can be used therefore is governed by the gap length of the heads and the tape velocity. It is not feasible to vary gap length, but it is possible to extend the high-frequency end of the recording spectrum by increasing the tape velocity. Within limits this is possible and is extensively done, but at the higher speeds head and tape wear become more pronounced and recording capacity of the tape is reduced.

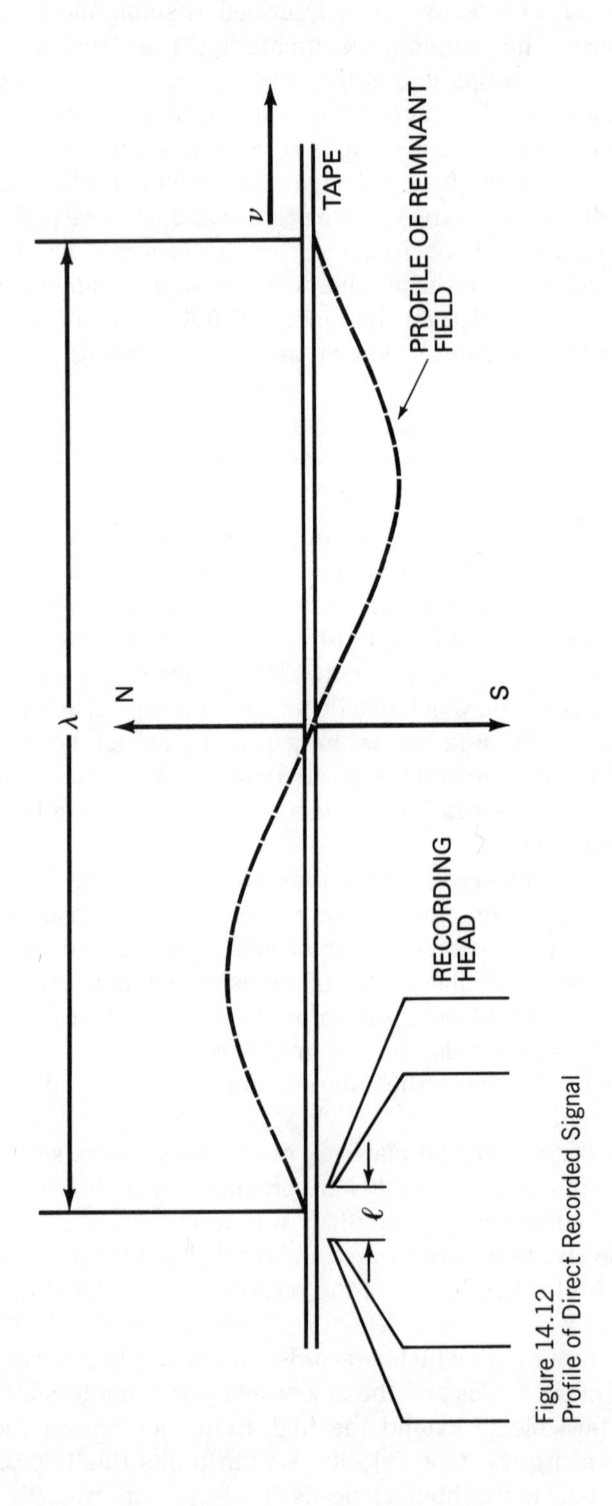

Figure 14.12
Profile of Direct Recorded Signal

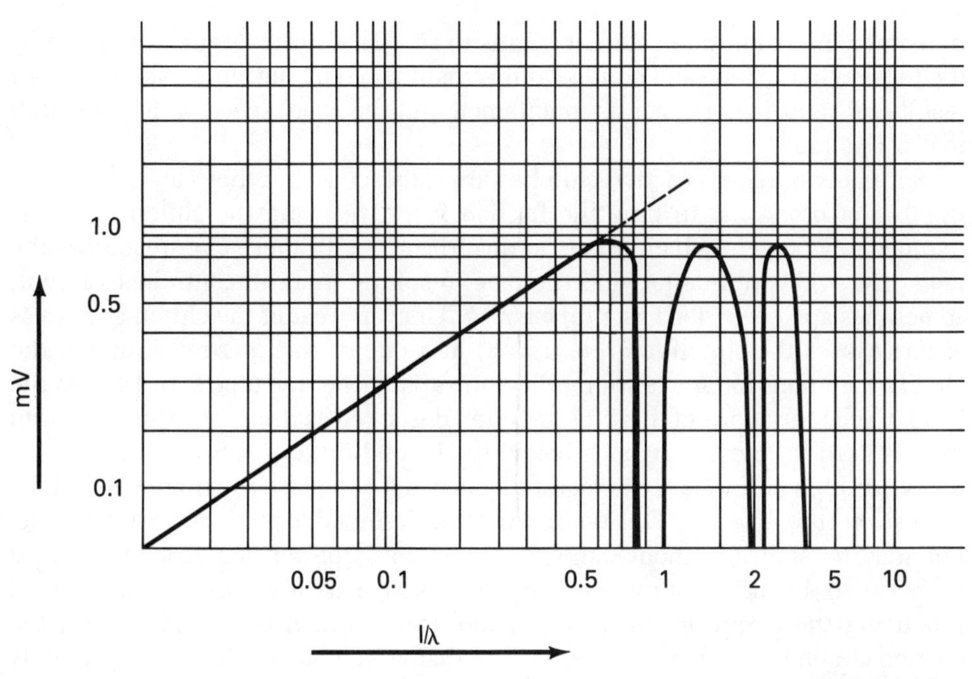

Figure 14.13
Playback Head Response

Tape Recorder Mechanical Considerations

If good reproduction of a signal is to be achieved, the mechanical design of a tape recorder is just as important as the electrical design. To attain high accuracy and consistent reproduction the initial design standard as well as the standard of maintenance must be high.

A tape transport has a number of functions: it must move the tape from one spool to a takeup spool without stretching the tape or allowing it to run wild; it must hold the tape in contact with the head under a pressure sufficient to provide good reproduction without producing excessive tape and head wear; and finally, it must provide the facilities for fast forward winding and rewinding of the tape between the spools.

Designers of tape transports face a number of problems in achieving these objectives, the first being that of obtaining constant tape speed. As tape runs from one spool to another the diameter of the tape reel on each spool is continually changing. This does not allow the use of a constant-speed motor on the takeup spool to provide constant tape speed. More important, however, is the converse; if the tape over the heads is being driven at constant speed, the speed of rotation of the takeup spool must be progressively reduced as the diameter of the tape reel increases.

Less expensive recorders use a simple motor and belt drive to move the tape and its spools. Usually the drive to the spool is a friction clutch that will only transmit a limited torque before it begins to slip. The motor torque is opposed by the tension in the tape and equilibrium is maintained by the clutch slipping. As a result, constant tape tension is maintained and the spool speed varies to match that of the tape.

For many purposes friction clutches are satisfactory but they can introduce a number of problems. In time the friction faces wear and the clutch torque is reduced to a point at which it fails to turn the spool. In these circumstances the tape drive will continue to wind the tape, resulting in an uncontrolled run out, sometimes aptly described as a *tape wreck*. Uneven wear of the clutch or ingress of dirt causes the clutch to *grab*, usually at a rate of once per revolution of the clutch. Variations in tape tension reflect in variations of the tape speed across the head. Cyclic variations of this type are often described as *wow* or *flutter*. The term wow is used for those variations below 10 Hz, and flutter for those above.

When high performance is required it is usual to replace the friction clutch by a *constant torque motor*. This type of motor is designed to produce a rated torque, but may be stalled without damage. When the tape tension reaches a value higher than the motor rating, the rotor *slips* relative to the rotating magnetic field producing the torque in the motor. The effect is much the same as with the friction clutch but with a magnetic rather than a friction coupling. Consequently the coupling is much more predictable and requires little maintenance. Rewind operation is similar to takeup and may be a friction clutch coupling or a separate motor. For cases in which fast forward or rewind operations are required, the tape is always lifted clear of the heads. Figure 14.14 illustrates the layout of a typical tape transport.

Tape running off the supply spool passes over a stabilizer that is a spring-loaded recessed pulley. The spring lightly holds the tape against the heads while the recess prevents lateral movement of the tape. After passing over the heads the tape passes through a constant-speed tape drive, then around a tensioning roller, and finally around the takeup spool. The constant-speed drive is more usually described as a *capstan drive*. This consists of two rollers, a rubber or plastic roller called a *pinch roller* driven at constant speed by a drive *roller*, called a *capstan*, running against its periphery. The tape is held firmly between the two rollers by pressure springs and is moved at constant speed by the friction of the pinch roller. Drive motors are coupled to flywheels to provide a stable drive speed, and some may be part of a servo system coupled to the takeup drive to maintain constant tape tension.

Signal Equalization

Below the point $l/\lambda = 0.5$ the output signal is proportional to the frequency of a constant amplitude signal. Since this is not the desired response it is necessary to correct the signal so that it is independent of frequency and proportional only to the amplitude of the input. Correction of tape recorder signals to eliminate the frequency factor is described as *equalization*.

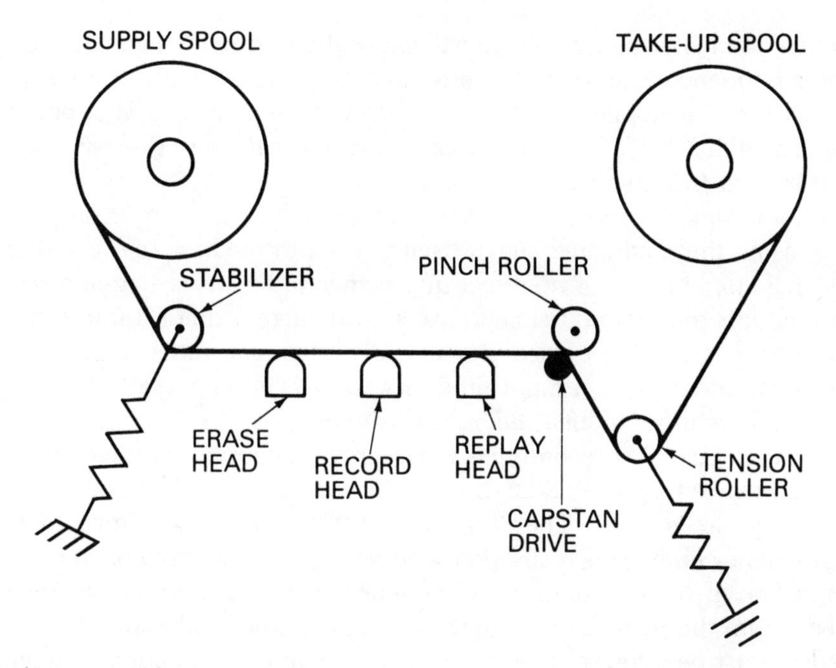

Figure 14.14
Schematic Diagram of Tape Transport

The response of the head is comparable to that of a velocity sensor under analogous circumstances (discussed in Chapter 13). The equalization circuits for the tape signals are similar to the integrators used with velocity sensors and have the same limitations with respect to the frequency range over which they may be adequately used.

The limitations of equalization circuits at low frequencies, and the low signal levels produced at these frequencies due to head response, limit the minimum frequency that can be reproduced on any recorder by direct recording methods; therefore, if dc levels or very low frequencies are to be recorded, some other techniques must be used.

FM RECORDING

The concept of *frequency modulation* has not previously been discussed. It is possible to design an electronic circuit described as a *voltage-to-frequency converter* that will output some predefined frequency signal with zero volts on the input. This frequency is known as the *center frequency*. A positive voltage applied to the input decreases this frequency and a negative voltage increases it. These changes in frequency are proportional to the applied voltage and consequently application of an alternating voltage will cause the frequency to modulate about its center frequency. The converse process is also possible and if a frequency-modulated signal is applied to a frequency-to-voltage converter, it is possible to recover the original signal. Such converters are generally described as *frequency modulators* and *frequency demodulators*.

When applying frequency modulation techniques to tape recording no bias signal is used; the modulated signal itself is used to produce the record on the tape. Signal amplitude is of no direct importance to the recording and recovery process and it is therefore possible to use a head current that saturates the tape. Although the resulting recorded signal is distorted, saturation reproduces the zero crossover points with certainty and these crossover points can be used in the playback process to reproduce the original signal.

The advantage of FM recording is the improved accuracy with which a signal may be reproduced, and in addition it provides the facility of extending the frequency range down to dc. The disadvantage is the restriction imposed on the upper frequency limit. This restriction occurs because the carrier of an FM signal is recorded on the tape. The highest frequency, therefore, that may be recorded, must lie within the recording range of the system, and it follows that the center frequency must be a lower value than this. In practice the center frequency is restricted to about 30% of the maximum that can be recorded, and the signal frequency to about ±20% of the center frequency. Thus the bandwidth of an FM system is less than that of a direct recorder by a factor of about 15.

MULTICHANNEL RECORDING

Instrumentation tape recorders are invariably multichannel. This allows the simultaneous recording of a number of signals in much the same way as a multichannel galvanometer provides equivalent paper records. A number of recording heads are *stacked* laterally, each provided with its own amplifier. The number of channels that may be recorded simultaneously depends on the tape width and head design. Four channels are common with quarter-inch tape, seven or nine with half-inch, and 16 with one-inch.

Head stacking must be done very accurately since any misalignment will cause simultaneous signals to be recorded at an angle to the lateral axis of the tape. This effect, described as *static skew*, introduces interchannel phasing errors when a play-back head other than the original recording head is used. In addition, even very small discrepancies between the position of heads on different recorders can cause substantial losses in playback signal.

The *stabilizer* is a tape guide designed to prevent lateral movement of the tape as it runs across the heads. If, for some reason, this is not effective, the tape will snake across the heads, producing a record that wanders across the width of the tape. This is described as *dynamic skew*. Recording channels of multichannel recorders are very narrow with small spacing; therefore, any variation in their lateral position will cause interference between adjacent channels during playback. This effect is known as *crosstalk.*

PULSE DURATION RECORDING

The number of channels normally recorded on a one-inch tape is 16, although the outer channels are often restricted to speech or marking functions. If high-frequency signals are to be recorded, it is necessary to use a channel for each signal, but if very-low-frequency or quasistatic signals are involved, multiplexing techniques may be used to increase the capacity. One method samples a signal, measures its amplitude, and presents a pulse to the recorder whose duration is proportional to the amplitude. It is possible to multiplex a series of signals sequentially in this way and present a train of pulses of varying duration to the recorder, each duration representing the amplitude of one signal. Such a pulse train is called a *pulse duration modulated signal* (PDM).

Multiplexers for PDM systems are often described as *commutators*, and the device that produces the pulses as a *coder* or *encoder*. During playback the pulses are *decoded*, that is, their durations are turned back into amplitude levels and a further commutator reassembles the original channels.

A typical recording channel may operate up to 900 pulses per second, but some of these will be used for synchronization and calibration purposes.

Commutators

For simplicity the commutators shown in Figure 14.15 are illustrated as mechanical switches, but in reality they are electronic. It is possible to build commutators in a number of ways, the most recent type being the microcomputer-controlled switch. To understand the fundamentals of commutation, however, it is better to consider a basic type shown schematically in Figure 14.16. This circuit utilizes a binary counter and a series of binary-to-decimal decoders described in Chapter 7.

The commutator illustrated multiplexes the signals from five sensors onto a common output line, and these signals are followed by three calibration signals. In operation, the clock provides a pulse train to the input of a binary counter. This input is summed and the output of the counter is applied to four *bus* lines. Also connected to the bus are the inputs of nine binary-to-decimal decoders. The first clock pulse of a sequence causes the binary-to-decimal-1 decoder to switch and this in turn is designed to switch sensor 1 onto an output line. A second clock pulse changes the output of the binary counter turning off the binary-to-decimal-1 decoder and turning on the decimal 2 decoder. The sequence continues at a speed determined by the clock frequency until the binary-to-decimal-9 decoder operates. This decoder switches a reset line to set the binary counter to zero so that the sequence may be repeated. As illustrated a series of voltage levels appear

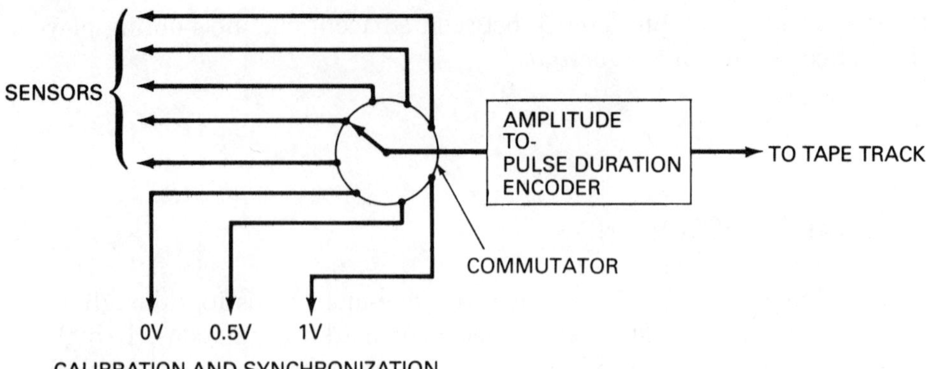

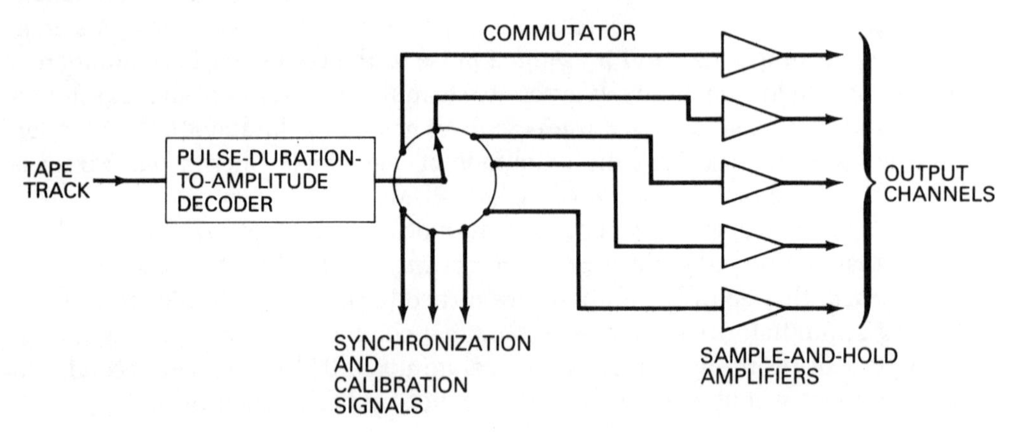

Figure 14.15
Pulse Duration Recording

on the output line, each voltage level being present for a time equal to the period of the clock pulse and representing sequentially the output of the sensors and calibration signals.

Amplitude-to-Pulse-Duration Encoder

Figure 14.17 illustrates the operation of one type of pulse-duration modulator.

The device operates from two commutator inputs, the multiplexed signals and the clock pulses. The commutator clock pulses are used to make the modulator operate synchronously with the commutator. A clock pulse transmitted to the output bistable turns it on and simultaneously starts a pulse generator that derives a *staircase generator*. This consists of a pulse counter operating into a digital-to-analog converter (DAC). The output of this DAC is a voltage that increases in a series of equal steps, each step being equivalent to an input pulse. The analog output from the commutator, and the output of the staircase generator are connected to the input of a comparator and when these become equal it

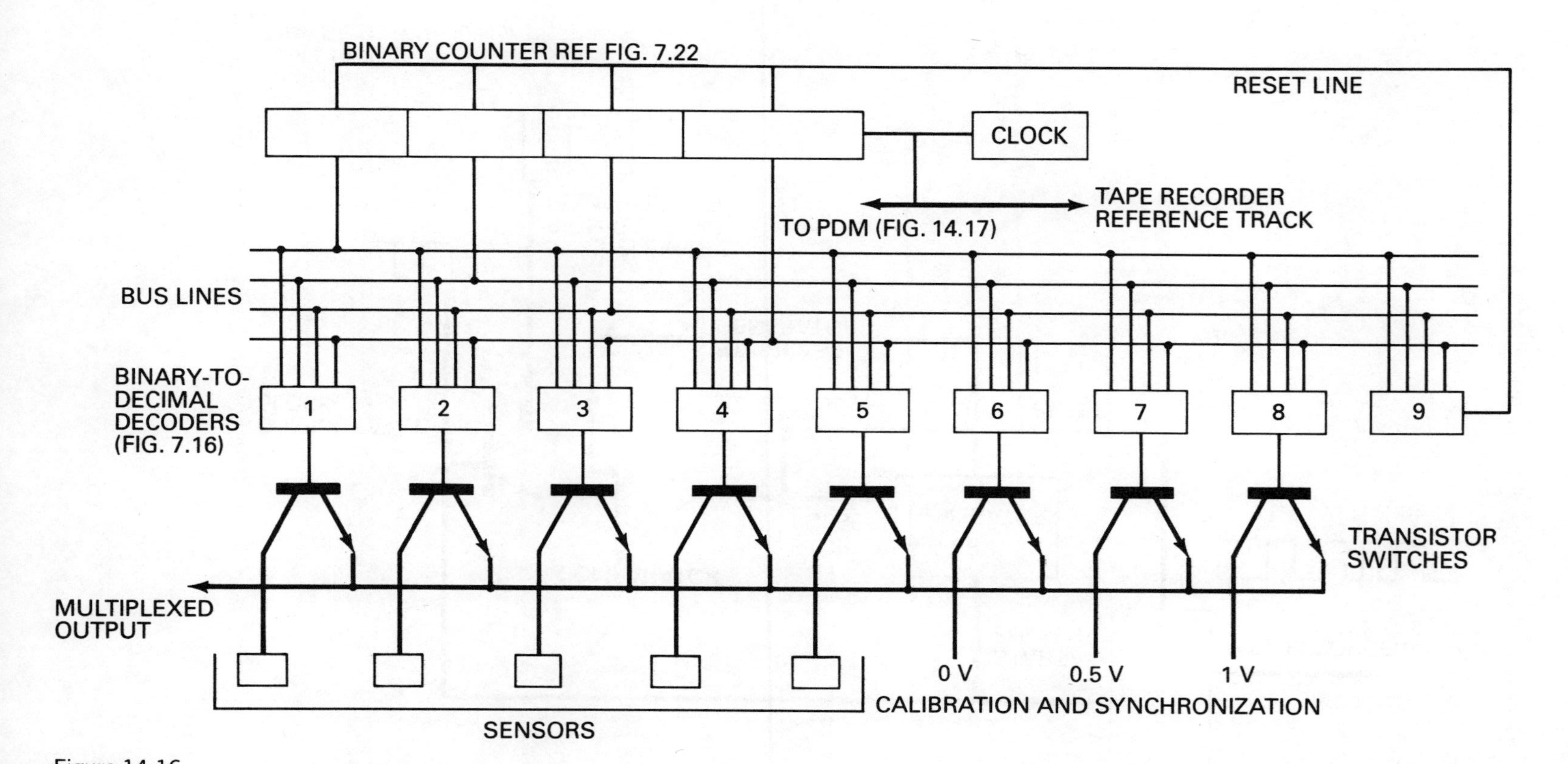

Figure 14.16
Basic Communtator Circuit

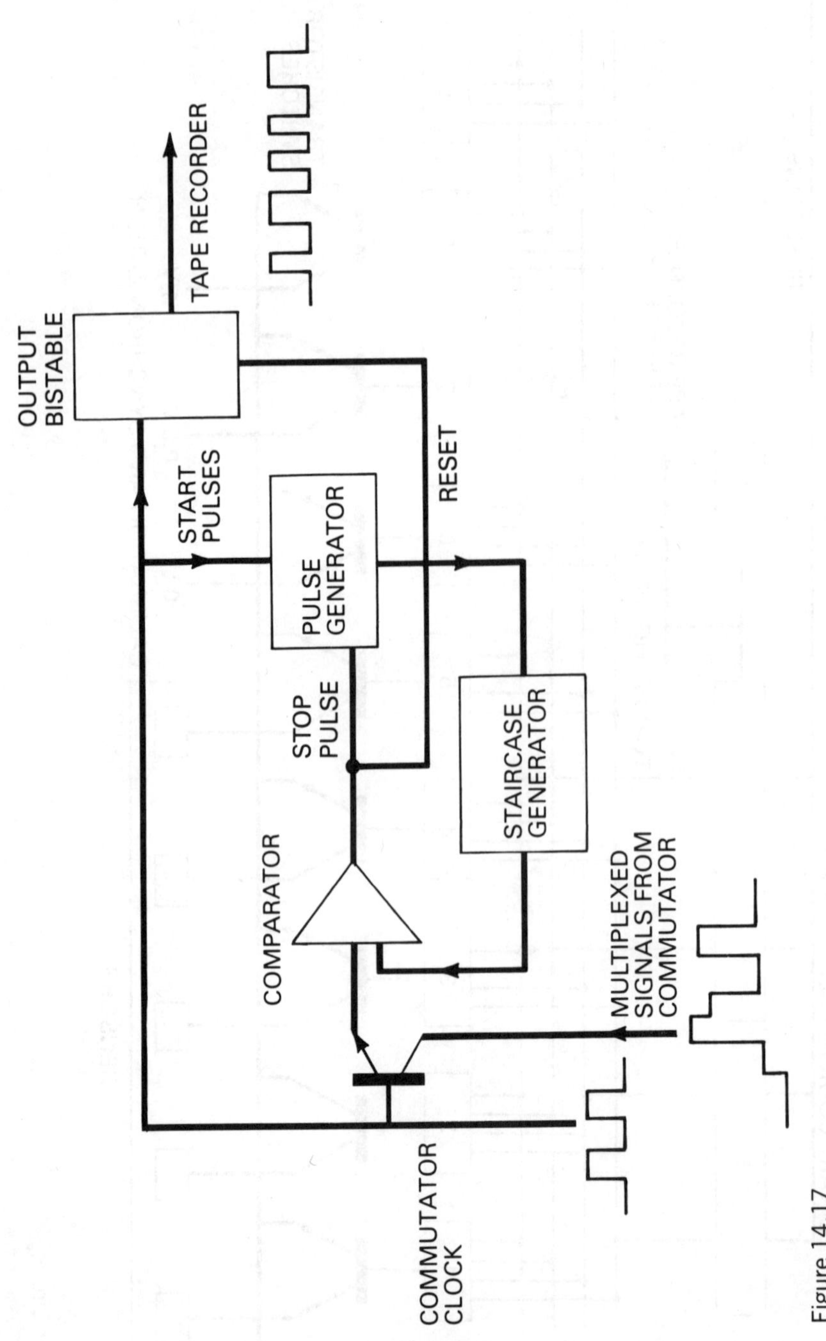

Figure 14.17
Pulse Duration Modulator

switches its state. In the process it stops the pulse generator and resets the output bistable. The pulse length sensed by the recorder circuit is determined by the time the output bistable is switched on, and this in turn is determined by the number of pulses required to make the staircase voltage equal to the input voltage. It follows, therefore, that the output pulse length is proportional to the input voltage.

The commutator clock determines the rate at which sensor signals are switched to the input of the comparator. In the time between one clock pulse and the next the measurement must be complete and the tape pulse returned to zero. The longest pulse that may be recorded, therefore, is something less than the length of the commutator clocking pulse. In addition, the accuracy of the measurement is limited by the resolution of the staircase generator. For example, to obtain 0.1% full-scale accuracy, the staircase generator will require a minimum of 1000 steps.

PDM Decoder

The change in voltage marking the beginning of a pulse is usually described as its *leading edge* and the change marking its edge is described as its *trailing edge*. The circuit outlined in Figure 14.18 uses a trigger circuit to modify PDM pulses so that their leading edges may be used to switch toggles to start a pulse generator and the trailing edges to stop it. Trailing edges are also used to switch bistable 1 to open the output gate of the decoding circuit, which is connected to a commutator. In the circuit illustrated, a second trigger circuit modifies recorded clock pulses so that their trailing edges may be used to close the output gate and reset the staircase generator. The operation of this circuit is the converse of the amplitude-to-pulse-duration encoder. The PDM pulses control the operation of a staircase generator to produce a series of voltage levels at the input of a commutator. The values of these voltage levels represent the recorded output voltages of the original sensors. Clock pulses are used to maintain the decoder and commutator in synchronism so that a sensor output always appears in the same commutator output line.

Signals reproduced by this type of circuit appear at the commutator input intermittently and will vary in duration depending on the elapsed time between the trailing edge of the PDM signal opening the output gate and the clock pulse closing it. This, in turn, depends on the length of PDM pulse, which is proportional to the signal producing it. Such signals may not be used directly to drive displays and analysis equipment, and it is necessary to use sample and hold techniques to provide a continuous analog signal.

Decoder commutators are effectively the reverse of that shown in Figure 14.16. The binary counter and the BCD decoder are driven by clocking pulses from the recorder clocking channel and the multiplexed output line becomes the input. Transistor switches shown in Figure 14.16 are reversed to become gates for sample and hold amplifiers.

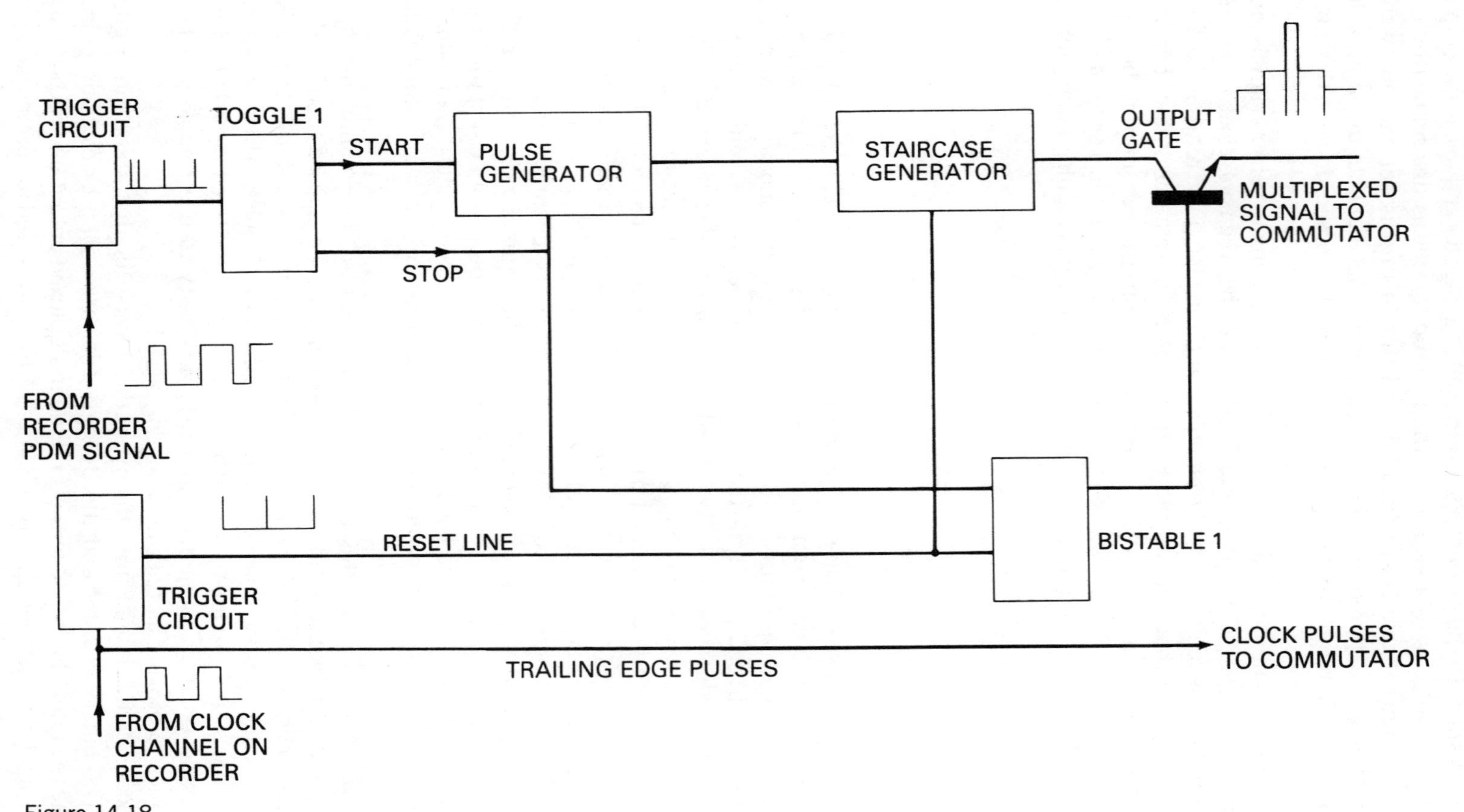

Figure 14.18
PDM Replay

Calibration and Synchronization

Between the input from the sensor and the output of the playback are a considerable number of circuits, any of which may be responsible for signal attenuation or amplification. It is necessary to provide some sort of calibration which provides a correction factor. In Figure 14.16 three levels are illustrated: 0, 0.5, and 1 V. It is possible to display these voltages on meters to provide a visual check that the calibration has not changed, or they may be used to correct the gain of the equipment to provide a corrected output. It is also possible for this type of circuit to lose synchronism. This happens only infrequently, but must be corrected immediately. One method is to use the calibration signals as synchronization signals. If the precise voltage levels fail to appear in the precise order at the correct time they may be made to operate a reset circuit when they do eventually arrive.

DIGITAL RECORDING

Another technique of recording is one using binary information recorded on magnetic tape. The binary signals may originate from a digital computer or they may originate from analog signals that have been converted into binary representations by an analog-to-digital converter.

Digital recorders are multichannel recorders whose heads are used to record a binary code across the width of the tape. The code can be one that uses the first channel to represent a value of 2^0, the next 2^1, and then in progression, to the final channel 2^n. With a ten-channel recorder, it is possible to represent a value with a precision of about 0.1%; the least significant binary digit (or bit) representing one part in 1024. An alternative method of recording uses a four-channel recorder and four consecutive records using a four-bit BCD code to represent a number to 0.1% accuracy. In practice four-bit codes are very restricted because they allow only the representation of numbers. Usually six- or seven-bit codes are used to represent numbers and alphabetic characters. One well-known code is the ASCII code (American Standard Code for Information Interchange, summarized in Appendix VI). This is a seven-bit code widely used as a standard in telecommunications and computer industries. Although the complete code allows 128 characters, usually a limited number are used for instrumentation work.

Binary digits may be recorded on a tape in a number of ways. The head may be energized to represent a binary 1 and de-energized to represent a binary 0. With this system the head is usually biased so that the signal returns to a bias level at binary 0. This is known as an *RB system* (return to bias).

Another system, the *RZ* (return to zero), represents one and zero bits by head currents moving in opposite directions, with a zero level inserted between each bit.

Two further systems known as *NRZB* and *NRZC* (non return to zero) operate with the current switching from one direction to the other without any zero pauses. NRZB systems change the direction of magnetization each time a binary

1 is experienced. NRZC systems change the direction of magnetization with each 0-to-1 change and each 1-to-0 change. The different systems are illustrated in Figure 14.19.

Analog signals may be converted and recorded digitally with a certain knowledge of their recorded precision. Their precision may be defined and their value reproduced exactly with each playback. This precision cannot be achieved with either FM or direct recording systems. In addition, digital recorders do not have the same sensitivity to wow and flutter as their analog counterparts. They still suffer dynamic skew, and gross errors may result in the extreme. If a tape runs obliquely across a head stack, it is possible that heads at one edge of the tape may be reading bits from an adjacent character. (See Figure 14.20.)

CHARACTER PRINTERS

The character printer is frequently used for data recording applications where limited and inexpensive printed output is required. It is often referred to as a *strip printer* and it prints a fixed number of characters across a 4-inch- (10 cm) wide paper. The characters are usually restricted to numbers and a few special symbols embossed round the periphery of a drum or *print head.* One set of characters is provided for each column of print, and the printer may provide for up to 14 columns of print. The drum rotates at constant speed, and as the number to be

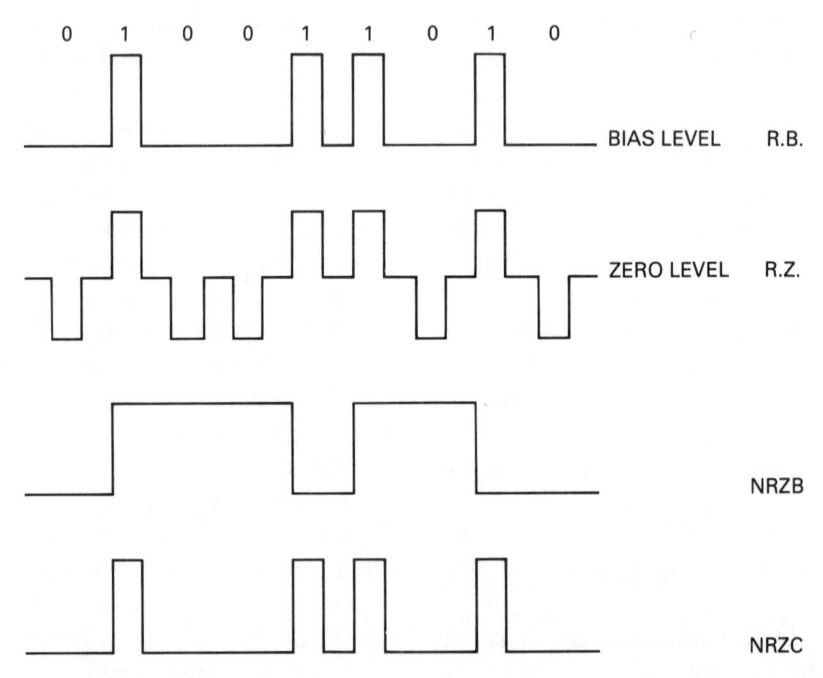

Figure 14.19
Digital Recording Techniques

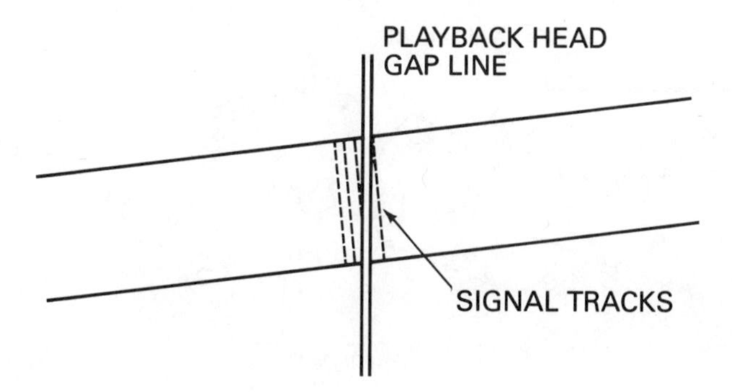

Figure 14.20
Dynamic Skew

printed passes above a hammer, a small area of printer paper is driven against the print head. A typewriter ribbon between the two causes the character to be printed.

While the principle of the strip printer is simple, its control is rather more complex. It is necessary for the control to recognize the character to be printed, and then actuate the hammer at the exact time the appropriate character on the drum is positioned over the hammer. Figure 14.21 illustrates a simplified schematic diagram of the control for the first three digits of a single column of printout.

The position of the print drum is often located by two coil and magnet sensors, one indicating a starting position on the drum, and another that pulses a position counter with a decimal output. At the end of each revolution, this counter is reset by a start pulse. One of the decimal output lines of a binary-to-decimal decoder is energized by input from an ADC. The decimal line in turn energizes one input of an AND gate whose second input is connected to an equivalent output of the position counter. When the drum rotates to its correct position, both inputs of the AND gate are momentarily energized, causing the hammer to operate.

DATA LOGGING

Data logging is a term often used to describe a special type of recording where a number of sensor signals are sampled, converted to digital signals, and recorded. The recorder may be a numeric printer, a tape punch, a magnetic recorder, or a computer. The multichannel servo chart recorder and PDM recorder are examples of equipment that may be described as data loggers but which do not involve a binary conversion.

A schematic diagram of a simple digital data logger is illustrated in Figure 14.22. Data logger designs are numerous and diverse. There is no standard way, therefore, that a date logger can be portrayed. Basically they consist of a multiplexer, an ADC, and a recorder whose functions are directed by a controller. This

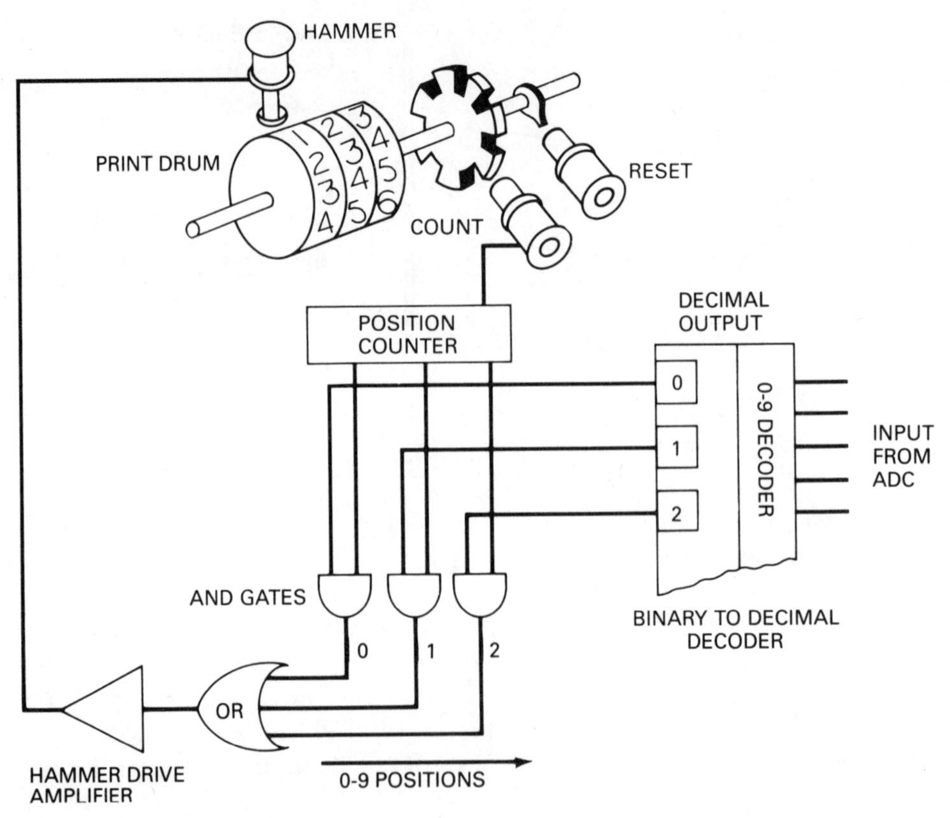

Figure 14.21
Drum Printer

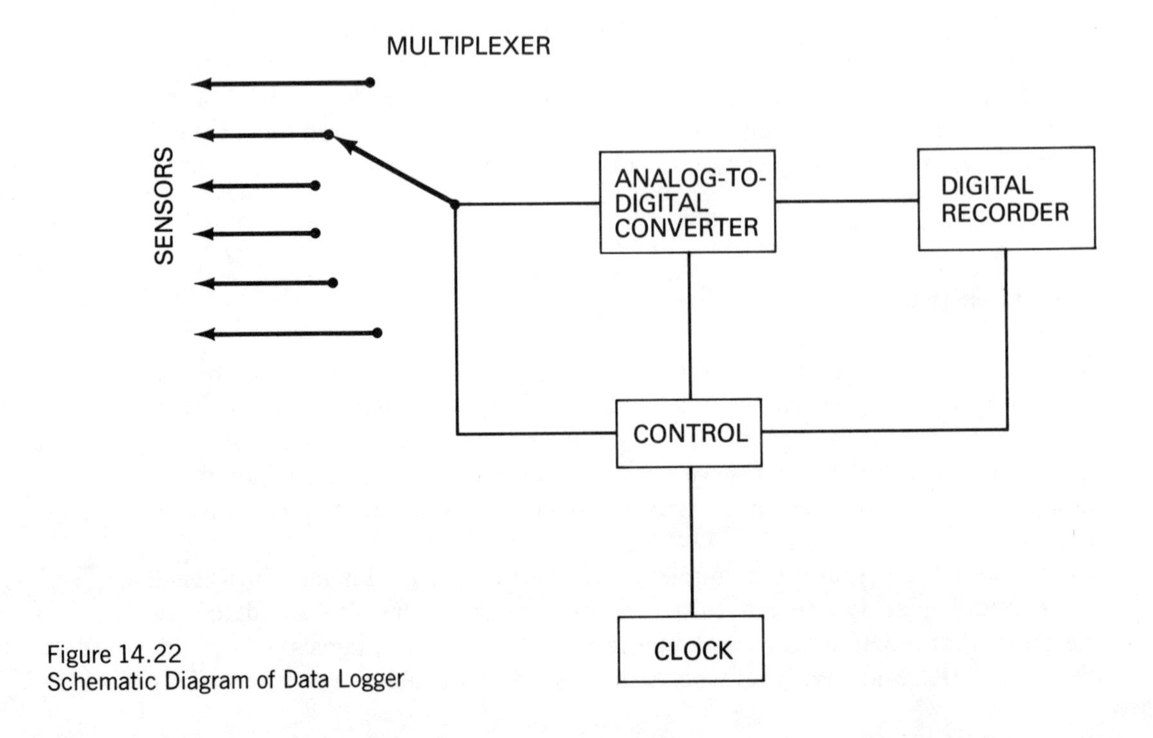

Figure 14.22
Schematic Diagram of Data Logger

controller may be a complex microcomputer-controlled unit, or it may be a simple sequencing control. In the latter case, it may do no more than advance the multiplexer one step and signal the start of signal conversion on a "time-out" basis. The recording sequence in this instance is under the control of the ADC, which automatically presents a series of BCD digits to the recorder in the time delay between one sample and the next.

The general case is more complex than this. The multiplexer is likely to be one that can be preprogrammed to some extent. Sometimes there is no restriction on the way individual switches are selected, but often access is to groups of switches rather than individuals. Where the recording sequence is concerned it is likely to be under the joint control of the ADC and the recorder. When the ADC has completed the conversion of a signal, the first digit is switched to the output lines. The ADC then activates a control line to signal to the recorder that a digit is present. The data are "read" by the recorder, which in turn activates a control line to "instruct" the ADC to present the next digit. The process continues until the ADC signals to the control that the last digit has been transmitted. The method of using control lines to maintain a sequence of operations is commonplace and is often described as *handshaking*. It has two functions: It ensures the integrity of the sequence of operation and it also ensures that the sequence is completed in the minimum time.

Standalone data loggers have a number of limitations, the worst being their form of output. If magnetic or punched tape is used as the recording medium, it requires off-line analysis before the operator has access to the data. If printed output is used, in an installation with a large number of sensors, the mass of paper produced makes comparison of results from different multiplexing scans extremely difficult. For this reason many data loggers are provided with communications facilities to allow their connection to a computer. Local recording is then primarily a protection against line or computer failure. Standalone data loggers, however, are frequently used for applications involving a few sensors that require infrequent scanning. Often these applications require only the simplest printed output for historical purposes and then an inexpensive data logger is an adequate means of recording.

PAPER TAPE PUNCHES

Another digital recording device is the paper tape punch that punches a binary code across the width of a paper tape. Paper tape punches are used less frequently now then in the past and will probably be superseded eventually by the cassette tape recorder. In simpler punches, tape is moved through the punch by a *sprocket*. This is a wheel with pointed teeth around its periphery. Sprocket teeth use a line of "sprocket holes" in the paper to draw the tape through the punch an increment, where it remains stationary until moved forward by the next tooth. During the stationary period, part of the tape rests under a line of punches operated by solenoid-activated hammers. Selected punches are driven through the paper, leaving a line of holes representing the code. Most punches generate "eight-hole" tape; seven are used for the code, and the eighth for a code check. This is described as a *parity check.*

PARITY CHECKS

The principle of parity checking is simple. If the number of one-bits in a binary group are counted, the number can always be made even or odd by the addition of a bit to the checking column. When the bits are made up to an even number, the group is said to have *even parity*. When the number is odd, the group has *odd parity*. Examples of odd and even parity are illustrated in the table below.

Byte	Bit Count	Parity Bit	Parity
10110011	5	1	Even
10010011	4	0	Even
10110011	5	0	Odd
10010011	4	1	Odd

Digital recorders and tape punches occasionally malfunction and for some reason do not record a complete code. The inclusion of a parity bit provides some protection against this error going undetected. During the reading operation each line of the code is checked, and if the count is no longer true, a failure is registered. Parity checking does not guard against the loss of two bits, but this is less likely to occur than the loss of one.

xy PLOTTERS

Another type of recorder, the *xy* or *flat bed plotter*, is used for automatic graph plotting. The pen in this type of recorder is free to move in both the x and y planes and is positioned by two servo systems similar to those illustrated in Figure 14.2. These recorders, though relatively simple in design, have good accuracy and are very reliable.

The servo recorder is an analog-input device, and while it is very useful in certain applications, it is not the best type of plotter for use with a computer. An alternative is the digital plotter operated by two *stepping motors*, one to drive the pen across the paper, the other to move the paper at right angles to the pen movement. Often the paper is a continuous length and the plotting table is the circumference of a drum driven by the second motor.

In addition to their use as plotter drive motors, stepping motors have many applications where precise control is required. The one shown in Figure 14.22 is a simple three-phase single-pole device that produces six steps per revolution. It is possible, however, with suitable gearing, and also by increasing the number of phases and magnetic poles, to make units with high resolution.

Suppose a permanent magnet rotor is fitted inside a case that has three field coils wound on three equally spaced pole pieces. If the coils are wired as shown and supplies are connected to the junctions of the field coils, a magnetic field will be set up with a resultant field lying at some angle to the windings. The magnet will rotate to a position of alignment with this field and, if the polarity of the supplies is changed, will turn to take up a new position of attraction.

Provided polarity of the windings is changed in a strict sequence, the magnet will rotate in six equally spaced discrete steps to execute a complete revolution. The table below illustrates a sequence of polarity change to obtain a complete rotation.

A	*B*	*C*
+	+	−
+	−	−
+	−	+
−	−	+
−	+	+
−	+	−

This stepping sequence is usually provided by manufacturer-supplied stepping-motor controllers that accept a serial pulse train input and translate it into a stepping sequence.

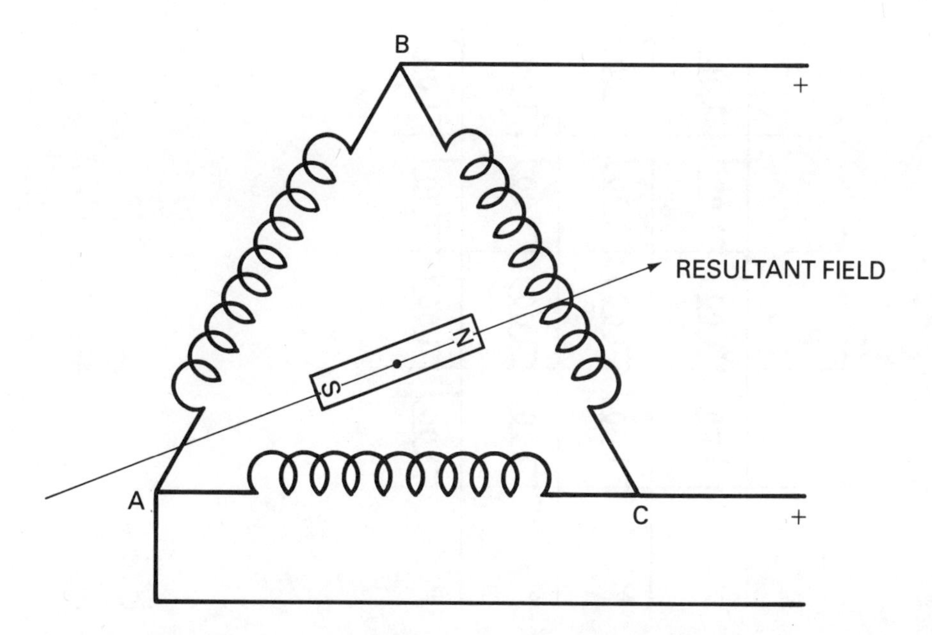

Figure 14.23
Stepping Motor

MULTIPLEXERS

The simplest multiplexer that can be used for data logging is the uniselector discussed earlier but because it is very limited in speed and is restricted to sequential scanning, it is seldom used. Where low-speed sampling is adequate, relay multiplexers, especially *mercury-wetted* types, are very reliable. For higher speeds, *solid-state switches* are used, but these have certain limitations that make their application in heavy industrial environments more difficult than relays.

Consider first a relay multiplexer. Provided only one relay is to be selected at a time it is possible to connect the relay coils in a simple matrix as illustrated in Figure 14.24. The advantage of this method of connection is realized with large arrays. The 4 × 4 matrix requires eight control lines to switch 16 relays, a 16 × 16 matrix requires 32 control lines and allows selection of 256 relays.

Matrix connection requires selection of two control lines to activate any relay. To select R_1, the lines 1 and A must be activated, and to select R_{10}, lines 2 and C. Switch selection may be by relay control, but solid-state circuits are much more reliable in their operation. It is possible to use an adaptation of the commutator circuit described in Figure 14.16 as a sequential controller. One such circuit is illustrated in Figure 14.25.

The control shown consists of two commutators. One steps the y control lines 1, 2, 3, and 4 and then resets. The reset pulse steps counter 2 controlling the x control lines from A to B, and then the cycle repeats itself. Using two four-stage

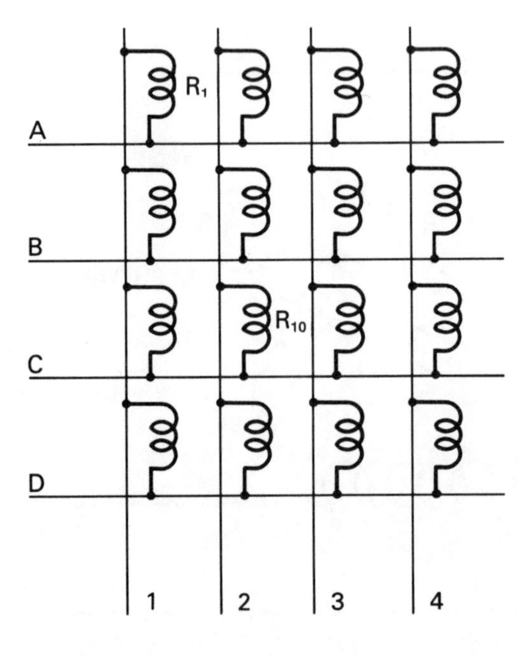

Figure 14.24
Matrix Multiplexer

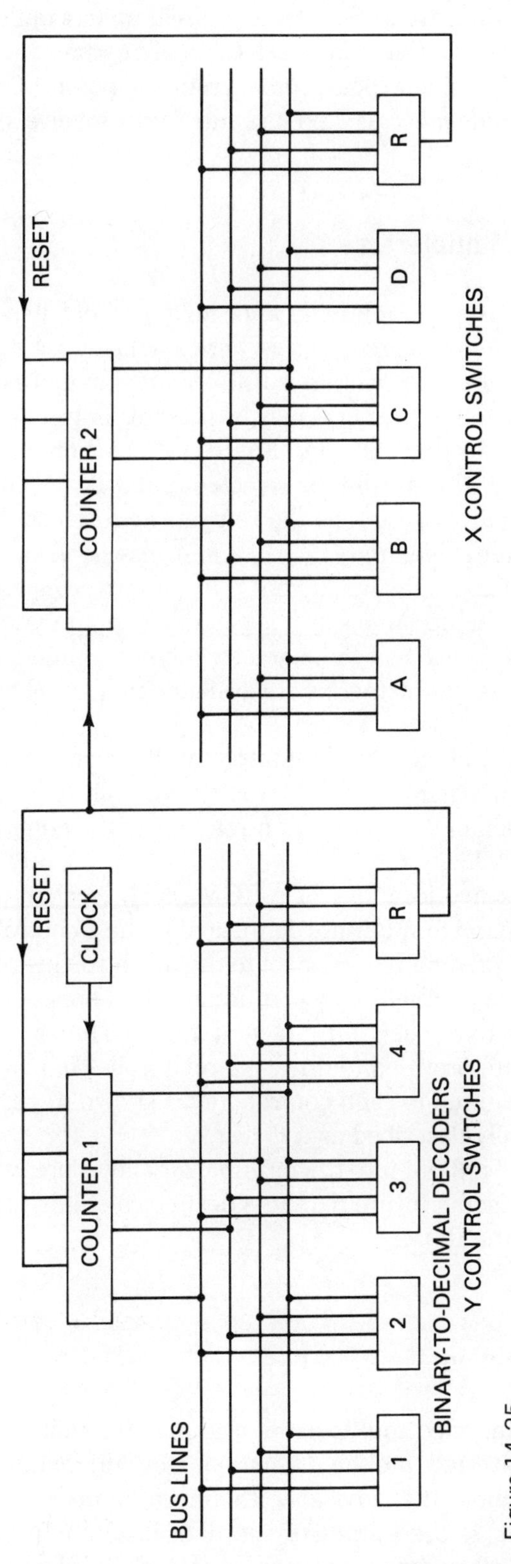

Figure 14.25
Matrix Multiplexer Control

counters, the control will drive a 16 × 16 sequential matrix multiplexer. A much improved multiplexer is one that allows selection of any point at will, and allows repetitive scanning of a single point, or a group of points. Such a system is described as having *random access*, and its control may involve a microcomputer.

Preprogrammed Multiplexers

Sometimes an infinitely programmable data logger is not necessary, although nonsequential selection may be required. Where a data logger is part of an installation, where the sequence of selection is unlikely to change frequently, a device known as a *ROM* (Read Only Memory) may be used to control the scanning cycle. Consider a matrix such as that shown in Figure 14.24, where the relay coils have been replaced by fuses. If the *A* line is selected and a heavy current allowed to flow through lines 2 and 4 using *A* as the return, the fuses connecting *A*, 2 and *A*, 4 will blow. At a later stage, lines 1, 2, 3, and 4 may be sensed through line *A* using a current less than the fusing current. The pattern registered will be that of a binary number 1010. In the same way it is possible to preprogram lines *B*, *C*, and *D* turning the matrix into a 4 × 4-bit character memory. Such a small memory is of little use, but much larger memories are available that are physically small and inexpensive.

A ROM can be introduced into a matrix multiplexer, to convert it into a preprogrammed system. Using the 4 × 4 matrix multiplexer as an example, it is possible to connect each of the x and y control lines to the output lines of a ROM as shown in Figure 14.26a.

This circuit shows five locations of a ROM, each location containing eight parallel bits. The left-hand four bits are connected to the x control lines of a multiplexer and the four remaining bits connect to the y control lines. Each location of the ROM is activated sequentially by a commutator that energizes relays depending upon which fuses have been left intact as drawn. The circuit does not work because no supplies are provided to drive the relay coils. To make it functional a transistor switch is required in each control line as shown in Figure 14.26b. This type of circuit is usually described as a *driver*.

The introduction of a ROM into the multiplexer allows preprogrammed selection of relays but if at some future date it is desired to change the program, it is necessary to replace the ROM.

Channel Identification

When using preprogrammed multiplexers it is possible that wrong information may be accidentally inserted into the memory or a circuit malfunction may cause incorrect switch selection. It is advisable, therefore, to maintain some means of channel identification. If the output of the data logger is printed, the line of printed characters is often extended to include a channel number. Where the

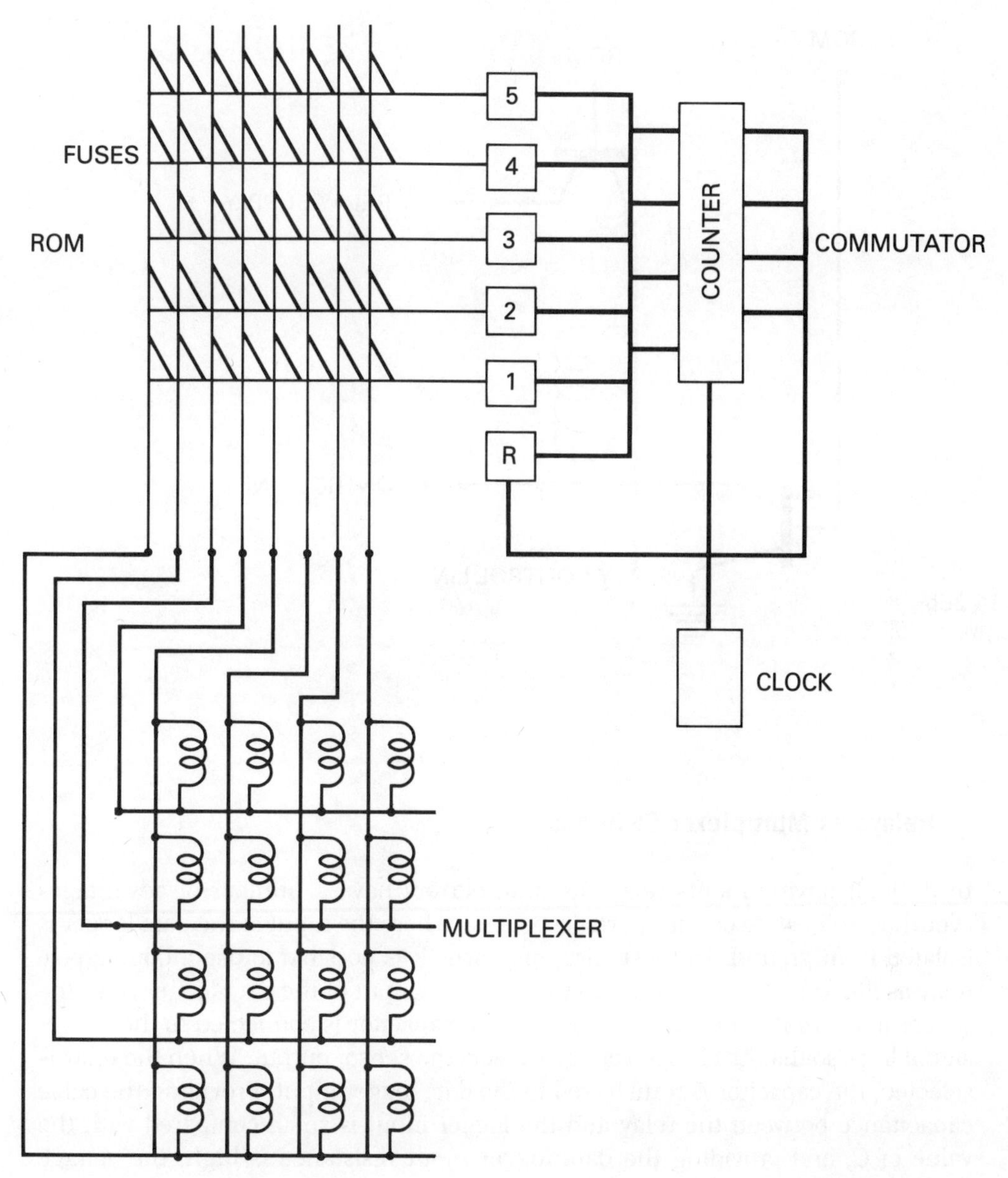

Figure 14.26a
Preprogrammed Multiplexer

output is a tape record, two sets of data are often recorded sequentially, with one representing the channel number, and the other representing the measurement.

One way of identifying a channel, illustrated in Figure 14.27a, uses a two-input AND gate to sense the coincident control lines of the matrix. The output of the AND gate is used to switch a multiplexer relay drive transistor, and also operate a channel identification line. This line may be one input of a BCD encoding circuit. Part of a two-decade encoder is illustrated in Figure 14.27b.

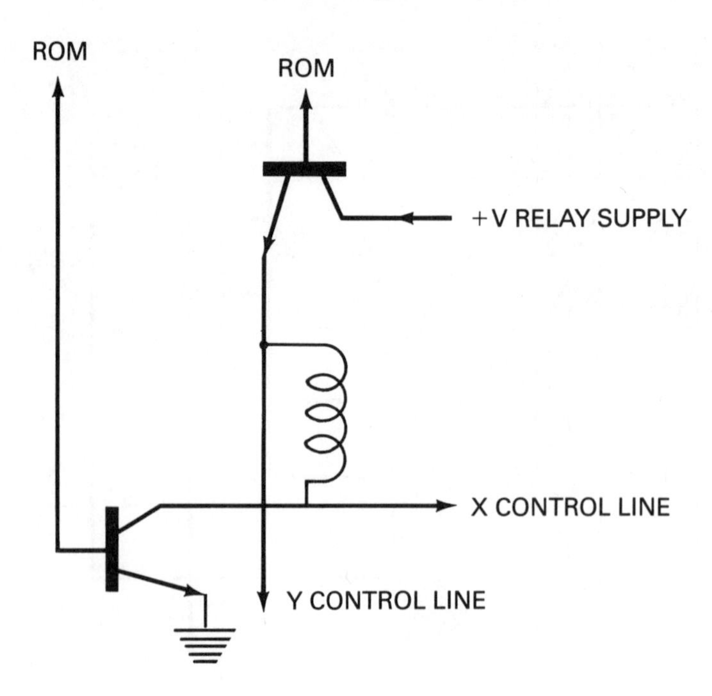

Figure 14.26b
Relay Driver

Relays as Multiplexer Switches

In difficult environments relays as multiplexers have a number of advantages over their solid-state counterparts, especially where the input of the data logger is isolated from ground. In these circumstances it is possible to use a double-pole relay as illustrated in Figure 14.28 to isolate the input of the data logger from the sensor lines. In its unenergized position the capacitor is connected to the sensor signal lines so that its charge voltage reflects the sensor output. When the relay is selected, the capacitor is transferred to the data logger input. Providing the cable capacitance between the relay and the logger input is small compared with the value of C, and providing the data logger input resistance is high, the voltage measured by the data logger will be that of the sensor. This type of multiplexer is sometimes described as a *capacitor store* multiplexer but probably is better known as a *flying capacitor* multiplexer.

These multiplexers provide the input of a data logger with a high degree of protection against common mode voltages, and spurious voltages generated by ground loops. In the event of a switch being destroyed by the application of high voltages, only one channel is affected, and not the entire system.

In addition to its function of transferring the sensor signal to the data logger input, the capacitor may be used as part of a filter network. Figure 14.29 illustrates how the addition of two series resistances turns the multiplexer into a simple balanced RC filter of the type discussed in Chapter 3. Typical values of R and C are 1000 ohms and 100 microfarads respectively. This will produce a reduc-

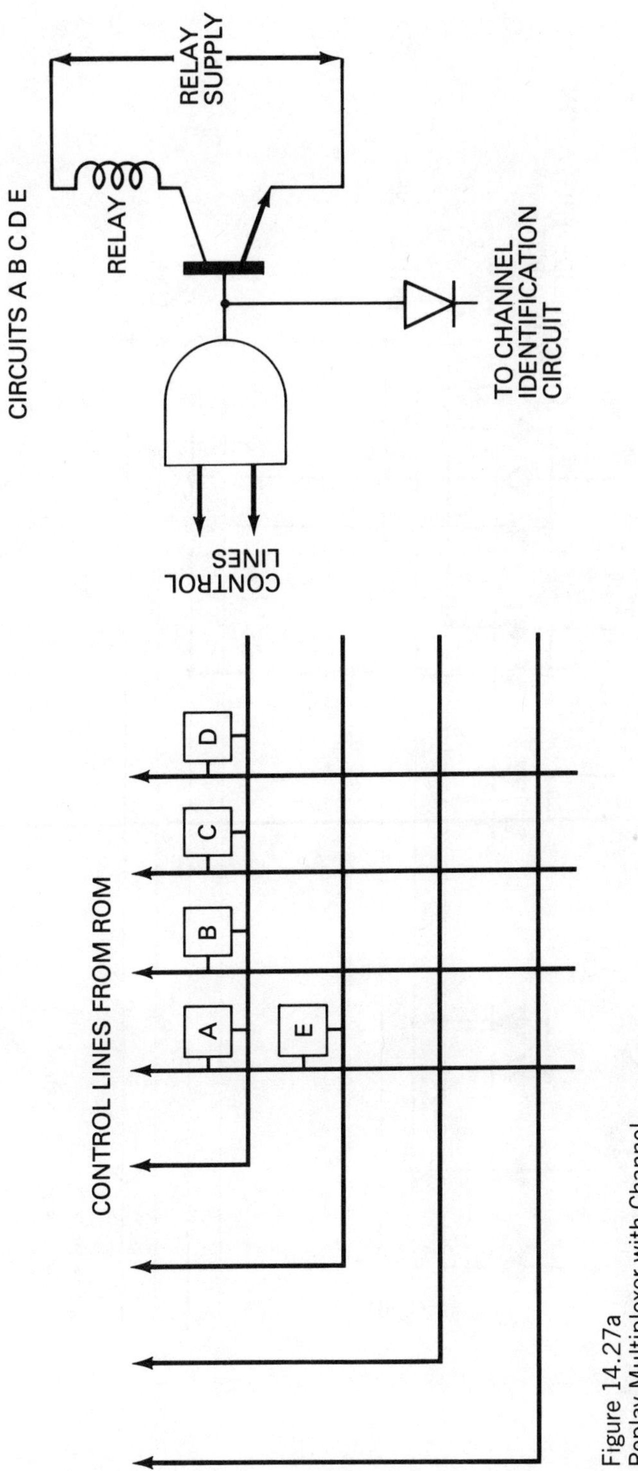

Figure 14.27a
Replay Multiplexer with Channel Identification Circuit

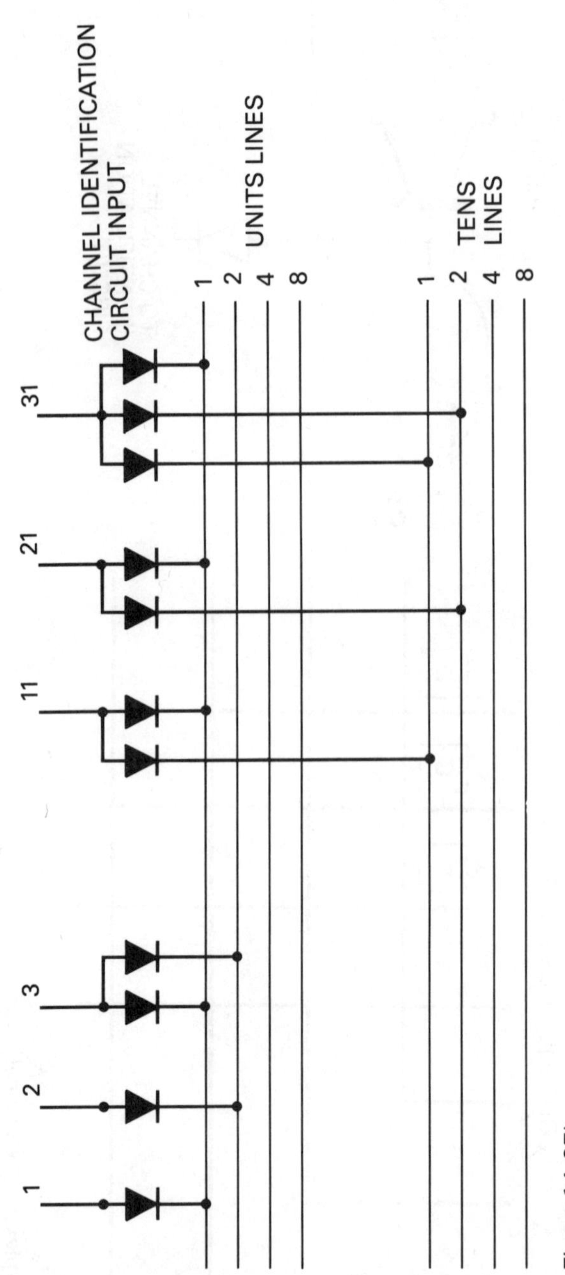

Figure 14.27b
Two Decade BCD Encoder

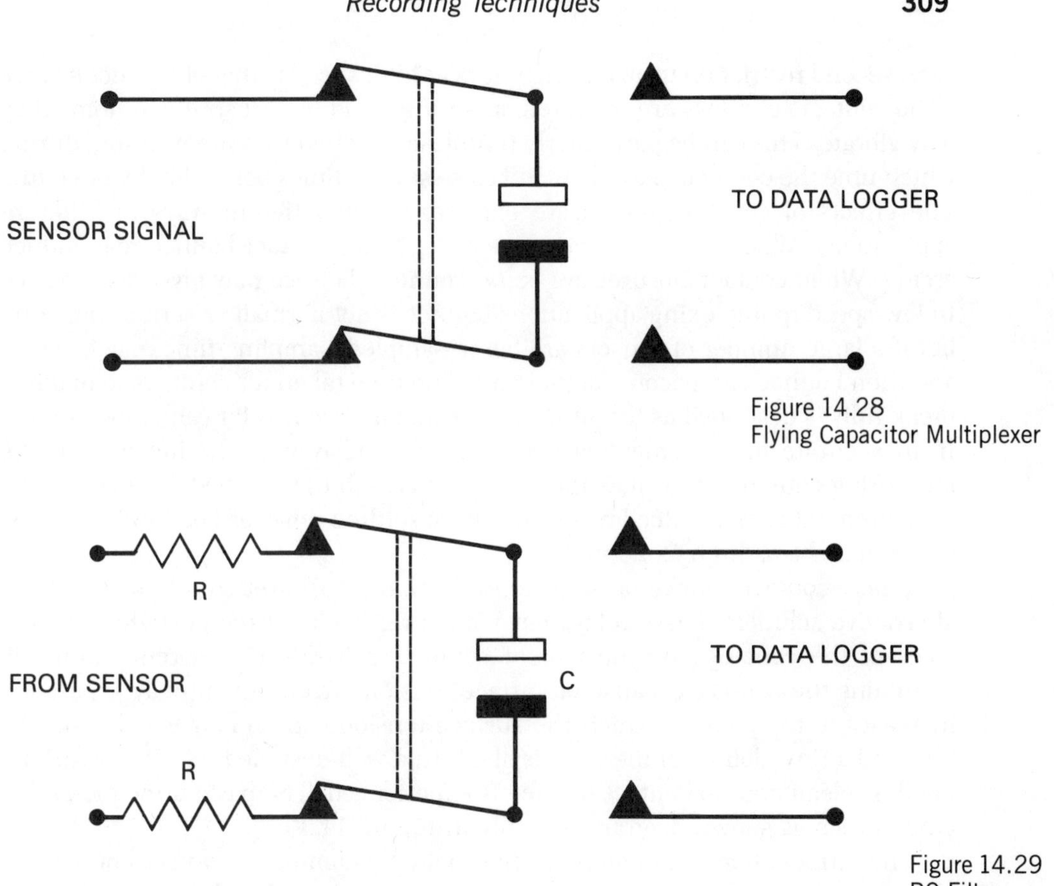

Figure 14.28
Flying Capacitor Multiplexer

Figure 14.29
RC Filter

tion in value of a 60-Hz signal frequency of about 100:1. If a settling time of 5 *RC* is assumed, the time constant will allow a sampling speed of one sample per second. Where relay multiplexers are required to operate at higher speeds, the capacitor and resistor values may be reduced with the sacrifice of noise rejection. This procedure is limited to a value of capacitance below which the loss of charge from the capacitor into the logger input circuit becomes significant.

There are two restrictions to switching relays at high speed, the first of which is relay life. Relays are designed to have a guaranteed life that can reach as high as 10^7 operations but is nevertheless finite. At one sample per second this would represent a life of 2800 hours continuous operation. While it is feasible to operate a relay at much higher speeds, its guaranteed life becomes disturbingly short. This being the case, relays are not the best devices to operate continuously at high sampling rates. It is a feature of relay multiplexers, however, that they may be run at relatively high sampling rates beyond their guaranteed life without failure. This is probably due to the fact that relay life is normally based on estimates where the contacts are switching their rated current. The currents involved in multiplexing are very small and the life of a multiplexer relay is probably governed more by its mechanical characteristics than its electrical ones.

A second restriction to switching relays at high speed is that of *contact bounce* The contacts of relays are mounted on springs, and like all spring systems they may vibrate. This can be particularly troublesome when they are opening, during which time the contacts may open and close several times before finally breaking. The effects of contact bounce are experienced in different ways in different applications. When a relay carries a heavy current, contact bounce may induce arcing. When contact closures are being counted, bounce may give false counts. In low-speed multiplexing applications bounce is not normally a serious problem, but if a large number of sensors are being sampled, sampling time may be short and then bounce can become significant. The time taken for contacts to make or break fully is described as the *settling time* and is specified for each type of relay. If an accurate measurement is being made, a delay must be imposed in the measuring equipment to allow the contacts to reach their settled state before the measurement is made. Reed relays may have settling times of 1 or 2 milliseconds, others may be as high as 20 ms.

Where contact bounce presents a problem, mercury-wetted relays provide an alternative solution. These relays have small reservoirs of mercury that wet the faces of the contacts. As the pair of contacts separate the mercury film still maintains the contact because of surface tension effects. Eventually separation increases to the point at which the surface tension can no longer maintain the film and a tiny globule of mercury is shed. In a well-designed relay the resulting break is clean and no contact bounce is experienced. The basic principle of this type of relay is shown diagrammatically in Figure 14.30.

The introduction of the mercury film not only eliminates contact bounce but also produces contacts having very low resistance when closed. In addition, mercury-wetted relays have a longer life than a reed relay with "dry" contacts.

Solid-State Switches

If a multiplexer switch is to be operated at a high scanning speed, at some point it becomes necessary to use solid-state switches. Relays are limited by their me-

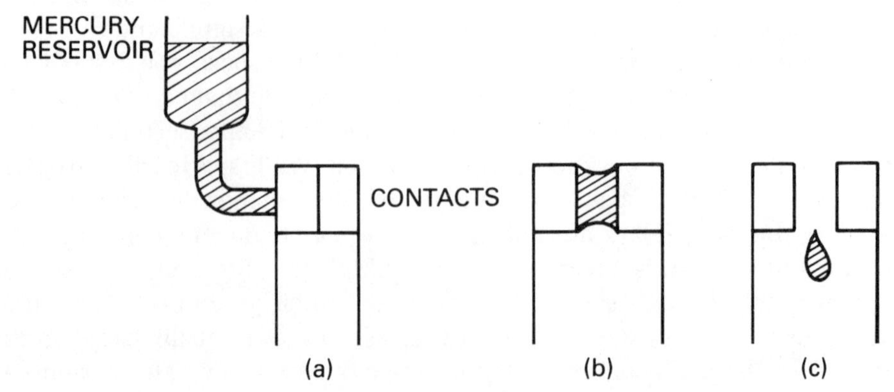

Figure 14.30
Principle of Mercury-Wetted Replay

chanical restrictions and contact bounce problems. While solid-state switches do not suffer these shortcomings, their application presents other problems. For example, they may not be exposed to the same common mode voltages as a relay, and they offer less protection to connected electronic equipment. It is possible to use any of the transistors previously discussed as multiplexer switches but their construction and mode of operation introduces errors that limits their value. One type of transistor not previously introduced is the *field effect transistor* (FET). These transistors are characterized by their high open-circuit resistance and their ability to switch currents flowing through them in either direction. They also do not suffer the limitations of the junction transistors. A typical FET multiplexer is illustrated in Figure 14.31.

Two FET transistors T_1 and T_2, are used in this switch. With no control voltage applied to their gates, both switches offer a very high resistance to the signal and act as a double-pole open switch. Application of a control voltage causes the resistance to fall to almost zero, and connects the signal to the data logger input circuit.

Because of the limited voltage that can be applied to such a switch without failure, it is desirable to protect it from excessive voltages that may appear on the lines. One method uses diodes connected to a low-impedance power supply. The two sets of diodes illustrated are described as *clamping diodes*. Provided voltages on the input lines are less than 12 V no current will flow through the clamping circuits, but any excursions of the voltage outside these limits will cause one or the other of the diodes will conduct to prevent the voltage from increasing more than a volt or two above the clamping voltage. The limit to this protection occurs when the current rises to a point at which the diodes fail.

ANALOG-TO-DIGITAL CONVERSION

A signal sampled by a multiplexer is presented to the input of a data logger for a limited period of time. During this sampling period, an analog-to-digital converter

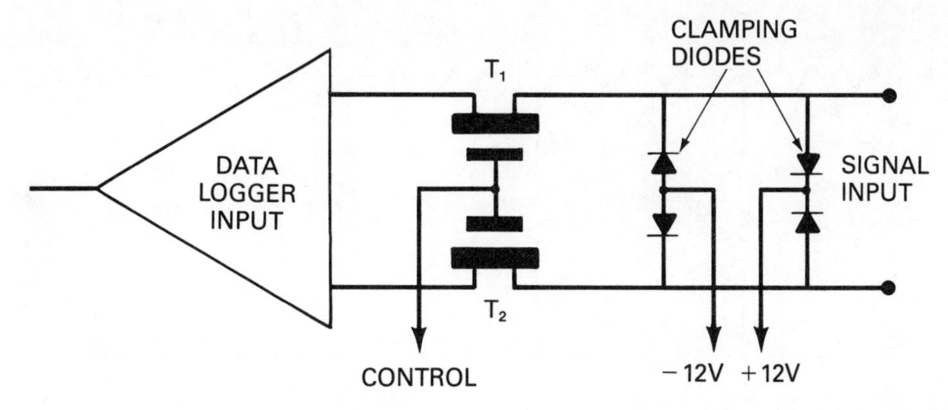

Figure 14.31
Solid-State Switch

must convert it into a digital representation, and then the converted signal must be recorded. Conversion time, therefore, must be appreciably shorter than sampling time. There is no problem in obtaining high-speed ADCs, many having microsecond conversion times. The problem is rather one of making a meaningful measurement of the sample that may have high-frequency noise superimposed. If the converter takes a "*snapshot*" of a low-frequency signal, only a few microseconds long, the instantaneous value recorded may bear no resemblance to the value of the signal being measured. Providing the sample is long compared with the conversion time, it is possible to make a number of measurements and average the results to obtain a true representation of the signal. Averaging in this way is effectively another method of filtering a signal.

The simplest type of ADC is the multicomparator type. This device uses a series of comparators to compare an unknown voltage with a series of reference voltages. If the unknown voltage is switched just to the comparator with the highest reference voltage and then to the others in a descending sequence, eventually a comparator will be switched whose reference is lower than the unknown voltage. This will cause the comparator to change state and indicate the completion of the operation. With such an arrangement it is theoretically possible to build up any required resolution by increasing the number of comparators and suitably dividing the reference voltage. Such converters are simple in principle, but in practice are expensive to construct. The limited resolution of comparators also restricts their accuracy.

Single-Comparator ADCs

Alternative methods of conversion use a single comparator and vary the reference voltage. Consider a resistance network shown in Figure 14.32a.

Consider the resistors *A*, *B*, *C*, and *D*, all of equal value *R* and the resistors *E*, *F*, *G*, and *H*, all of equal value 2*R*. Suppose the network is terminated by a resistor *J*

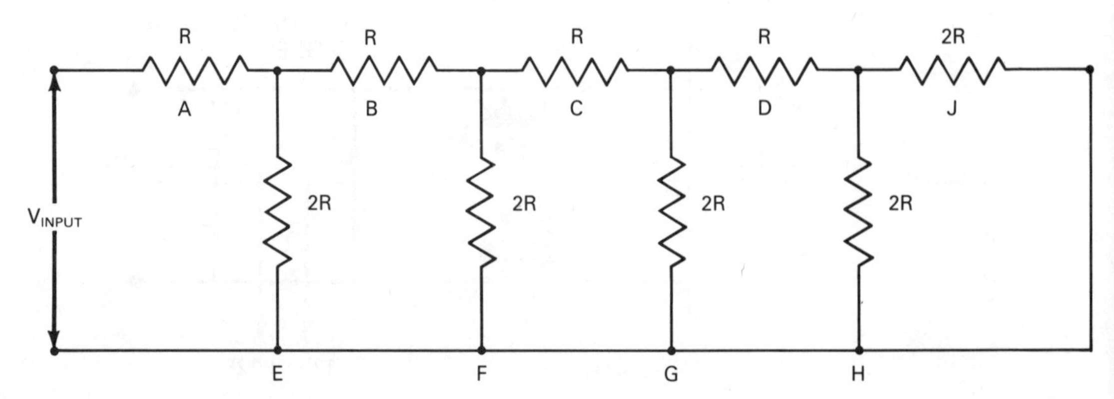

Figure 14.32a
Binary Ladder

with a value 2R, the effective value of $H = H^1$ and is that of H and J in parallel (i.e., R.).

The resistors D and H^1 form a potential divider with equal resistance values. It follows, therefore, that a voltage appearing across H must be half that of a voltage across G.

Consider now the effective resistance value of G. This is effectively the resistor G in parallel with two series resistors D and H^1. Since D and H^1 each has a value of R, the effective value of $G = G^1$ must also equal R. It follows, therefore, that a voltage appearing across G must be half that of a voltage across F.

Continuing this reasoning, it can be shown that the voltage across F is half that of a voltage across E, and the voltage across E is half that of the input voltage. This progression is a binary one and the network is described as a binary ladder. Such networks are adapted to provide reference voltages for the comparator of single-comparator A to D converters as illustrated in Figure 14.32b.

At the beginning of a conversion, a control circuit sets a register to a predetermined value. Usually this sets the divider network to half the maximum output of the comparator. This reference voltage is compared with the unknown input and the comparator generates a signal to the control indicating whether the reference voltage is too high or too low. The control then resets the register to a new value based on the polarity of the previous setting. Carried out in a circuit portrayed in Figure 4.33, this process of successive approximation continues until a balance is reached when the control issues a Conversion Complete signal.

Single-Ramp ADCs

A variation of this method of conversion uses a ramp generator as a D-A converter incremented from zero by a clock. When the output of the generator and the unknown signal are equal, the clock is stopped. The contents of the counter at

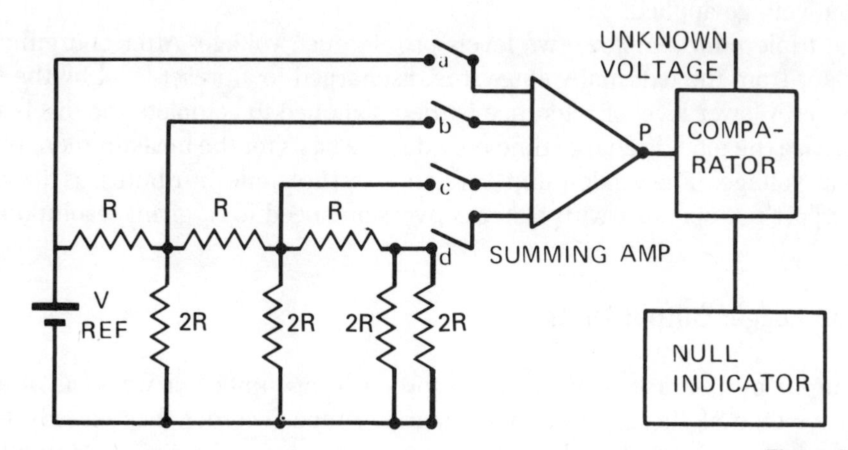

Figure 14.32b
Single Comparator ADC

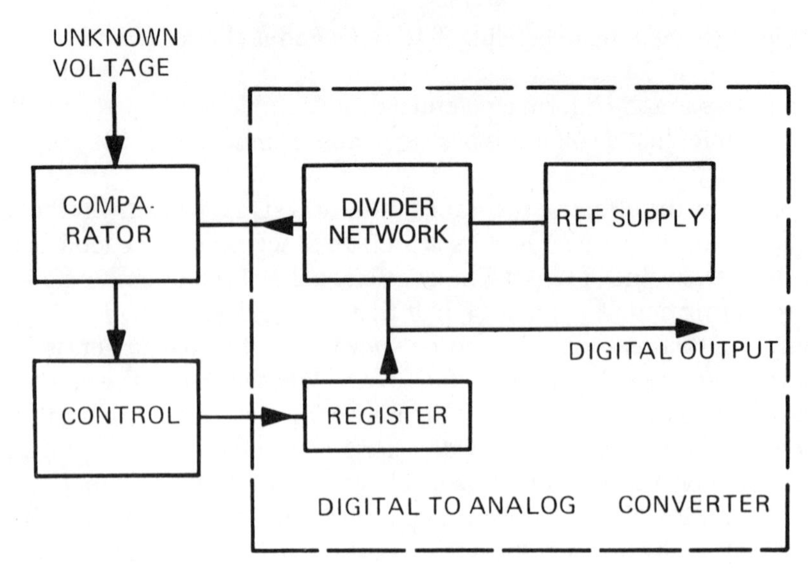

Figure 14.33
Successive Approximation ADC
Block Diagram

this point provide a measurement of the external voltage. A single-ramp ADC is illustrated in Figure 14.34.

Double- and Triple-Ramp ADCs

An improvement on the single ramp is to be found in the dual- and triple-ramp ADCs. In the dual ramp an integrator is exposed to the external voltage for a fixed period of time and then the integrator is switched to a reference source of opposite polarity. The time taken for the integrator to return to zero is proportional to the external voltage applied.

The triple ramp employs two levels of reference voltage. After charging the integrator from the external voltage it is discharged to a preset level by the first reference. A lower level of reference is then switched to complete the discharge. In this case the total discharge time provides the basis for the measurement of the external voltage. The triple-ramp version has the same attributes as its dual counterpart but operates with faster conversion speed for a given resolution.

Data Logger Output Units

The output of a data logger can be recorded in a number of ways, as already indicated. One of these, the digital magnetic tape recorder, has already been discussed in some detail, and another, the computer, will be discussed in a later chapter.

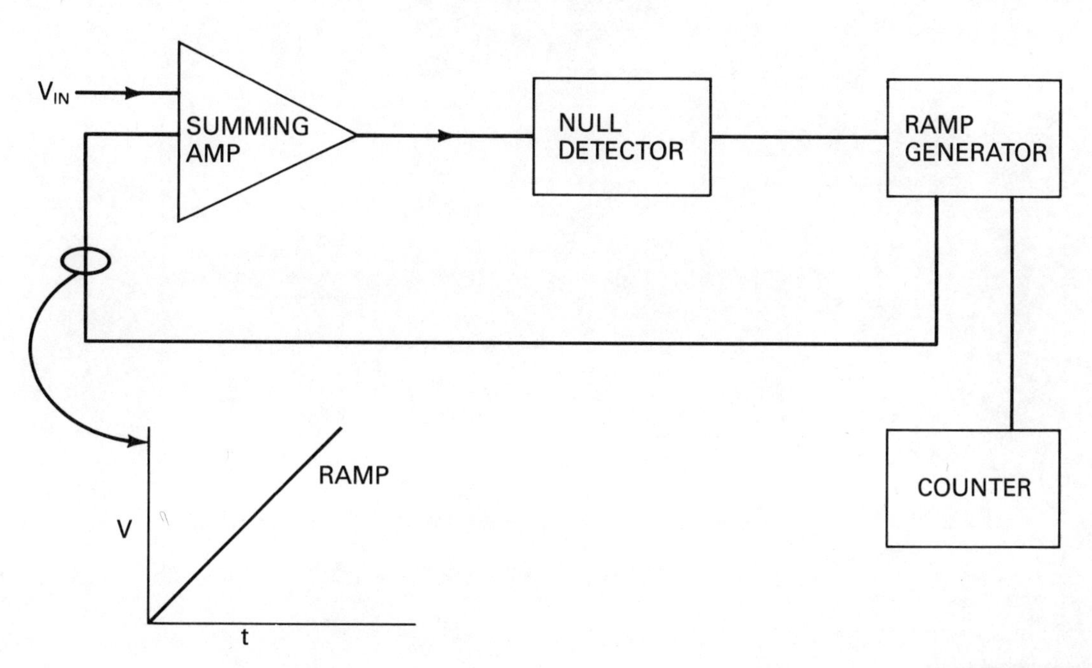

Figure 14.34
Single-Ramp ADC

CHAPTER 15

CALIBRATION TECHNIQUES

It is intended in this chapter to confine the discussion to the calibration of equipment used for making measurements with an accuracy of 1% FS or less. These represent a large proportion of all common measurements. Where more exacting measurements are required, the reader is recommended to the appropriate ISA Standard & Practice.

The attitude toward providing calibration facilities varies widely depending on local attitudes toward instrumentation. Some firms are prepared to spend considerable sums to provide and maintain calibration laboratories while others resist providing even minimum standards. The majority of firms, however, are prepared to provide a calibration room where conditions are relatively stable, and a modest range of calibration standards for their instrumentation engineers.

Industries requiring very high and consistent standards of measurement accept the overhead of expensive calibration laboratories and specialized staff. In these situations the instrumentation engineer often draws equipment from an equipment room and either returns it at the end of a test, or on a regular basis so that it can be checked and recalibrated before reissue. The recommended environment for high-quality calibration laboratories is specified in ISA Standard 52.1. At the opposite extreme are firms opposed to providing calibration equipment. Such firms may solve their problems by returning equipment to the manufacturer periodically, or by employing the services of an independent calibration laboratory. Some firms may not even own their measuring equipment, but rent it from firms specializing in equipment rental.

The more common situation in small or medium-size firms is one in which the instrumentation engineer is responsible for calibrating equipment as well as making on-site measurements. In many ways this is a very satisfactory arrangement as it allows the flexibility to calibrate equipment in the way best suited to a particular circumstance. Such an arrangement, however, considerably extends the responsibilities of an instrumentation engineer and increases the possibility of standards being neglected because of the pressure to make on-site measurements. The solution to this problem is one of management. In the final analysis someone must be accountable for the maintenance and periodic calibration of local standards.

The quality and the amount of calibration equipment carried by a firm obviously depend on the types of measurement, and ultimately the accuracy required. Because it represents a significant capital investment, the purchase of equipment should be considered rationally. For example, some high-quality calibration equipment gives optimum accuracy only in environments of high stability. If the measurement requires only a few percent accuracy, it may be possible to use a lower accuracy standard, less sensitive to its environment and at a much reduced cost.

Improvements in electronic equipment in recent years have brought about a change in attitude toward the quality of equipment. The introduction of integrated circuits has made possible the design of robust equipment suitable for on-site measurements whose performance rivals that of laboratory standards of a few years ago. The problem with such equipment is that of calibration. Many firms cannot afford the standards to equip a facility. The best they can achieve is an identical unit in the calibration room against which they can compare their site equipment. In general, the accuracy of such equipment is so much higher than the sensor with which it will be used that small deviations from the true value are of little consequence. Gross errors, of course, can be recognized and the equipment returned to the manufacturer for recalibration or repair.

If on-site equipment has accuracies only an order or two better than the sensors, the high-quality site equipment provides an adequate performance for calibration. Typical of equipment that can be used this way are high-quality digital voltmeters, digital frequency meters, counters, and timers.

CALIBRATION OF SITE EQUIPMENT

A typical measurement channel consists of a sensor, a balancing network, an amplifier, and a recorder. For best accuracy it is desirable to calibrate the channel as an entity but unfortunately this is not always possible. There are many systems where measuring equipment is built into a larger installation remote from the sensor calibration facilities. In these circumstances, removal of the equipment each time a sensor is calibrated can be an unreasonable undertaking. To overcome this, it is often necessary to calibrate the amplifier and recorder *in situ*, while the sensor is removed for calibration at the central facility. When the sensor is reinstalled, the results of the two calibrations are combined mathematically.

The procedures for calibrating a measurement channel are similar for most sensors and the parameters they measure, and for the convenience of this discussion a pressure measurement is assumed.

At the commencement of a calibration, it is common practice to balance the equipment, and adjust the output of the recorder to zero at the datum of the sensor. Following this procedure, maximum calibration pressure is applied to the sensor so that the amplifier gain can be adjusted to provide a convenient recorder deflection without overloading the amplifier. Pressure is now released and the recorder checked to establish that the channel has returned to its correct zero state.

Between the time that the equipment was calibrated and a measurement is made there is no guarantee that the channel gain will remain constant. It is necessary, therefore, to perform some sort of equipment calibration that may be repeated at the time of the measurement to provide a correction factor for any changes that may occur. One method requires the operator to switch a calibration resistance across one arm of the input bridge network. This provides a recorder output that can be used as a reference to correct the gain at the time of a measurement. Such a procedure is sometimes described as an *internal calibration.*

With the completion of preliminary procedures it is now possible to determine what is sometimes described as the *channel sensitivity*, or alternatively the *calibration factor*. This is the ratio of the recorder output/applied pressure and is obtained by applying to the sensor a source of pressure whose value is accurately known. At the time of making a measurement the calibration factor is divided by the recorder output resulting from the application of the measured pressure to obtain its value.

When performing a calibration it is exceptional to base a calibration factor on a single input. It is more usual to base it on a number of points, obtained by applying pressures in a series of discrete steps from the datum to the maximum and back to the datum. The channel sensitivity may be determined from these points either mathematically or graphically.

In the majority of instances, output of a pressure sensor will be linear with pressure and the slope of the line drawn through the points will provide a calibration factor. Normally points on this graph will lie close to the line, their scatter representing a measurement of the channel accuracy, but occasionally one or two points may plot well outside the normal distribution. It is probable that these are due to errors either in setting the pressure, reading the output of the recorder, or plotting the point. Providing the points do not repeat in subsequent pressure cycles they may be ignored, but if they do, they may indicate a faulty sensor, or a fault in the calibration equipment.

In addition to providing a calibration factor, it is often necessary to calibrate a sensor to determine the effects of nonrelated parameters on its performance. It is easy in the stability of a calibration laboratory to overestimate the true accuracy of a system. Pressure sensors may introduce errors when subjected to temperature changes, for example, or a recorder may introduce errors when subjected to vibration. Manufacturers of these types of equipment provide data to cover most eventualities, but if there is any reason to suspect that the equipment is not operating within its specification, then a calibration will be necessary.

PRESSURE STANDARDS

Essential to every pressure calibration is the availability of a pressure standard. A desirable standard is one whose accuracy is ten times the accuracy of the sensor being calibrated. If the discussions are restricted to the calibration of systems with an accuracy of 1% FS or less, the desirable accuracy of the standard is 0.1% FS or better. There are, in fact, many moderately priced standards whose accuracies are higher than this, and which will operate satisfactorily in any laboratory where the temperature is relatively stable and the atmosphere is relatively free from contaminants.

Pressure standards are divided into two types; one produces an accurately known pressure, the other measures an arbitrary pressure accurately. The dead-weight tester shown in Figures 15.1 and 15.2 is one of the former. This device produces an accurate pressure in a fluid by the application of weights to a close-tolerance-area piston. The fluid is usually oil, although air is sometimes used for lower ranges.

The sensor to be calibrated is connected loosely to the test connection point, the priming valve is opened and the screw press opened to its maximum position. The tester is then purged by forcing oil through the system with the priming pump. After purging, the test connection is tightened and the priming valve closed. Weights are added to the piston and the pressure raised by compressing the oil in the system with the screw press. When the pressure reaches a value equal to the applied weight divided by the piston area, the piston rises from its rest position on its lower stop and floats between this and an upper stop, which re-

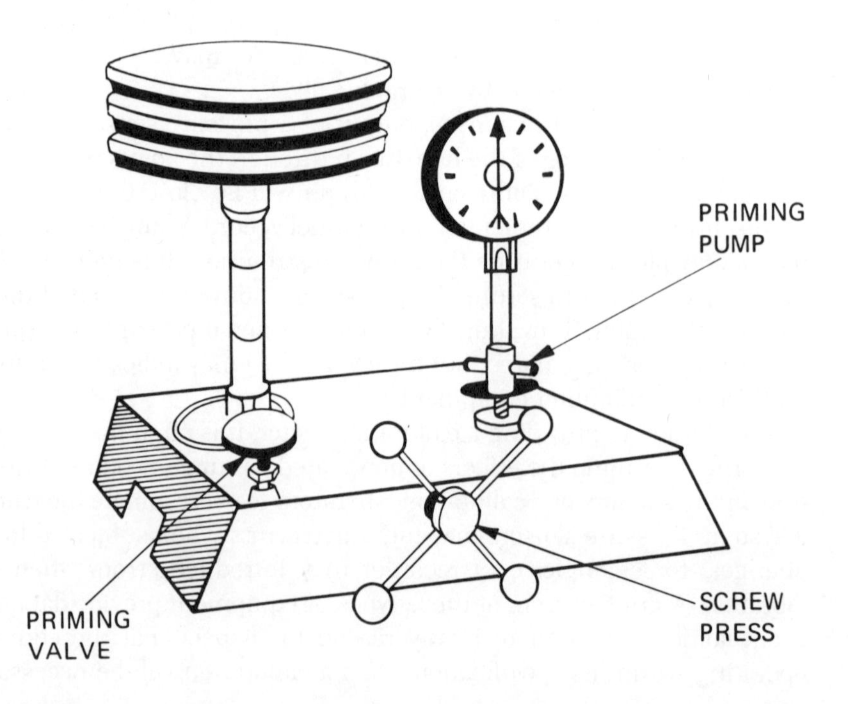

Figure 15.1
Dead Weight Tester

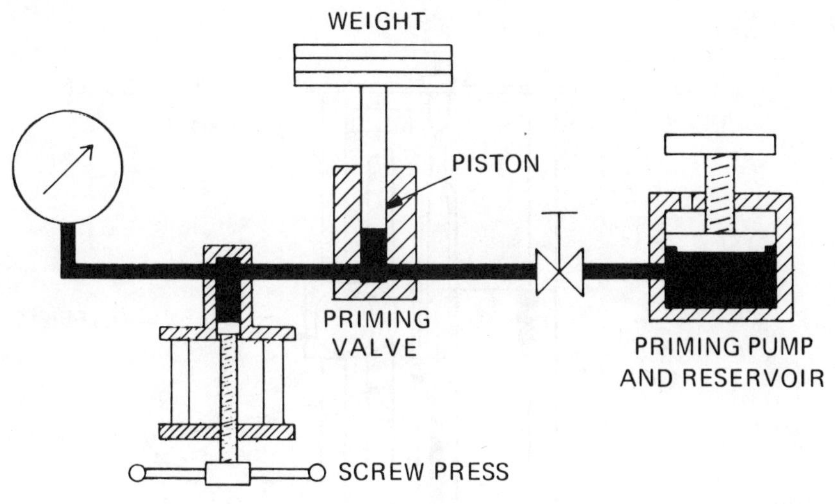

Figure 15.2
Principle of Operation of Dead Weight Tester

stricts the limits of the piston travel. With the piston floating an increase in pressure drives it against the upper stop, a decrease allows it to fall on the lower. The difference in pressure to move the piston between the two stops is partly due to friction of the piston in the cylinder and it can be reduced to a minimum by spinning the weight and the piston during a measurement.

The accuracy of the tester depends on the tolerances of the piston area and weights, but ±0.03% of the measured value is typical. Oil-operated dead-weight testers may be used safely up to very high pressures, as the volume of oil is small and any failure results in an immediate drop in pressure. Testers with ranges up to 100,000 psi (7000 kg/cm^2) are available.

Manometers

For low-pressure ranges, up to about 15 psi (1.0 kg/cm^2), various types of manometers are used as standards. The simplest form is the precision standard manometer shown in Figure 15.3. These manometers are usually supplied with sighting aids to assist in reading the level of the mercury, and a vernier that allows accurate subdivision of the scale.

The manometer tube has a large diameter, is precision bored, and is mounted on a solid base containing a well for the mercury. The legs are adjustable and the base is levelled against a built-in bubble level. The tube is supported by steel columns that also act as supports for the scale and sighting mechanism. Resolutions of 1 part in 6000 are possible with this type of instrument, and accuracies of ±0.02% FS.

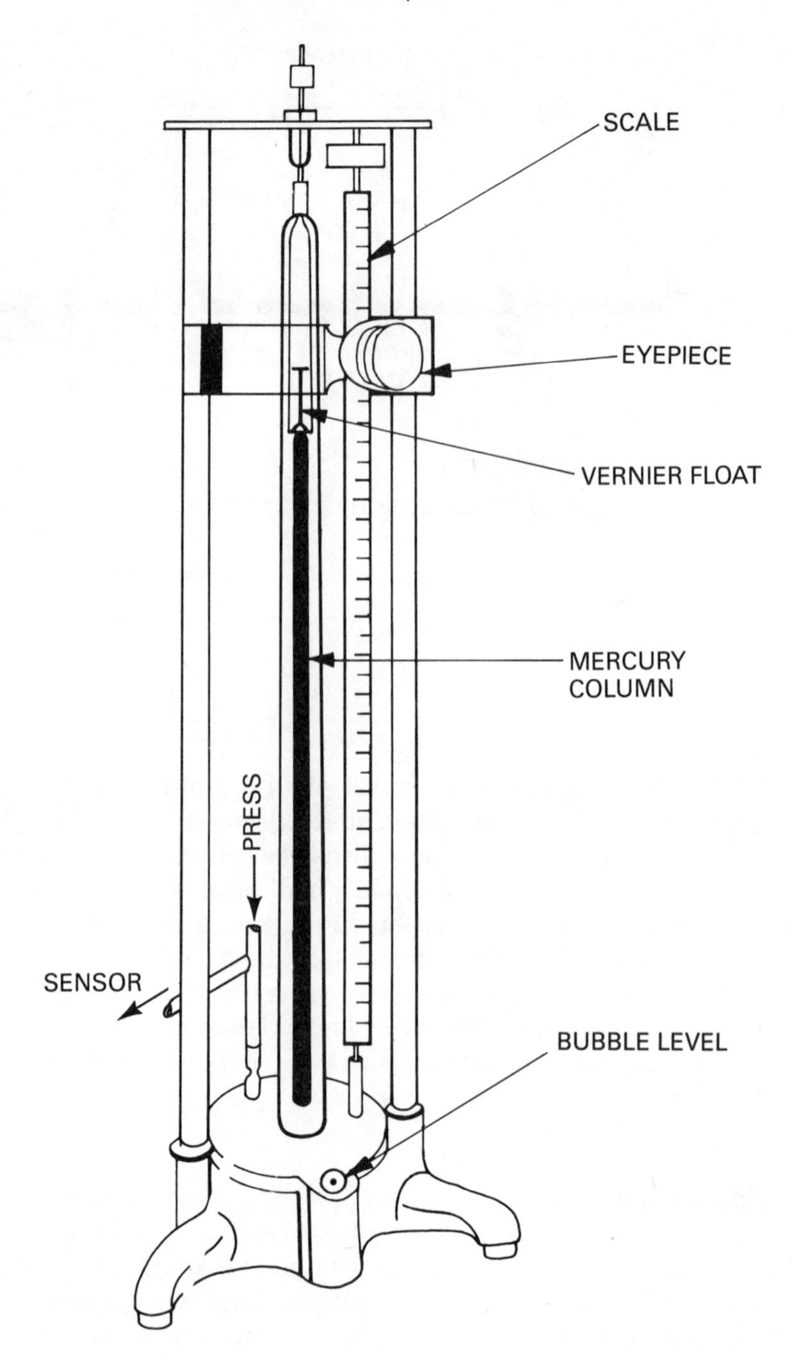

Figure 15.3
Mercury Manometer
(Wallace and Tiernan)

In the type of manometer shown in Figure 15.4 a LVDT is used as a level-sensing device. The core of the transformer floats on the surface of the mercury column while the body is free to move along the length of the leg. A perforated tape connected to the transformer body and wound around two sprocket wheels allows the body to be driven by a servo motor to a position of electrical balance. The servo position is a measurement of the difference in the height of the mercury.

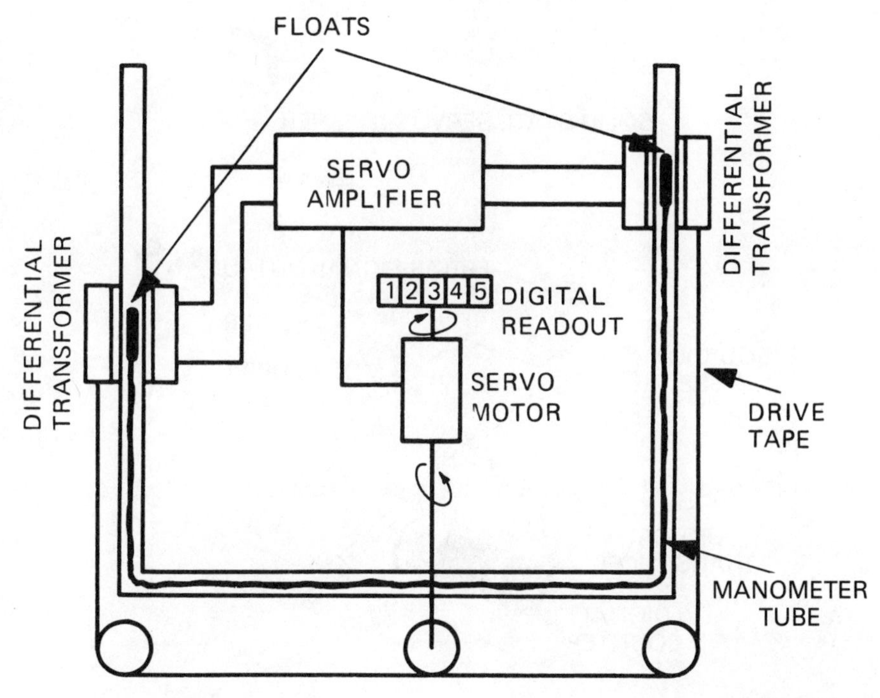

Figure 15.4
Self-Balancing Manometer
(Exactel Instrument Co.)

Bourdon Tube Pressure Calibrators

Spiral Bourdon tubes made of quartz and certain nickel alloys display excellent spring characteristics with low hysteresis and thermal coefficients. Quartz tubes make ideal transducers up to about 500 psi (35 kg/cm^2), and nickel alloys up to about 10,000 psi (700 kg/cm^2). The characteristics of these tubes make possible the design of a simple pressure calibration device with an accuracy in the region of 0.05% FS. Bourdon tubes are often designed with mirrors attached to their free ends that can be used as part of an optical servo system. This allows the angle of rotation to be measured without restricting the movement of the tube.

Consider the schematic diagram of the calibrator shown in Figure 15.5. Pressure is admitted at the pressure port causing the tube and the suspended mirror to rotate. As the mirror rotates the servo motor rotates the geared disk to maintain equal light on a matched pair of photocells in a photocell assembly. The balance is very fine and capable of a resolution of the order of one part in 10,000.

It is possible to obtain calibrators that use two concentric Bourdon tubes, one of which is used as a reference tube for absolute or differential pressure measurements. It is also possible to use some units as precision pressure regulators using the Bourdon tube as a sensing element in control servos. These devices make very useful standards; they have an accuracy considerably higher than the average sensor, and the ability either to measure an unknown pressure, or to provide an accurate source from an unregulated supply.

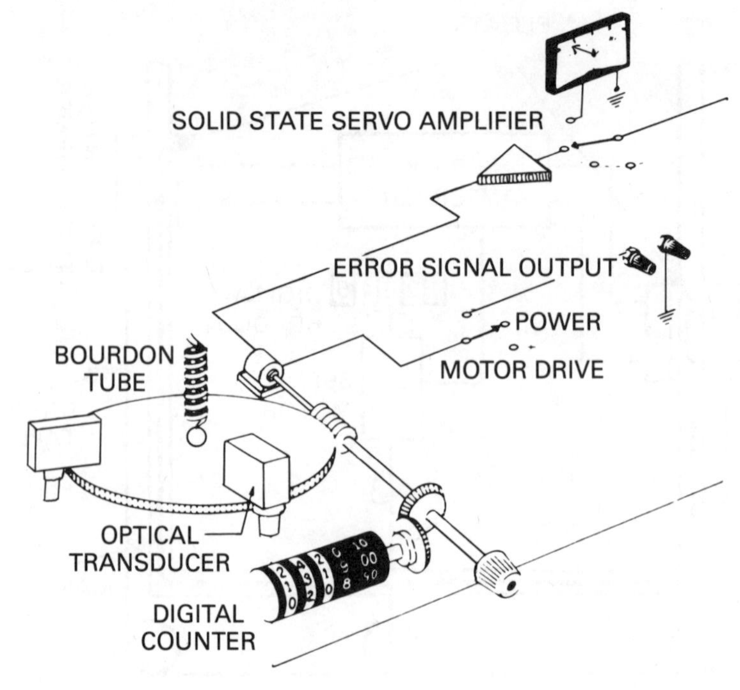

Figure 15.5
Bourdon Tube Pressure Calibrator
(Texas Instruments)

SENSITIVITY CONSIDERATIONS

When making a measurement, the accuracy of the measurement is no better than the accuracy of the entire measurement channel. Variations in the sensitivity of a sensor, its amplifier, the recorder, and, often, variations of bridge excitation voltages can all affect the measurement. Considering a linear bridge-type pressure sensor, the output is proportional to the pressure applied and also the bridge excitation voltage. If a pressure P is applied to a sensor whose output is balanced to zero at its datum and the corresponding output of the sensor is V volts, then the sensitivity of the sensor is V/P. This value of sensitivity is dependent on the bridge voltage and any subsequent variation will affect the accuracy of the measurement. Suppose the value of the bridge voltage at the time of calibration is v_1, the factor V/Pv_1, provides a value of sensitivity of the sensor with 1 V applied to the bridge. If this factor is multiplied by the bridge voltage applied at the time a measurement is made, a corrected value of sensitivity is obtained which is applicable to the measurement.

In those instances where the sensor and the recording equipment are calibrated separately, the calibration of the amplifier and recorder can be accomplished by applying an accurately known voltage to the amplifier input. If the voltage is V_{in} and the corresponding recorder deflection is d_1, providing the amplifier and recorder are balanced at zero, the recording circuit sensitivity is:

d_1/V_{in}

The amplifier output is dependent on its gain factor at the time of calibration, and if this is used to reduce the factor to a value of sensitivity per unit gain, the expression becomes:

$d_1/V_{in}G_1$

where G_1 is the amplifier gain at the time of calibration.

From the two sensitivity factors it is possible to derive a *composite factor* whose value is:

$$\frac{V}{Pv_1} \cdot \frac{d_1}{V_{in}\, G_1}$$

If the sensor and amplifier are connected and a measurement made that produces a recorder deflection of d_2 the pressure that produces this deflection is:

$$P = \frac{d_2}{\text{composite factor } v_2\, G_2}$$

where

v_2 is the bridge voltage at the time of measurement,
G_2 is the amplifier gain at the time of measurement.

The values of V/P and d_1/V_{in} are not normally obtained from single measurements, but are calculated from values obtained over the operating range.

Consider now the case where the entire channel may be calibrated as an entity. It is possible in these circumstances to use a different technique to provide simpler calibration factors. Most amplifying equipment intended for instrumentation measurement is provided with internal calibration resistors that can be switched across one arm of the measurement bridge. If sensors are four-arm bridges, the resistor is switched across one of the sensor arms. If the sensor is of the two-arm variety, the bridge is completed by two internal arms and the calibration resistor is switched across one of these. The value of the calibration resistor is high compared with the resistance of the arm. A ratio of about 1000:1 is typical. Its actual value, however, is only significant if the measurement is to be made on a different channel from which the sensor is calibrated.

At the time of the calibration, the recorder, amplifier, and bridge are adjusted to zero balance. The calibration resistance is switched across one arm of the bridge and the resultant off-balance of the recorder noted. At the time of measurement this internal calibration procedure is repeated. Differences in the recorder response to the change caused by the calibration resistor provide a measure of the changes in sensitivity of the equipment and the bridge excitation voltage.

Suppose a linear sensor produces a deflection of d_1 when a pressure P is applied. The sensitivity of the system at the time of calibration is d_1/P. If the amplifier gain is G_1, then the sensitivity per unit gain is:

d_1/PG_1

Suppose the recorder deflection due to the calibration resistor is d_2. Then a calibration factor can be obtained

$$\frac{d_2}{\text{sensitivity per unit gain}}$$

which is sometimes described as the *equivalent calibration.*

At the time a measurement is made, this equivalent calibration may be used to calculate the value of a pressure that gives a recorder deflection as follows:

$$\text{Pressure} = \frac{\text{equivalent calibration}}{d_3} \times \frac{d_4}{G_2}$$

where

d_3 is the recorder deflection due to the calibration resistor at the time of measurement,
G_2 is the amplifier gain at the time of measurement,
d_4 is the recorder deflection due to the measurement.

This method of using a calibration resistance provides a means of correcting changes in bridge voltage, amplifier sensitivity, and recorder sensitivity that may take place between the calibration and the time of making the measurement. It does not take into account any changes in the sensor itself. Changes occurring during the actual measurement may not be corrected but a second internal calibration performed after the measurement gives some indication if a change has occurred and whether it is acceptable.

Instrumentation amplifiers often have switched gain controls. When applying internal calibration resistors the amplifier gain control should be set at the time of calibration and at the time of measurement to the same value. This allows an easy direct comparison to be made without introducing calculations to account for gain factors.

Measuring Sensitivity

The accuracy of a calibration depends on the accuracy of the calibration equipment and the stability of the environment in which the calibrations are being made. For the measurement accuracies under discussion this does not usually demand more than reasonable precautions in maintaining a steady room temperature. When higher accuracy is required (better than about 1% FS), precautions become increasingly stringent.

If a pressure sensor is to be calibrated using oil as the pressurizing fluid, before starting the calibration it is necessary to purge the system to ensure that all air is removed from the pressurization system. The sensor is then cycled several times over its calibration range, a process known as *exercising* the sensor. This effectively settles the equipment and the sensor so that no irregular readings are experienced around the zero point. During this operation the measurement channel gain is adjusted so that convenient recorder deflections are obtained without overloading the amplifier and the system is checked for leaks.

Following these preliminary procedures, pressure is applied from the calibration datum point to the maximum in a suitable number of incremental steps. For convenience these increments are usually made equal, but it is often considered desirable to increase the number of measurements around the operating point of the sensor. At the beginning of the depressurization procedure the first point is sometimes made half of the normal increment so that the remaining decremental points fall halfway between the incremental ones.

During the calibration pressure is normally incremented and decremented in a continuous operation. If a point is overshot, no attempt is made to return to that point; either the settled value is recorded or the change is continued to the next increment. This procedure makes possible a measurement of the hysteresis of the sensor which would be obscured if local looping were allowed at points on the curve. To complete the calibration it is necessary to obtain a measure of the sensor's ability to reproduce an output with repeated applications of input. This is known as the repeatability of the sensor and its value is obtained by repeating the measure cycle several times.

Before each calibration cycle an internal calibration procedure should be performed. If the channel gain changes significantly, or if the zero base drifts, consideration must be given as to the validity of that run. If conditions vary by an unacceptable amount, it follows that either the equipment has not been stabilized before attempting the calibration, or that the accuracy being demanded is too high for the quality of the equipment being used.

Calibration Factors

The results of calibrations for measurements not involving a computer are frequently presented graphically, and estimations of the best straight line through the points are made by eye. Where a computer is part of the installation, mathematical techniques may be used to determine the calibration factor. One method is illustrated later in the chapter. Most sensors are linear within the limits of accuracy of the measurement, displaying only small amounts of hysteresis. Any gross deviation from the mean line may indicate a faulty sensor, equipment malfunction, or signal overload. Smaller deviations include nonlinearity, hysteresis, or equipment adjustment errors, and provide an indication of the accuracy with which a measurement may be made. This is done by drawing two further lines parallel to the calibration line to include all the scattered points. The distance between these lines is quoted as a percentage of the full scale output and provides a measure of the certainty with which a measurement can be made. The value is usually described as *the accuracy of a sensor quoted as a percentage of full-scale output.* This is illustrated in Figure 15.6.

TEMPERATURE EFFECTS ON SENSITIVITY AND ZERO

Pressure sensors can change both their zero datum and, to a lesser extent, their sensitivity with changes of temperature. Manufacturers often adjust their sen-

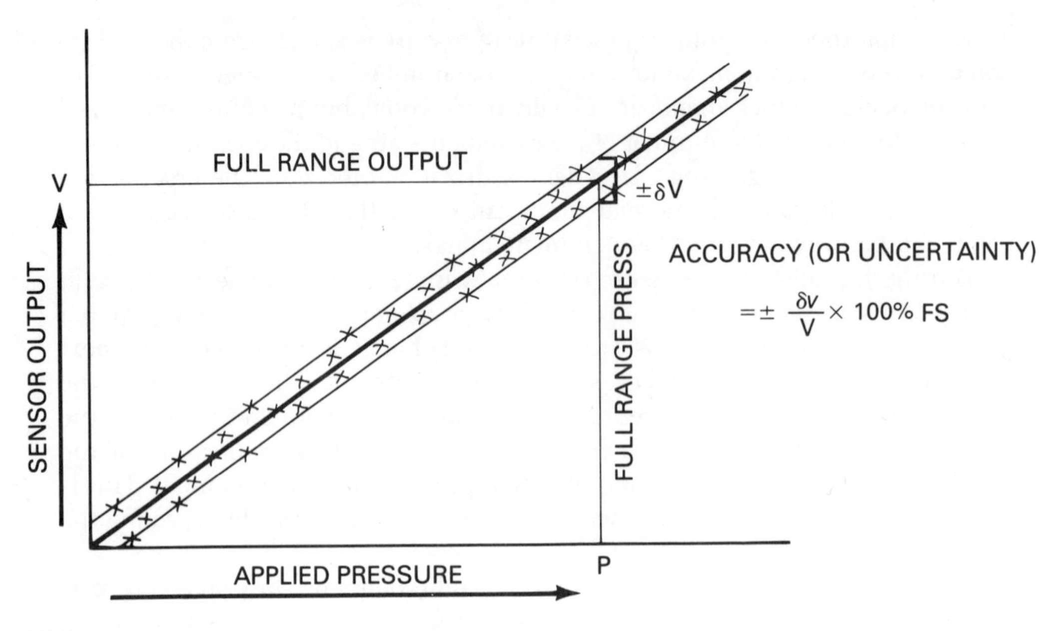

Figure 15.6
Typical Calibration Curve for
Pressure Sensor

sors by introducing internal balancing components to minimize these changes and usually quote the limits in performance data supplied with each unit. When a sensor has been in service for some time, especially if it has been operating at the limits of its specification, it may be prudent to recalibrate it over its temperature range. If there is reason to believe the specification has been exceeded, a recalibration is certainly required. One way of doing this is to change the temperature over the range incrementally and measure the output at the datum and maximum pressure for each increment of temperature. This will provide data to calculate the change in sensitivity with temperature, as well as the amount of zero drift. The major time element in such a calibration involves stabilizing the calibration chamber temperature. In comparison, the time for making the pressure measurements is usually small. Consequently, intermediate pressure points may be recorded at the different temperature increments to obtain a comprehensive calibration of the sensor's performance without incurring an unreasonable time penalty.

Because manufacturers' calibrations are normally carried out at a steady temperature, there is often no indication of what may happen to a sensor output if it suffers a sudden change. It is possible, in some instances, that fluid in contact with the diaphragm can suddenly change its temperature while the body of the sensor is held constant. Alternatively the body can change while the diaphragm remains at constant temperature. Both conditions are difficult to simulate, partly because of the difficulty in defining a "sudden change" which is realistic, and partly because of the difficulty in reproducing that change.

Some precautions can be taken to prevent the worst effects of violent temperature changes. It is possible to isolate the diaphragm of most pressure sensors by mounting the sensor at the end of a tube which connects the diaphragm to the pressure source. Cases may also be protected by lagging. The isolation of other types of sensor, however, may be much more difficult. When it is not possible to protect a sensor all that can be accomplished is a best estimate of the errors that may occur.

In principle, the simplest way of producing a sudden change on the case of a sensor is to protect the diaphragm from coming into contact with a fluid and then plunge it into an oil bath. A galvanometer recorder is generally adequate for the measurement of the response that will show a curve similar to the one in Figure 15.7.

While dipping a sensor into an oil bath may be the easiest way of producing a step temperature change there are practical difficulties that limit its value. Sensor cables and attachments make the time required to immerse the unit something greater than is desirable to reproduce a step function. Also at high temperature the oil bath is likely to make such an operation hazardous.

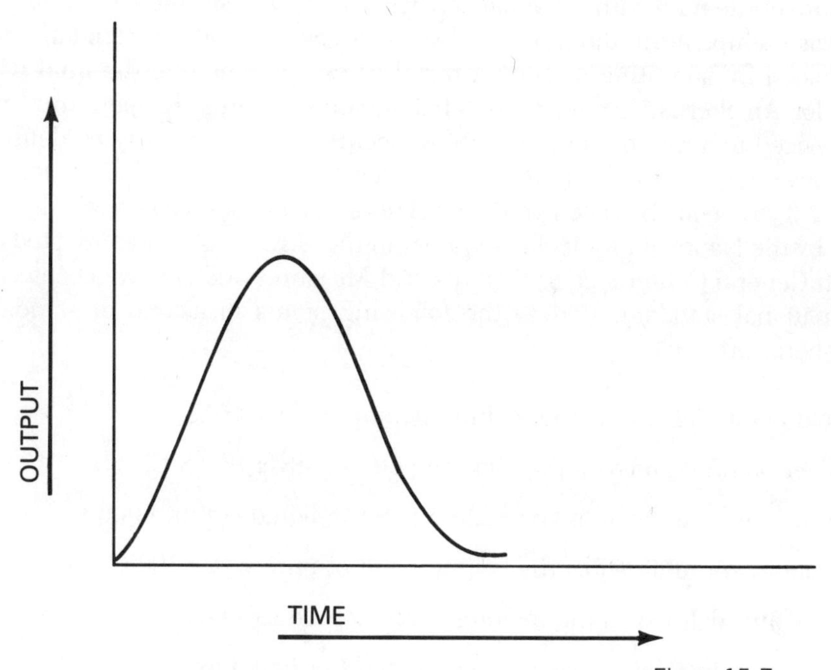

Figure 15.7
Sensor Output Caused by a Step Change in Temperature

CALIBRATION OF OTHER SENSORS

The procedures for calibrating other sensors of the bridge type are similar to those described. Load and displacement measurements are easily understood, and standards are well known and readily obtainable. Two types of calibration that warrant special consideration, however, are temperature and vibration sensors. Both are important, and both present their own problems.

TEMPERATURE CALIBRATIONS

It is unusual for on-site temperature measurements to be required to the same precision as those in a laboratory. With care, on-site measurements to within 2 or 3°C of a value are achievable, and for the most part, are considered adequate.

By far the most common temperature sensor is the thermocouple. Provided the thermocouples used are produced in accordance with national standard specifications, it is usually necessary to check only one point during calibration. The remaining points may be assumed to conform with the tables developed for the particular thermocouple. In the United States these standards are authorized by the American National Standards Institute (ANSI).

To obtain an accurate single-point calibration measurement one of two methods may be used. The first is to immerse the thermocouple in a fluid maintained at a temperature around its operating temperature and to measure the temperature of the fluid with a standard thermocouple or resistance thermometer. In this case temperature difference between the couple and the standard may be eliminated by mounting both on a metal block and agitating the fluid with an impellor. An alternative method uses fixed-point standards. Temperatures may be reproduced to a much higher degree of accuracy than they may be defined. To take advantage of this, a temperature scale based on a number of accurately known fixed points has been produced. This scale, based on one originally formulated by the National Physical Laboratory in the U.K. in 1927, was adopted by the Ninth General Conference of Weights and Measures and is now accepted as an international standard. It uses the following points measured at standard atmospheric pressure.

1. Gold point: defined as the melting point of gold 1063°C.
2. Silver point: defined as the melting point of silver 960.8°C.
3. Sulfur point: defined as the boiling point of liquid sulfur 444.6°C.
4. Steam point: defined as the boiling point of pure water 100°C.*
5. Ice point: defined as the melting point of pure ice 0°C.*
6. Oxygen point: defined as the boiling point of liquid oxygen −182.97°C.

*Fundamental fixed points.

In addition to the international scale of fixed points, a number of secondary points are also internationally accepted. These include:

Freezing point of copper in a reducing atmosphere	1,083°C
Freezing point of zinc	419.4°C
Freezing point of lead	327.3°C
Freezing point of tin	231.8°C
Freezing point of mercury	−38.87°C
Temperature at which solid CO_2 changes into the gaseous state at normal atmospheric pressure	−78.50°C

This is not a comprehensive list but materials such as tungsten, palladium, and antimony are not readily available to the average instrumentation engineer.

If the freezing point of a substance is used as a fixed point, it is a feature of these substances that their cooling is *arrested* at their freezing point until the latent heat is dissipated. This *arrest point* is reproducible to a high degree of accuracy and is easily recognizable. Suppose a thermocouple is used to measure the temperature of a pure metal that has been heated to a temperature above its melting point. If the metal is allowed to cool under stable conditions, the initial rate of cooling will be proportional to time. Suppose the time taken for the metal to cool 1°C is noted and this time is considered to be one measurement interval. The time taken to fall the next 1°C will also be one measurement interval. This will continue until the metal approaches its freezing point when it will take something longer than one interval to cool 1°C, reaching a maximum at the arrest point. These measurements allow a graph in the form of Figure 15.8 to be plotted from which can be obtained an accurate measurement of thermocouple output at the freezing point of the metal.

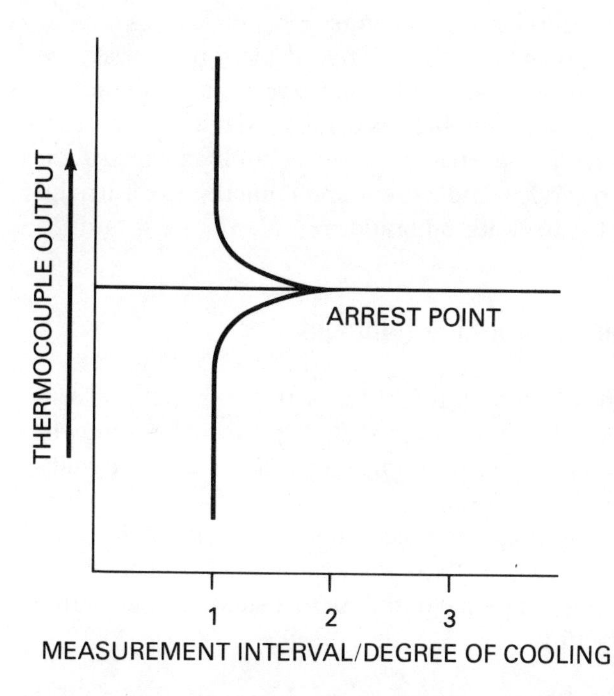

Figure 15.8
Arrest Point Method of Determining the Freezing Point of a Substance

Temperature Standards

The international scale not only defines the values of the fixed points but also how the values between the fixed points should be interpolated.

Between 0°C and 630.5°C a standard platinum resistance thermometer is to be used as the measuring device and the temperature calculated from the expression:

$$R_t = R_o(1 + \alpha t + \beta t^2)$$

where R_t is the resistance at a temperature between the defined points, R_o is the resistance at 0°C, and α and β are obtained by measuring the resistance of the thermometer at the ice point, the steam point and the sulfur point.

Between −190°C and 0°C a standard platinum resistance thermometer is to be used and the temperature calculated from the expression:

$$R_t = R_o(1 + \alpha t + \beta t^2 + C(t - 100)t^3)$$

The constant C is determined by measuring the resistance of the thermometer at the oxygen point.

Above 630°C and below the gold point, the measuring device is a standard platinum/platinum rhodium thermocouple, one junction being maintained at 0° while the other measures the temperature. The output of this thermocouple is given by the expression:

$$e = U + Vt + Wt^2$$

where e is the output of the thermocouple.

U, V and W are constants obtained from measurements made at the gold and silver points. The third point chosen is antimony, which has a freezing point of 630.5°C. Antimony is one of the secondary fixed points.

It is obvious that producing and maintaining standard platinum resistance thermometers and thermocouples is not practical in the average instrumentation department. Fortunately, it is possible to purchase secondary standards at a reasonable cost that are adequate for most purposes. Manufacturers of these standards usually provide a calibration service, and it is common practice to return the secondary standard periodically to check its calibration.

Calibration of Thermocouple Measuring Equipment

Assuming that a thermocouple has been manufactured with wires that conform to the standard specifications, the calibration of associated measuring equipment is most easily achieved indirectly using thermocouple tables and a standard millivolt source. A number of firms provide accurate millivolt sources for this purpose, most based on the principle of the standardized potentiometer. Some of the portable versions reproduce a voltage to about 0.1% of the stated range and are small enough to be carried in a tool bag. A simplified circuit of one such voltage source is illustrated in Figures 15.9 and 15.10.

To operate this equipment the amplifier and the supply are switched on and the equipment left for a few minutes to stabilize. The switch S1B is turned to the position illustrated in Figure 15.9 and the standardization resistor adjusted until the meter indicates a null. At this point the voltages developed across the ballast resistor by the operational battery and the standard cell are equal. In this condi-

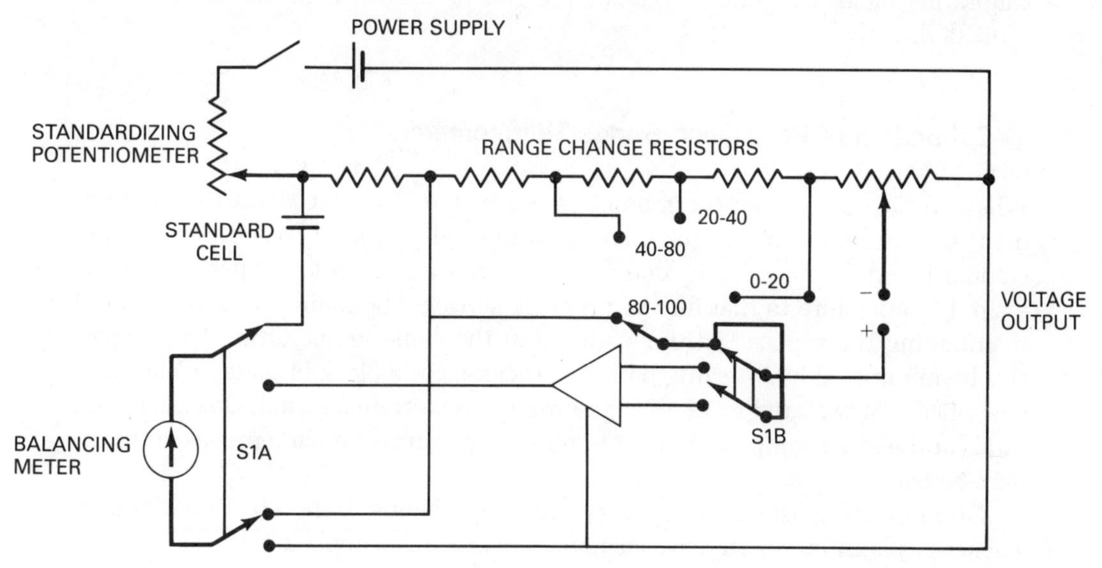

Figure 15.9
Standard Millivolt Source
(Voltage Source)

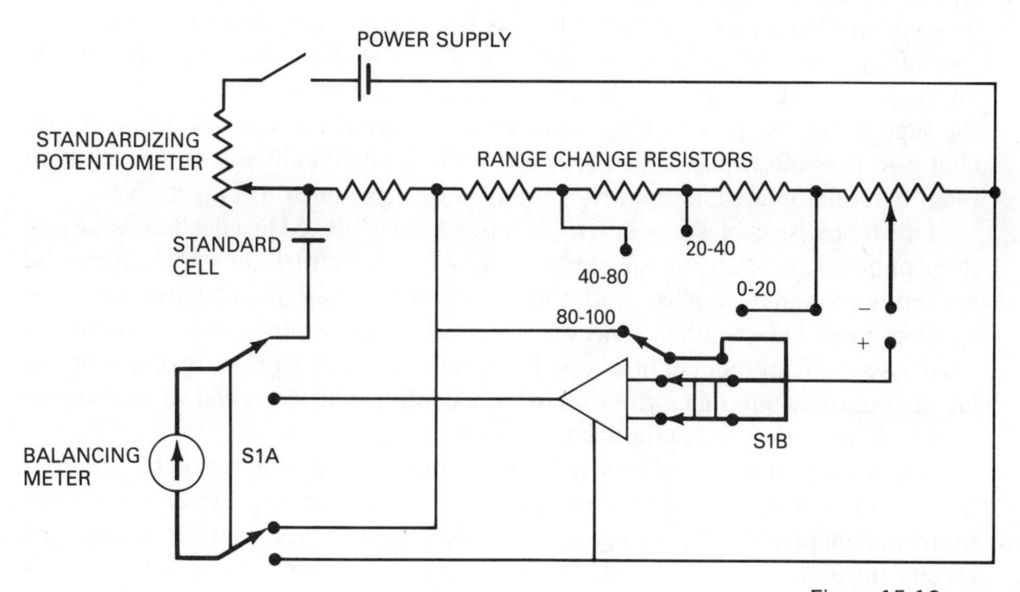

Figure 15.10
Standard Millivolt Source
Thermocouple I/P

tion the source chain is designed to provide a voltage source with scale output across the instrument terminals. With the switch S1B in its alternative position, shown in Figure 15.10, an unknown input voltage applied to the instrument terminals is balanced against the potentiometer until the meter indicates null. The value of this measured voltage is read directly from the scale.

Potentiometers are extremely useful instruments. They may be used either for calibrating measuring equipment or they may be used to measure the output of sensors directly.

Calibration of Resistance Bridge Thermometers

Where resistance bridge thermometers are in direct contact with the environment they are measuring, they may be immersed in a fluid with a temperature standard, and their sensitivity determined by raising the fluid temperature using a similar procedure to that for the pressure sensor. The main problem is that of maintaining the sensor and the standard at the same temperature. Differences can be minimized by inserting both into recesses in a block of metal, sometimes described as a *sink* and by gently agitating the heat transfer fluid. Oils are useful fluids at the lower temperatures; at higher temperatures, molten metals and salts may be used.

One type of sensor that presents special problems is the *stick-on* element. These are platinum resistance thermometers made and applied in the same manner as strain gages. Sometimes out of necessity they must be adhered to the outside of a container to measure the temperature inside. Pipes that cannot be removed from the line, or that cannot be drilled to accept an insertion sensor are typical of such a situation. Usually pipes where temperatures are to be measured by stick-on elements are lagged, but this lagging is a compromise between the cost of lagging and the cost of heat loss. There is therefore a case for locally improving the lagging over the sensor to reduce the temperature gradients across the pipe walls and protect the sensor from the effects of external temperature changes. It is often possible to reproduce these site conditions in a calibration room by building a calibration rig similar to that sketched in Figure 15.11.

This rig consists of a header tank where a fluid is heated to a desired value and then pumped through the specimen pipe. At a constant fluid temperature the system is allowed to stabilize until the sensors measuring the inlet and the outlet temperatures have settled. The difference in temperature (which should be small) is a measurement of heat loss through the lagging. Enclosure of the pipe in an environment chamber through which conditioned air is blown allows the effects of ambient site conditions to be estimated.

When investigations of this nature are undertaken it is usual to adhere a number of stick-on elements to the pipe so that the variation in performance of individual elements within a batch can be investigated. Even so, when the final on-site measurements are made there is no absolute certainty that the results obtained will fall within the limits obtained from the test rig. They will, however, represent the best measurements that can be made under the circumstances.

The use of test rigs to simulate site conditions and measure the response of a

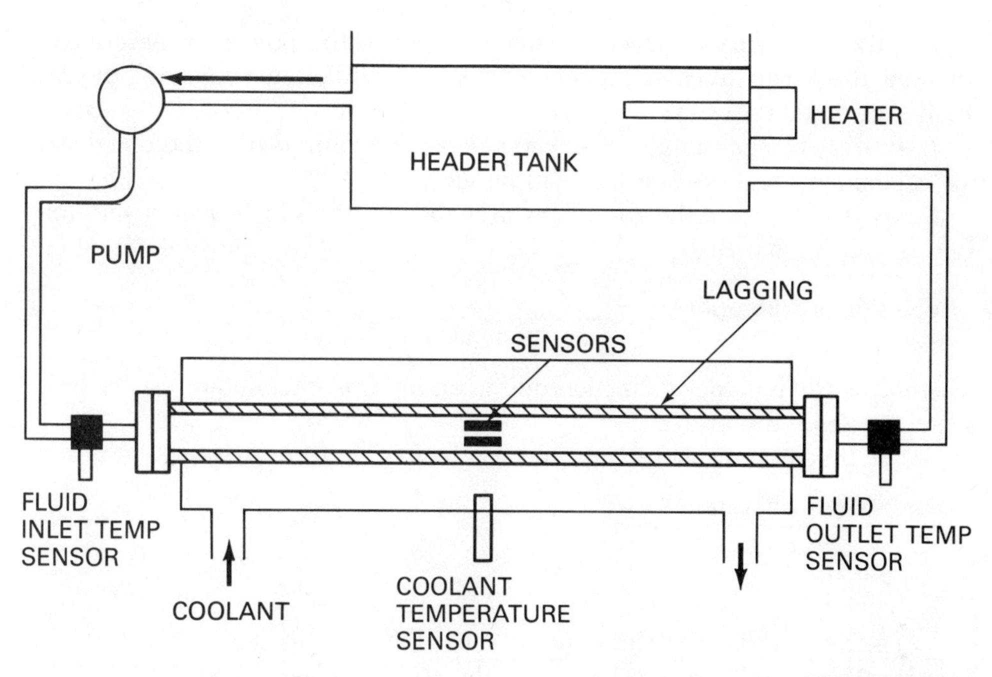

Figure 15.11
Calibration of "Stick On" Temperature Sensors

sensor when subjected to those conditions is a common instrumentation practice. Strain gages, for example, are frequently tested on a batch basis to check their gage factor and temperature response factors.

CALIBRATION OF VIBRATION SENSORS

The most common vibration sensor with which the instrumentation engineer will be involved is the accelerometer.

Where spring-mass accelerometers are used their sensitivity can be measured by a calibration recording their response to the effects of gravity. An accelerometer to be calibrated is laid on a flat surface with its movement axis at right angles to the surface. The measuring equipment is balanced and a record made of its zero position. The accelerometer is then turned onto its end so that the force of gravity acting along its axis causes the mass to deflect against the spring. A measurement is made of the output of the accelerometer which is then rotated 180° so that the mass deflects in the opposite direction. A calibration of this type is sometimes described as a "2 *g*" calibration because the total output of the sensor will be equivalent to one measuring an acceleration of two times the gravity constant (i.e., 64.4 ft/sec^2). Vibration engineers usually measure vibration in units of "*g*" obtained by dividing the value of acceleration in ft/sec^2 by the gravity constant. The 2 *g* calibration, therefore, provides a convenient method of determining the sensitivity of accelerometers and when the procedure described is followed, the gravity force acting on the mass is equivalent to subjecting the case to an acceleration of $\pm 1g$.

If the trace of a galvanometer recorder deflects $\pm d_1$ mms from the zero during the calibration, then the sensitivity in mm/g is $\pm d_1$

Suppose the gain of the amplifying system is G_1, then the sensitivity at unity gain is $\pm d_1/G_1$, and the

$$\text{equivalent calibration} = \pm \frac{d_2}{\text{sensitivity at unity gain}}$$

where d_2 is the recorder deflection due to an internal calibration resistor being switched across the bridge.

At the time of a measurement, the acceleration in terms of g will be:

$$\frac{\text{equivalent calibration}}{d_3} \times \frac{d_4}{G_2}$$

where

d_3 is the recorder deflection due to the calibration resistor
d_4 is the recorded deflection due to the accelerometer output
G_2 is the amplifier gain at the time of the test

In principle, the accelerometer output over its designed frequency range should be proportional to acceleration and the static 2 g calibration should be the only one necessary. Unfortunately, accelerometers do not always function according to principle; the effects of a loss of damping fluid, for example, or a damaged spring may not be apparent during a static calibration. If malfunction is suspected, it is necessary to carry out a calibration over the sensor's frequency range to prove its integrity.

ACCELEROMETER CALIBRATION RIGS

It has been demonstrated in Chapter 13 that the output of an accelerometer is proportional to the square of the signal frequency. This implies that any calibration rig must have good frequency stability and be provided with means of measuring frequency accurately. In addition it is desirable to calibrate the accelerometer over its operating range, and it also follows that the rig must be provided with easily adjustable amplitude settings.

A number of types of calibration rigs are used, each one suitable for a particular application.

Centrifuges

Where a sensor is to be used to measure static g forces, the centrifuge allows the extension of the ± 1 g calibration over the working range. The centrifuge is a rotating table on which accelerometers are mounted with their operating axis at right angles to the periphery. Signals and bridge excitation power supplies are connected through slip-rings. The table may be driven by a variable-speed electric motor or alternatively through a variable gear box, preferably one of the "gearless" varieties which allow stepless ratio adjustment without the necessity to stop the rig during speed changes. A typical centrifuge is sketched in Figure 15.12.

Consider a centrifuge where R is the radius of the center of the accelerometer from the center of the table, and ω is the angular velocity of the table. The acceleration along the axis of the accelerometer is

$$\omega^2 R = 4\pi^2 f^2 R$$

where f is the table speed in revolutions per second.

Expressing this in terms of g:

$$\text{acceleration} = \frac{\omega^2 R}{g}$$

where g is the gravity constant.

It is obvious from this expression that due to the frequency squared term small errors in the measurement of frequency introduce substantial errors when calculating acceleration. Fortunately the accuracy of modern digital frequency meters is such that the problems of frequency measurement are no longer acute. When these meters are used in conjunction with a frequency multiplier, errors are virtually eliminated. The basic input to the frequency meter may be an optical sensor or a coil and magnet sensor activated by the rotation of the table. A slot cut in the edge of the table, for example, can be used to allow light to fall on a photocell to produce a once-per-revolution pulse. The number of pulses per revolution may be multiplied by spacing further slots round the periphery of the table.

An alternative method of frequency measurement is one that may be used when an accurate variable-frequency oscillator is available. In this instance an ac tachometer is used to provide the frequency signal that is compared against the oscillator. The tachometer output is connected to the x plates of an oscilloscope, and the oscillator to the y plates. When the two signals are equal in frequency and amplitude a circular trace is produced on the screen described as a *Lissajous figure.* This trace is very sensitive to differences in frequency. Any small discrepancies will cause it to vary from a circle through an ellipse to a straight line and then to a circle again. More complex Lissajous figures are produced when the signals on one plate of the oscilloscope are multiples of those on the other. However, only when the two are equal will a circle be produced.

Vibration Tables

There are basically two types of vibration tables. Those using electromagnetic excitation may be employed at high frequencies, but have a restricted amplitude. This limits the accelerations they can produce at low frequencies. Mechanical vibrators are capable of producing large amplitudes, but their frequency range is

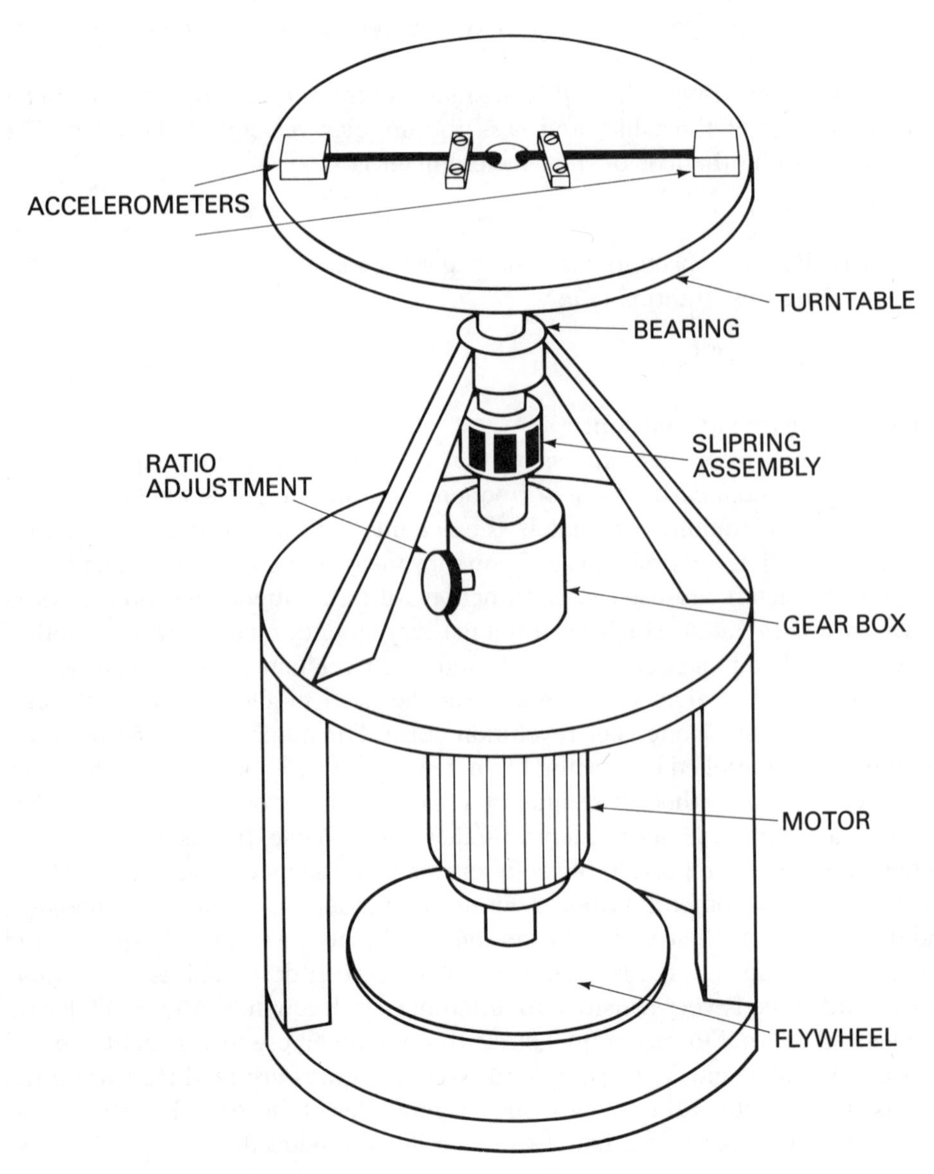

Figure 15.12
Accelerometer Calibration Centrifuge

restricted. Consequently, when calibrating some accelerometers it may be necessary to change from one type of table to another. The principle of a mechanical table is illustrated in Figure 15.13

A variable-speed motor in this design drives an eccentric which in turn drives a spring-loaded follower. In practice, the design of the oscillating system is more complex than illustrated. Plain bearing surfaces, for example, wear quickly and are prone to seizure. It is usual therefore, to use follower guides made with roller bearings.

The problem with this type of table is that of amplitude adjustment. Normally, to obtain an accurate amplitude setting the movement of the table is measured with a distance gage while the motor is being rotated by hand. This makes calibration at a constant acceleration level impractical. Consequently it is necessary to calibrate over a range of frequencies at each amplitude setting and then rationalize the accelerometer output for amplitude and frequency. The response curve shown in Figure 15.14 illustrates this procedure.

To obtain this curve, the exciter table amplitude is set and a series of acceleration measurements are made over the frequency range. Suppose in this instance a galvanometer recorder is used to measure the accelerometer output. Suppose also that at some frequency, the trace deflection of a measured point when corrected to unity gain is d. If this value is divided by the table amplitude a and the

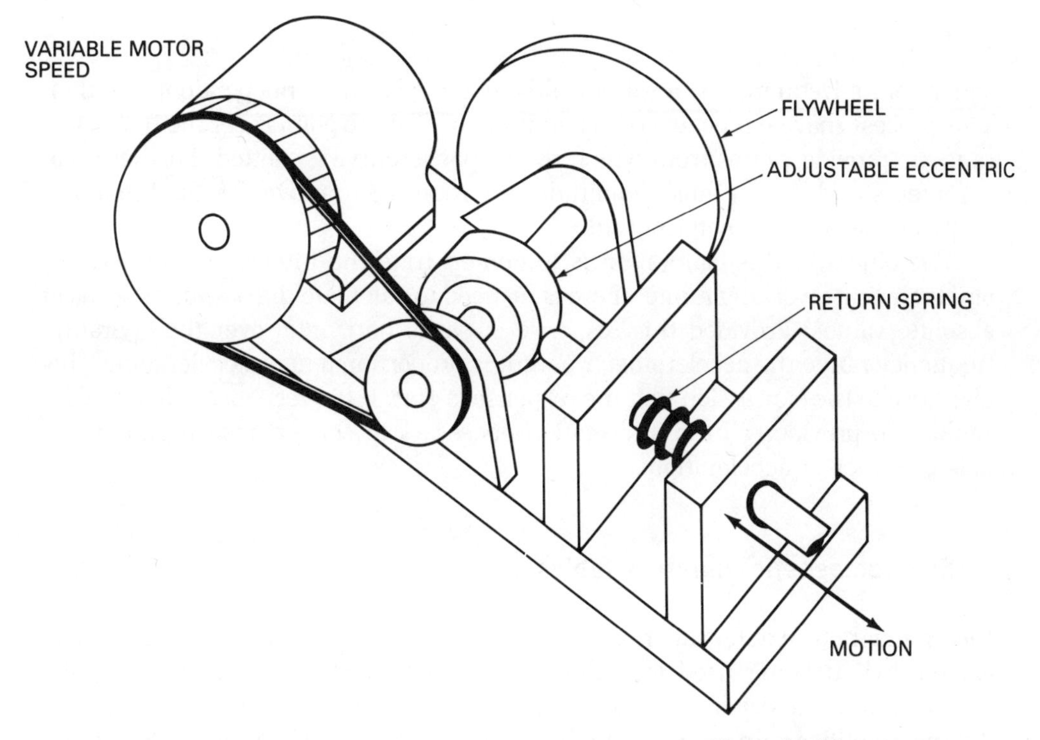

Figure 15.13
Accelerometer Linear Vibration Table

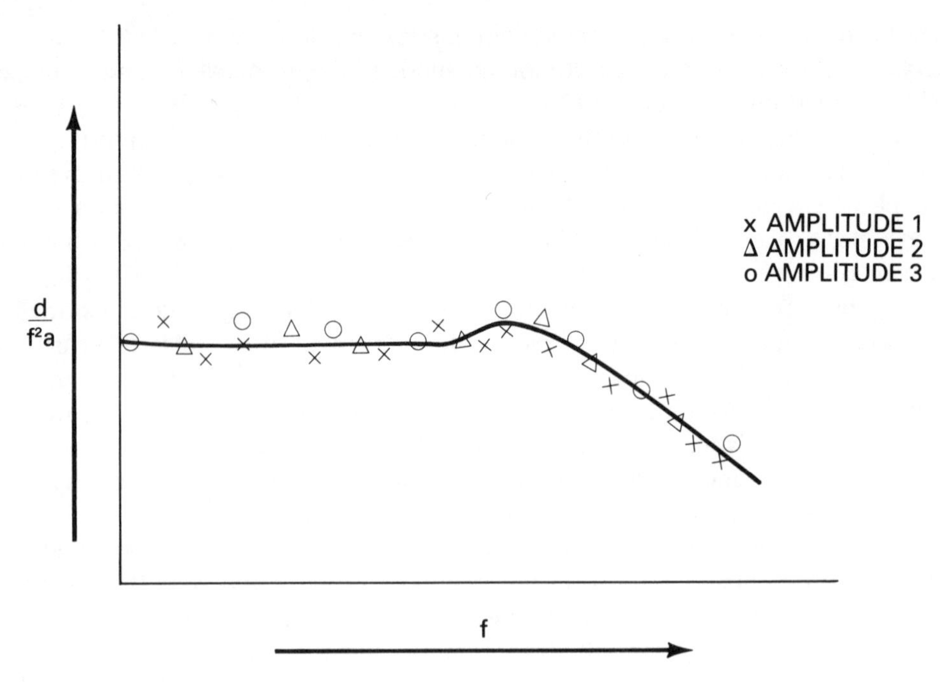

Figure 15.14
Typical Accelerometer Calibration Curve

square of its frequency f, a value will be obtained that is independent of either. This process may be repeated with a number of discrete points over the frequency range to provide values from which the response curve is plotted. Further measurements at different table amplitudes are required to prove the amplitude linearity of the accelerometer output.

The object of this calibration is to demonstrate the integrity of an accelerometer over its working range; there is no need to calculate the output in terms of absolute values. Provided this response curve is horizontal over the operating frequency range the accelerometer output is proportional to its acceleration. This gives rise to the expression that the response is *flat*. If the response is flat, the 2 g calibration provides all the information necessary to convert measurement signals to values of acceleration.

Electromagnetic Vibration Tables

Electromagnetic exciters are much more flexible in their control than mechanical exciters when being used for calibration purposes. Their frequency faithfully follows that of the driving oscillator, and the amplitude is a direct function of the driving amplifier gain, both of which are easily adjusted. The problem lies in the difficulty of measuring the table amplitude accurately.

One method uses a standard accelerometer that allows chosen values of acceleration to be set up at any frequency. Capactive accelerometers with a stated accuracy of about 0.05% *FS*, and piezoelectric devices with an accuracy of 1% *FS* are both used for this purpose.

If standard accelerometers are not available, it is necessary to resort to measurements of amplitude and frequency to calculate acceleration. Differential transformers provide one method of measuring amplitude, but the frequency over which they may operate is limited by their carrier frequency, and strain-gaged devices are often limited by their natural frequency.

Various optical methods are also employed. The simplest uses a stroboscope and a travelling microscope lined up on the edge of the table. A slowly moving image of the table edge produced by the stroboscope when tuned to a frequency slightly different from that of the table provides a measurement of its amplitude. Such a sighting mark is not always easy to see. An improvement on this technique uses a piece of carborundum cloth stuck to the edge of the table as a sighting aid. A light directed at an angle reflects bright spots from the grit. It is usually possible to sight one spot fine enough to be used to measure the spread of light with reasonable accuracy.

Pendulum Calibrators

Seismic sensors and velocity sensors that are required to work at low frequencies can often be calibrated on a long pendulum. The pendulum rod is made of light tube for rigidity and is pivoted on conical bearings at the suspension point. The vibration table is part of the heavy bob at the lower end designed to allow mounting of sensors with their movement axes at right angles to the rod. Sensor wiring is laid along the pendulum rod and taken off at the pivot.

The period of a simple pendulum T is given by:

$$T = 2\pi\sqrt{\frac{l}{g}} \qquad (1)$$

where l is its length and g is the gravity constant.

The frequency f of this pendulum is

$$f = \frac{\sqrt{g}}{2\pi\sqrt{l}} \qquad (2)$$

Provided the angle of swing is small compared with its length, the acceleration of the pendulum bob is simple harmonic and has a value of $\pm\omega^2 x$ where x is the displacement of the sensor in the direction of swing from its rest position.

But

$$\omega = 2\pi f$$

Therefore the acceleration of the bob

$$= \pm(2\pi f)^2 x \qquad (3)$$

Substituting from (2)

$$\text{Acceleration} = \pm \left(\frac{2\pi\sqrt{g}}{2\pi\sqrt{l}}\right)^2 x$$

$$= \pm \frac{gx}{l}$$

The acceleration acting on the sensor expressed in terms of g is

$$\pm \frac{x}{l}.$$

When using a pendulum as a calibrator it is usual to provide a sensor such as a potentiometer mounted at the pivot to measure the angle of swing, and to record its output with that of the accelerometer. Ultraviolet recorders are very convenient for this purpose. A graph of acceleration versus accelerometer amplitude is plotted for the preset frequency and plots at other frequencies are obtained by adjusting the pendulum length.

CALCULATING CALIBRATION FACTORS MATHEMATICALLY

There are occasions when a large number of sensors must be calibrated periodically. In these instances mathematical techniques are useful. This is particularly true when a computer is available to do the calculations, or when the sensor is operating directly into a computer. A number of mathematical methods are available for determining the slope of a line, the most common being the method of *least squares*. This is described below.

Consider a line $y = mx + c$. Suppose points on the line are x_1, y_1, x_2, y_2, x_3, y_3, and x_4, y_4. The slope of the line is obtained in the following way:

1) Average the y coordinates
2) Average the x coordinates
3) Subtract y_1, y_2, y_3 and y_4 in turn from the average y coordinate
4) Subtract x_1, x_2, x_3 and x_4 in turn for the average x coordinate
5) Multiply 3) by 4) for each pair of coordinates
6) Sum the products of 5)
7) Square each difference in 4)
8) Sum the squares of 7)
9) Slope $= \frac{6)}{8)}$
10) c is obtained by substituting the mean value of 1) in the equation. The procedure is summarized in the table below.

Step 1 $\overline{y}$	Step 2 $\overline{x}$	Step 3 $\overline{y} - y_i$	Step 4 $\overline{x} - x_i$	Steps 5 and 6 $\sum (\overline{y} - y_i)(\overline{x} - x_i)$	Steps 7 and 8 $\sum(\overline{x} - x_i)^2$
y_1	x_1	$\overline{y} - y_1$	$\overline{x} - x_1$	$(\overline{y} - y_1)(\overline{x} - x_1)$	$(\overline{x} - x_1)^1$
y_2	x_2	$\overline{y} - y_2$	$\overline{x} - x_2$	$(\overline{y} - y_2)(\overline{x} - x_2)$	$(\overline{x} - x_2)^2$
y_3	x_3	$\overline{y} - y_3$	$\overline{x} - x_3$	$(\overline{y} - y_3)(\overline{x} - x_3)$	$(\overline{x} - x_3)^3$
y_4	x_4	$\overline{y} - y_4$	$\overline{x} - x_4$	$(\overline{y} - y_4)(\overline{x} - x_4)$	$(\overline{x} - x_4)^4$
$y_1 + y_2 + y_3 + y_4$	$x_1 + y_2 + x_3 + x_4$			$\sum_{i=1}^{4} (\overline{y} - y_i)(\overline{x} - x_i)$	$\sum_{i=1}^{4} (\overline{x} - x_i)^2$
$\overline{y} = \frac{1}{4}(y_1 + y_2 + y_3 + y_4)$	$\overline{x} = \frac{1}{4}(x_1 + x_2 + x_3 + x_4)$				
Step 9 $m = \frac{\sum_{i=1}^{4} (\overline{y} - y_i)(\overline{x} - x_i)}{\sum_{i=1}^{4} (\overline{x} - x_i)^2}$			Step 10 $c = \overline{y} - m\overline{x}$		

CHAPTER 16

THE COMPUTER

Computers may be considered to have two basic roles in industrial operations involving instrumentation. They may be used to monitor the operation, or they may be required, in addition, to produce control responses. The use of computers in these roles relieves control engineers of those tasks for which they are not well suited. The computer, for example, can be programmed to monitor large numbers of sensors and recognize trends without oversight or error. It can also calculate quickly complex control responses and generate control signals, unaffected by fatigue or misjudgment. Even in those instances where final control is given to the engineer, the computer can still be used to monitor the response to control actions and provide guidance messages. In addition to its monitoring and control functions, however, the large storage capacity of the computer provides the engineer with access to large amounts of historical data which may be used for fault tracking, maintenance, and many other operational functions.

Where a computer is required to fulfill a monitoring role it may be provided with data-logging facilities mounted in a common rack with the processor. These modules, when provided by the computer manufacturer, are often easier to program than alternatives, but they require that all sensor wiring be terminated at the computer. Where sensors are scattered over a large industrial site, the cost of wiring can be high. In these circumstances, it is often more economical to use several data loggers installed at convenient remote locations and connect them to the central system by inexpensive telephone-type transmission lines.

Where a computer is required to provide control as well as monitoring facilities, it is possible to use it as a central controller, with all sensor and control wires terminated at the central point. Not many years ago this would have been the only economically feasible choice. Now it is often more economic to use a distributed system using a relatively small central system connected to remote computers by communications lines. In such a complex the tasks of monitoring and control are shared between the peripheral computers, while the central system coordinates the operation of the peripherals.

It is a feature of computer application that not many years ago their application was an expensive undertaking, requiring careful economic justification. Now, most instrumentation systems include at least one digital computer, if not more. The application of computers has, in fact, become so commonplace that, in many instances, preprogrammed microprocessors are embedded into traditional equipment without the user being aware of their existence. There also appears to be no limit to new applications. Each week, it seems, produces new innovations which force a reappraisal of established concepts. Indeed, the only certain feature in the changing world of computer application is the certainty that any discussion will be outdated before it is published. In this environment, every student must start somewhere and this chapter is designed to provide an introduction to the basics of computing and the application of computers to simple industrial operations.

BINARY NOTATION AND ARITHMETIC

In common with all digital equipment, digital computers operate with binary representation of numbers, that is, numbers expressed as powers of 2. For example:

$1 = 2^0$
$2 = 2^1$
$3 = 2^1 + 2^0$ etc.

If the least significant binary digit is always kept to the right-hand side, and digits of increasing value progress to the left, then the form may be abbreviated to a series of ones and zeros. One, in a column, indicates that the power of 2 that would normally occupy that column is present. Zero will indicate its absence. For example:

2^5	2^4	2^3	2^2	2^1	2^0
1	1	1	0	0	1

would represent

$$2^5 + 2^4 + 2^3 + 2^0 = 57$$

When using binary notation it is possible to add, subtract, multiply, and divide in binary, although the rules are somewhat different from decimal arithmetic.

Adding rules:

$0 + 0 = 0$
$0 + 1 = 1$
$1 + 1 = 0$ carry 1

Example

$$\begin{array}{r} 1011 \\ +\ 1001 \\ \hline 10100 \end{array}$$

Subtraction rules:

$1 - 0 = 1$
$1 - 1 = 0$
$0 - 1 = 1$ borrow 1

Example

$$\begin{array}{r} 1101 \\ -0110 \\ \hline 0111 \end{array}$$

Multiplication rules:

$0 \times 0 = 0$
$0 \times 1 = 0$
$1 \times 0 = 0$
$1 \times 1 = 1$

Example

$$\begin{array}{r} 101 \\ \times\ 101 \\ \hline 101 \\ 000\ \ \\ 101\ \ \ \ \\ \hline 11001 \end{array}$$

Division:

Subtract and borrow as with decimal division

Example

$$\begin{array}{r} 1011 \\ 101\,\overline{)\,110111} \\ 101\ \ \ \ \\ \hline 00111 \\ 101\ \ \\ \hline 0101 \\ 101 \\ \hline 000 \end{array}$$

A group of eight ones and zeros in the form described is often referred to as a *byte*, and one or more bytes make up a binary word. Computers operate with fixed length binary words; a common length is one containing 16 *bits* (BInary digiTS). This is not universal, however, and different computers may use different length words. It is usual to reserve the most significant bit as a sign indicator. If the most significant bit (left-hand column) is 1, then the number is negative; if the bit is 0, the number is positive.

The use of negative numbers in binary arithmetic presents a problem. Suppose it were necessary to add +1 and −1 in binary, the answer should be 0. Using the notation described above, however,

$$\begin{array}{rl} 0000\ 0000\ 0000\ 0001 = & +1 \\ +\ \ 1000\ 0000\ 0000\ 0001 = & -1 \\ \hline 1000\ 0000\ 0000\ 0010 = & -2 \end{array}$$

If the sign bit is included in this operation the answer is not zero; thus an alternative method representing negative numbers is required. The method is described as *twos complementing* and is that number which, when added to the original number, produces all zeros. For example:

```
   45 =    0000 0000 0010 1101
  -45 =  + 1111 1111 1101 0011
           -------------------
           0000 0000 0000 0000
```

The carry bit generated by the left-hand column is ignored.

The rule for finding the twos complement of a binary number is to invert all the bits in the number and add 1 to the least significant (right most) bit. This process is illustrated below:

```
 + 45   0000 0000 0010 1101
Invert  1111 1111 1101 0010
Add 1   1111 1111 1101 0011 = -45
```

The largest positive number in a 16-bit signed word is 32,767. Any attempt to add to this will cause an error. For example:

```
32,767 = 0111 1111 1111 1111
         0000 0000 0000 0001
Add      1000 0000 0000 0000 = -32,768
```

This is described as an *overflow situation* and in a computer, is detected by a special circuit that switches on a single bit called an *overflow indicator*.

BINARY ADDER

It is not possible to discuss all the circuits required for binary arithmetic, some of which are complex. The binary adder is an exception.

The circuit can be based on two types of logic gate; the *NAND gate* and the *exclusive OR gate*. The NAND gate has been discussed in an earlier chapter and the operating conditions for a two-input gate are summarized as follows:

One or no input present	Output voltage present	(1 state)
Two inputs present	No output voltage present	(0 state)

The exclusive OR gate (Fig. 16.1a) has not been previously discussed. Its operating conditions are:

With a signal on one and only one input	Output is in a 1 state
With no input signal or with a signal on two inputs	Output is in an 0 state

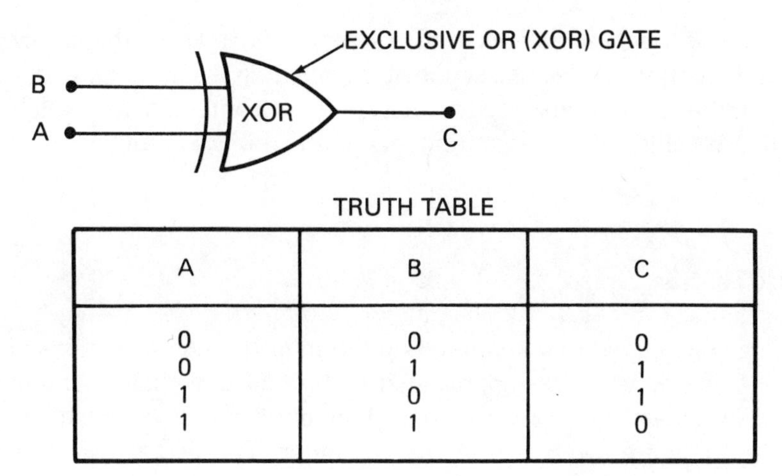

TRUTH TABLE

A	B	C
0 0 1 1	0 1 0 1	0 1 1 0

Figure 16.1a
Truth Table for Exclusive OR (XOR) Gate

The truth table is shown in the figure

Two exclusive OR gates and three NAND gates make up the adding circuit shown in Figure 16.1b

Assume that, initially, the input and output states are all as shown in the figure. Suppose now input *A* is switched on. The output of XOR1 becomes 1 and the output of XOR2 becomes 1. There is no change in the state of NAND1 and NAND2. If now an input is applied to *A* and *B* the output of XOR1 becomes 0 and the output of XOR2 becomes 0. The output of NAND2, however, is changed by

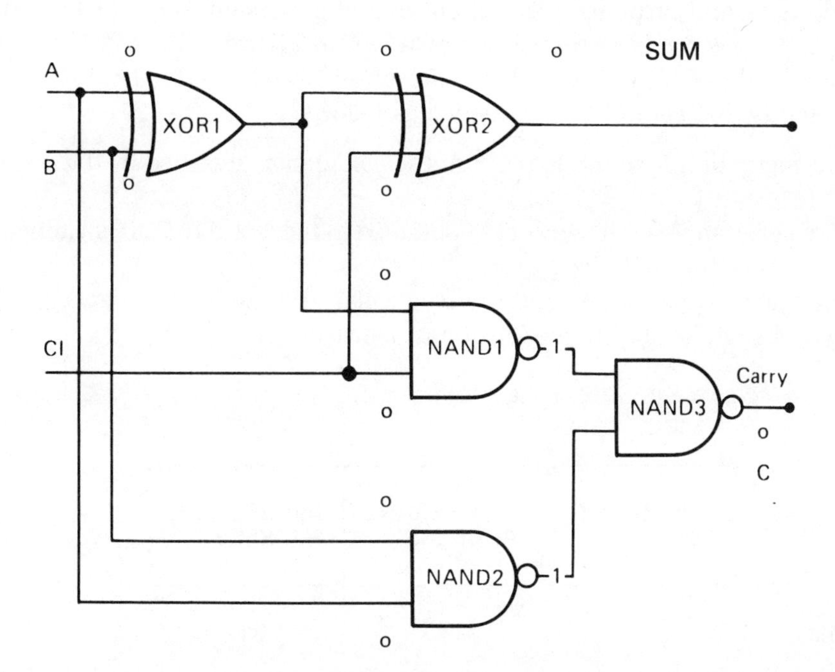

Figure 16.1b
The Binary Adder

signals A and B to its input. This in turn switches NAND3 so that a carry signal appears at its output, and satisfies one of the conditions for binary adding. If the operation of the circuit is analyzed in all its possible combinations it will be found, in fact, that it fulfills all the requirements of an adding circuit.

INSTRUCTIONS

It is widely appreciated that computers perform arithmetic operations on binary numbers. What is often less appreciated is the fact that their operation is also controlled by binary coded instructions. This implies that two distinct types of binary information are retained in the computer. The first is called *data*. The second type, *instructions*, cause the computer to perform defined functions.

To understand the operation of an instruction it is necessary to understand something of the operation of the computer and, in the simplest case, three basic units may be considered.

Computer Storage: The computer storage holds both data and instructions. Groups of circuits capable of holding one byte or a binary word each has its own unique access code, or *address*.

Arithmetic Unit: Is capable of performing a specified number of calculations, each of which may be selected by the application of a particular binary instruction code to its selector circuits. The binary adder is one example.

Arithmetic Unit Control: Effectively causes a controlled series of switching actions depending on the instruction applied to its input code.

The instruction itself may be divided into several sections called fields. The particular fields and their arrangement in the instruction vary between both computer designs, and particular instructions in a single design. One type known as the *two-address format* instruction is illustrated in Figure 16.2.

A number of bits comprise each field, representing a code for that field. These are interpreted by the control in a sequence as follows:

a. The arithmetic unit is set up to perform the calculation specified by the code in field 1.
b. Two words of storage, with addresses contained in fields 2 and 3, are switched to the inputs of the arithmetic unit.
c. The calculation is performed and the result placed in an output register.
d. The control switches to the next sequential instruction.

OPERATION CODE	1st OPERAND ADDRESS	2nd OPERAND ADDRESS
1	2	3

Figure 16.2
Two-Address Instruction

e. The cycle is repeated using the next instruction.

Where a number of instructions are grouped together to perform a specific function they are known as a *routine.* A set of routines and instructions that allow a computer to perform its function is called the *computer program.* Once started, the sequence of operation is automatic and will continue until the last instruction has been operated, or *executed.*

COMPUTER FUNCTIONAL UNITS

It is convenient to summarize the basic units of a computer in the form of a block diagram. Figure 16.3 provides one interpretation.

Arithmetic-Logic Unit and Control

In the previous section reference was made to the arithmetic unit. In fact this unit performs a number of logical operations, and is more often described as the

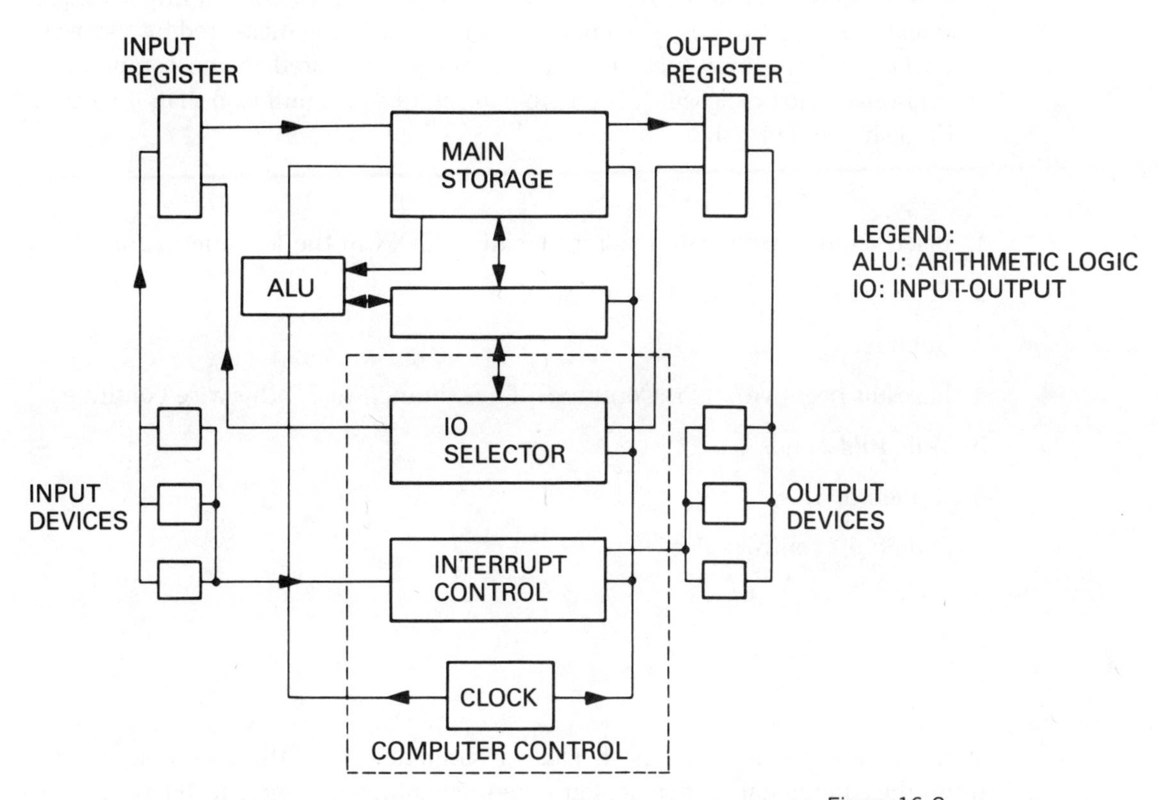

Figure 16.3
Computer Block Diagram

arithmetic-logic unit (ALU). The logical functions performed by the ALU usually include COMPARE AND, OR and EXclusive OR. The operation of the ALU is directed by the ALU control, which, among its functions provides one that causes the computer to branch from one part of a program to another.

In the instance of a two-address instruction, the next instruction to be executed will be the next of a sequence. A record of the *next instruction address* is maintained in a register described as the *instruction address register* and this is incremented by one each time an instruction is executed. If the address record is modified by the program, the computer will continue its operation from a new point in the sequence. This is called *branching* and is initiated by a special instruction, predictably described as a *branch instruction.*

There are two types of branching operations. A branch instruction appearing at the end of a programming sequence will cause the computer to branch to another part of the program. This is called an *unconditional branch.* A second type of branch is caused by a test on the result of an arithmetic operation when a branch occurs if the test condition is satisfied. For example, if the result of an arithmetic operation is negative, the branch instruction will cause the computer to branch to a new specified point in the program. If the result is zero or positive the computer is instructed to continue its programming sequence. Such an operation is known as a *conditional branch* and provides the computer with the ability to make decisions.

The technique of branching may best be illustrated by considering a simple practical example. Suppose the pressure in a tank is being measured by a sensor, and when the pressure exceeds a certain set point, stored in the computer, a control valve is to be closed. It is possible to set up a program loop that, if written in English, would read:

1. Place setpoint value from storage location XXXX in the arithmetic unit.
2. Place sensor output in the arithmetic unit.
3. Subtract.
4. Is result negative? (No: Continue) If yes, branch to 7, otherwise continue.
5. Wait 10 seconds.
6. Branch to 1.
7. Switch off control valve.
8. . . . . . .
9. . . . . . .

While the above description is obviously a simplification of the operation, it illustrates the conditional branch at step 4, an unconditional branch at step 6, and the operation of a program loop between steps 6 and 1.

The Computer Control

The computer control is a set of units that sychronize the operation of the computer. Although the names of units differ in different computers, the control circuits usually include the following:

- *Computer Clock:* The computer consists of a complex series of electronic switching circuits that have two stable states. Switching sequences occur at high speed many times during the execution of an instruction and they require a periodic *forcing signal* to synchronize their operation. The part of the control that produces these forcing signals is known as the *computer clock.* A pulse generator that typically runs at a frequency of 5 MHz provides the basic clocking signal. The clock itself usually operates at a subdivision of this frequency and the time between successive clock pulses is known as the *machine cycle time.* Depending on the computer, a number of machine cycles are required to produce a *storage cycle.* This represents the number of machine cycles required to retrieve data from storage.
- *Signal Routing Circuits:* These are switching circuits that route signals and control data flow through the computer. They are controlled from a large number of sources, one of them being the ALU control.
- *Input/Output (I/O) Selector:* If the full operating speed of the computer is to be utilized, the computing function must not be forced to wait for slow input and output devices. An output device may take as long as 100 ms to print a character and during that time the computer may execute 100,000 instructions. To overcome delay problems that would be imposed by slow I/O devices, *buffering* techniques are used that allow independent operation of device circuits.

  It is usual to connect a number of devices to a common *I/O bus* consisting of a set of data input/output lines with associated address and control lines. Each device is provided with an individual address code and is selected by a unit called the *I/O selector.* A character to be printed is placed in an output register, and simultaneously the I/O selector is instructed which device is to be addressed. The circuits are so arranged that the printing function is carried out without further intervention by the computer. When the printing operation is complete, the device issues a *device interrupt* to signal to the computer that it is ready to accept the next character.
- *Interrupt Control:* All interrupts are controlled by the interrupt control which services them according to a preallocated priority structure. The use of interrupt control of devices not only allows optimum use of the computer but allows optimum use of the device within the constraints of I/O channel loading.

Computer Storage

Not many years ago practically all main storage systems in a computer were described as *core store* but due to developments in integrated circuit design their use is rapidly diminishing. Although the term *core* is often used for any type of computer main storage, true core stores consist of thousands of tiny ferrite rings, each one capable of being magnetized. Two sets of wires passing through the rings are used to address individual rings in a *matrix* arrangement as shown in Figure 16.4. When current flowing through two coincident X and Y wires causes a magnetic change in a ring, the change is signalled by a voltage induced in a third wire Z, which passes though all the rings of a group. Even using very small rings, large magnetic core stores tend to be bulky and their access time is limited to about 1 microsecond.

During recent years the progressive reduction in the cost of electronic devices has made them an attractive alternative to the core store. Integrated circuit techniques have also reduced their size to the point at which hundreds of bistable circuits can be created on a single silicon substrate, making them much smaller than their magnetic counterparts. Electronic stores are described as *volatile*; that is, they lose their information when power is switched off. Their advantage lies in their price and size and also that they operate at speeds appreciably faster than core store.

INPUT AND OUTPUT UNITS

There is a wide variety of input and output units associated with the computer, ranging from data lights and switches on the computer console, to other computers. Between these extremes lies a range of electronic and electromagnetic devices often referred to as *data processing I/O units*, used for input and output of both data and programs. Where small computers are concerned, probably the most frequently used I/O units are magnetic disks and tape, serial printers, video screens, and keyboards. Larger installations also may include high-speed line printers, and, sometimes, magnetic drums.

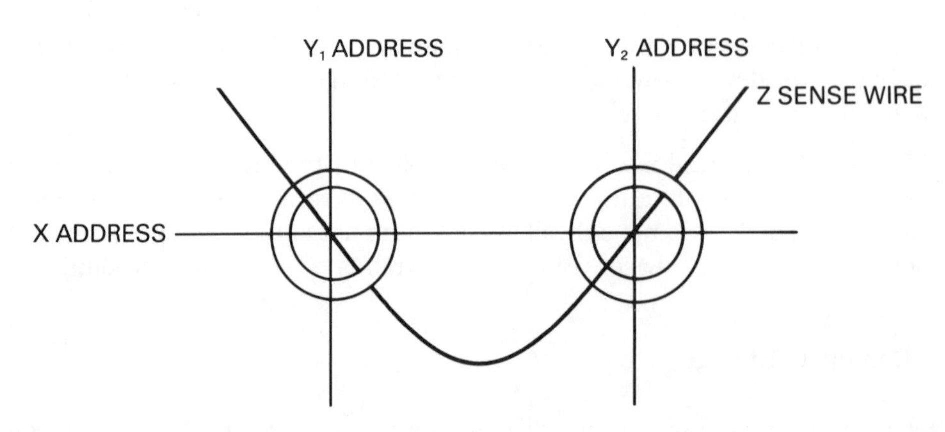

Figure 16.4
Core Store

Paper Tape Punches

Paper tape punches have already been discussed in the section on data loggers. They are still used on small systems but they are now being superseded in many instances by magnetic storage devices. A number of codes are used for paper tape, one of which is illustrated in Figure 16.5.

Paper Tape Readers

When paper tape is presented to a reader, the line of punched holes is positioned by a sprocket that picks up existing sprocket holes. Early readers used a line of electrical contacts to detect the presence or absence of holes. Readers now employ photoelectric or capacitive reading heads.

Punched Cards

In some respects punched cards are superior to punched tape. Individual cards are easier to replace if a punching error is found or if program modification is required. Card punches and readers, however, are considerably more expensive than their tape equivalents. In addition, a stack of cards required to hold a program is considerably more bulky and expensive than an equivalent tape reel. Storage and handling can also be more difficult. A typical 80-column punched card with its code is illustrated in Figure 16.6.

Teletypewriters

Small computers are often provided with a unit comprising a keyboard and serial character printer that is similar in appearance to a typewriter. Many of these are

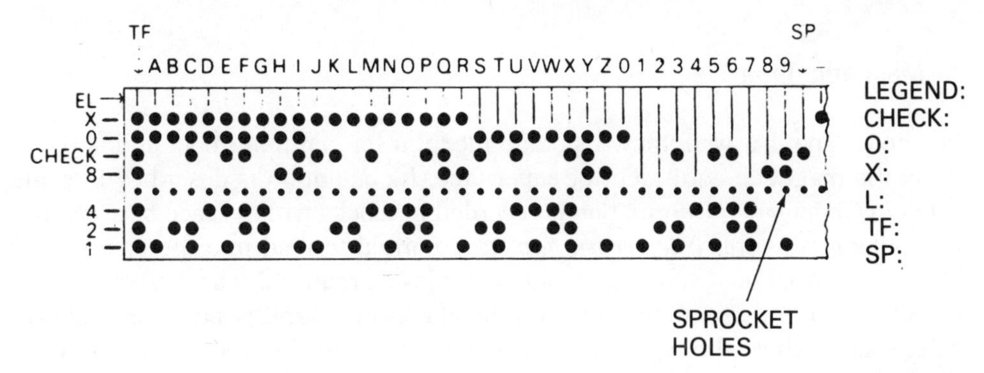

Figure 16.5
Paper Tape Code

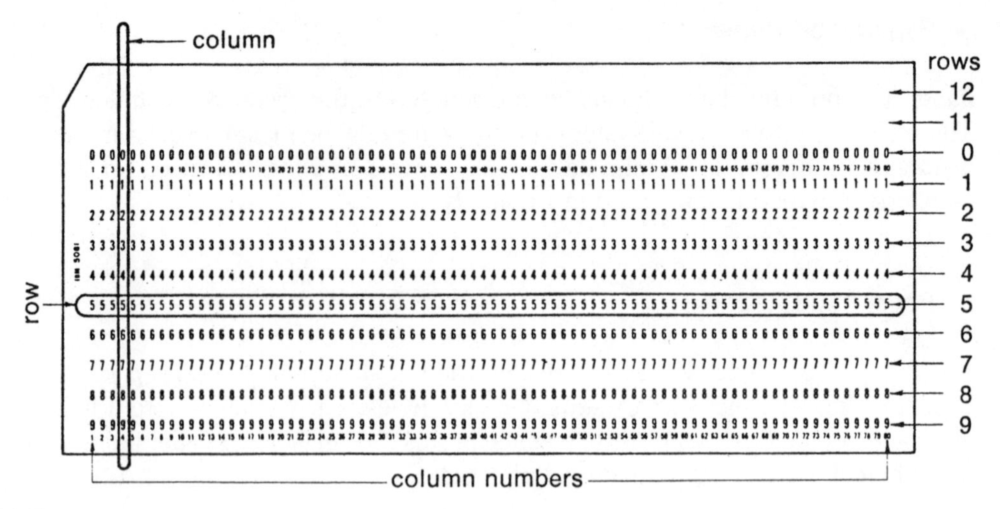

Figure 16.6
80-column Punched Card

derivatives of those used in Telex systems and connect to the computer through a serial communication line, often up to a mile (1600 meters) long. They provide a convenient and inexpensive form of interaction between an operator and a remote computer where data volumes are small and where printed output described as *hard copy* is required:

Video Display Terminals

Most modern computers are provided with at least one display terminal. These are units provided with a video screen and an alphanumeric keyboard. They operate at much higher speeds than teletypewriters and are much more convenient to use, as well as being more reliable. They do not, however, provide hard copy. Where this is required, a separate printer must be provided.

Magnetic Tape

Magnetic tape is a well-known storage medium for computer data and is effectively the magnetic equivalent of paper tape. Its operation is described in some detail in Chapter 14. Information is recorded in blocks with a space between the blocks known as the *interrecord gap*. It is normal to read into the computer a block of information, even though only part may be required. The interrecord gap provides a natural separation between the blocks for counting purposes and provides spaces where the tape may be stopped and started between readings. Magnetic tapes are useful when large quantities of data are to be stored and access time is of no serious consequence.

Cassette Recorders

Some small computers are provided with input and output circuits to allow connection of standard ¼-inch tape cassette recorders. In many instances these recorders are useful for storing and loading comparatively small programs. They are relatively inexpensive and consequently are ideal for use by maintenance engineers.

Magnetic Drums

These are drums coated with magnetic oxide. The drum is rotated at high speed and records are made around the circumference. Occasionally a single record head is used which is moved along the drum's axis by a stepping motor, although it is more usual to have one head for each track.

Magnetic Disks

The magnetic tape combines high storage capacity with a long access time; the magnetic drum has comparatively low capacity with a fast access time. Magnetic disks fall between the two. A metal disk covered with magnetic oxide on both surfaces is spun by a constant-speed electric motor. Data may be read from the top and bottom surfaces of the disk by heads which combine the functions of reading and writing in one unit. These heads are mounted on arms that can be moved by a servo system across the face of the disk from its center to its edge. With the heads in any stationary position a complete circle of information may be accessed from both surfaces. The circles are known as *tracks* and are often divided into a number of sectors, the first sector being identified by some type of sensor. Although the information on a disk is serial, it is converted to parallel format by the disk controller, before it is transferred to the computer I/O channel. Disk packs vary in size from a single disk, about 6 inches (150 mm) diameter, to multiple disks about 16 inches (400 mm) with storage capacities up to 150 megabytes or more. Some disk drives are provided with interchangeable disks but many are fixed, and contained in sealed compartments to improve their reliability. Also in many instances, a number of disks are stacked and driven by a common drive motor.

Diskettes

Another version of the disk is a small flexible disk contained in a protective dust cover. These devices, variously called "diskettes," "floppy disks," or just "floppies," are about 6 inches (150 mm) in diameter and interchangeable. They are designed principally to hold programs and data being transferred from one system to another and also to "back up" the main disk store. Diskettes are intended to provide a comparatively inexpensive form of storage and consequently their design is much simpler than that of the main disk. As a result of this, and as a result of the handling they receive in normal operation, diskettes can be expected to have a finite life. This should be considered when they are used to store data and programs which are particularly difficult to retrieve. In these circumstances, diskettes are sufficiently inexpensive to allow more than one copy to be made.

PROGRAMMING THE COMPUTER

When a computer has been connected to sensors and controls it must still be programmed before it becomes an operable installation. For convenience programs may be considered in two separate categories; the *operating system* and *application programs*. The operating system is required to control the internal operation of the computer while application programs direct the operating system to perform defined tasks.

The Operating System

Modern computers have a modular construction. This means that circuits may be added or removed to provide the most economic assembly to suit an application. The IBM Series/1 illustrated in Figure 16.7 is a typical general-purpose computer built on the principle of modularity. Putting together circuit modules in the most economic way is a skilled process often described as *configuring the system*.

If it was necessary to write a new operating system each time the configuration of such a system is changed, the cost of computing would be very high and time scales for their implementation long. This problem has been overcome by designing operating systems which, like the circuits, are modular. Program modules are selected to match a configuration and then the modules are joined or *linked* to form the operating system.

The preparation of operating systems used to be considered to be the province of programming specialists, engineers having neither the time nor the inclination to acquire the skills. Developments in computing, however, have changed this situation. Now operating systems are available that can be tailored by engineers without an unreasonable investment of time to acquire the skills. Most computer manufacturers provide operating systems for their computers and since they represent a considerable investment, a license fee is often charged for their use.

Application Programs

Application programs are written in one of the available computer languages, many of which approximate a normal written language. These are referred to as *high-level languages*.

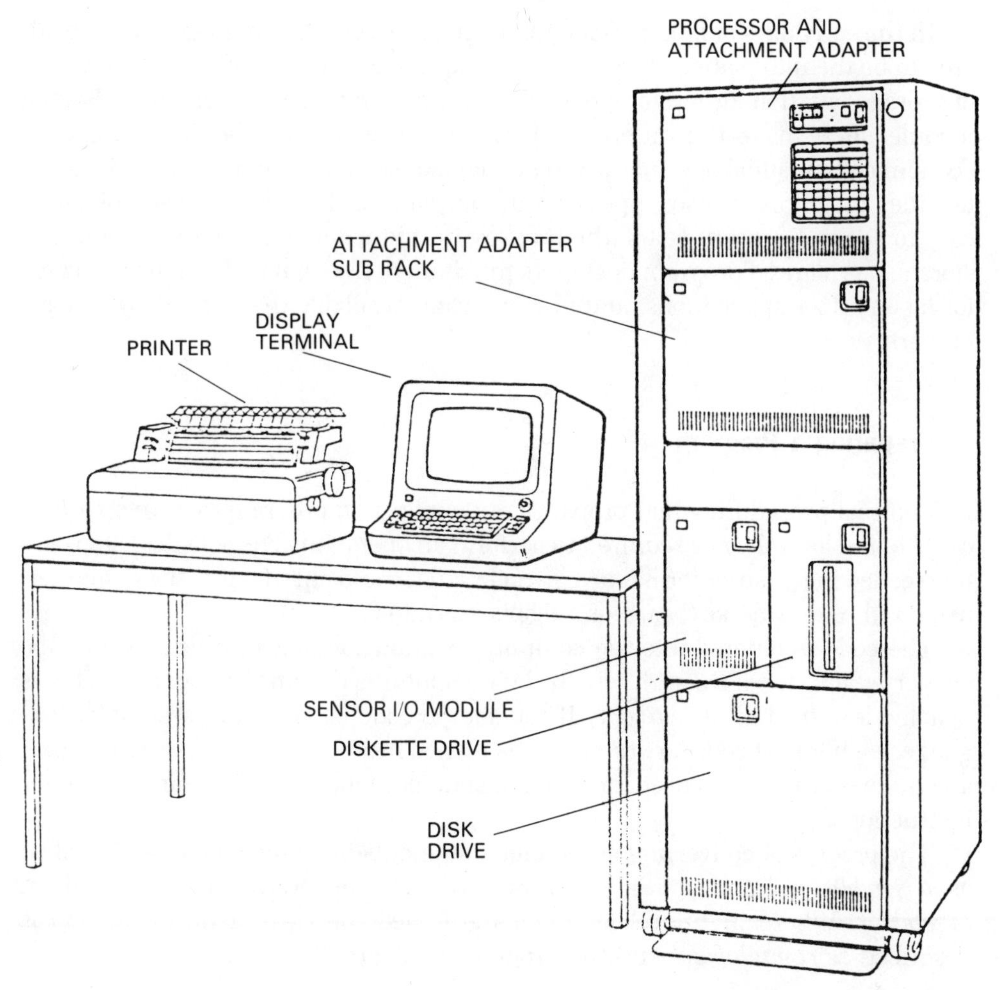

Figure 16.7
General-Purpose Computer (IBM Corp.)

The most basic language in general use is called *assembler language*. This is used by skilled programmers when operating speed and storage space are of prime concern. Each assembler statement converts directly into an equivalent binary instruction, and each program consists of a sequential list of instructions. Because computers differ in design, assembler languages are peculiar to a computer type.

High-level languages tend to be designed for specialized areas of application. *COBOL* (Common Business Oriented Language), for example, is designed for commercial applications, while *FORTRAN* (Formular Translation) is designed for scientific calculations. Some of these languages have been standardized and can be used with any computer provided with the necessary conversion programs.

In the early days of factory floor applications, assembler language was considered to be the most satisfactory application language because of its economic use of storage. At that time storage was expensive and the cost of an additional module often made the difference between an application being regarded as economically justifiable or not. Commercial and scientific languages also had no facilities to address sensor input and output, and in this respect assembler was easy to adapt. In recent years, the progressive reduction in the cost of computer storage has allowed program designers much more freedom to develop languages for factory floor applications, and now many are available that are relatively easy to learn.

Preparing a Program

It is normal procedure when preparing an application program to develop a *flow chart* of the logical steps required to perform an operation. These logical steps are then coded in the strict format of a given language and entered into the computer, usually through the keyboard of a display terminal.

The code as entered into the computer is known as *source code* and must be converted into the form with which the computer is designed to operate. This is usually described as *object code*. If the source code is written in assembler language, each line of code, or *statement*, is turned into one computer instruction. If a high-level language is used, one source statement may produce many computer instructions.

The process of converting assembler statements into object code is described as *assembly* and the special program used for the conversion is called an *assembler*. Where high-level languages are used the equivalent conversion is described as *compilation*, and the program as a *compiler*.

APPLYING COMPUTERS TO INDUSTRIAL OPERATIONS

There are three ways of using a computer to assist industrial operations. If the time required to perform an operation is considerable, it is possible to collect and enter data into a computer manually. The computer is used to calculate a control response and then a manual adjustment is made based on that response. This is described as *off-line* control. As the time cycle of an operation is reduced, there comes a point where the interpretation of manually collected data is too slow to be effective. Then data collection must be automatic. This is often described as *on-line monitoring*. Finally, when an operation cycle becomes too fast for an operator to intepret analyzed data and provide timely response, the response, too, must be automatic. Then the system may be described as an *on-line control system*.

The way in which computers may be used in their monitoring and control roles is most easily demonstrated by describing two simple applications.

CYCLIC MACHINE MONITORING

A considerable number of industrial operations involve the use of automatic or semiautomatic cyclic machines. Textile looms, plastic injection molding machines, automatic piece part manufacturing machines, and component assembly machines are all commonplace examples.

Cyclic machinery is designed to work at an optimum speed at which maximum output can be maintained without either frequent breakdown, or the production of an unacceptable quantity of scrap. The cycle time of a machine running at this speed is described as the *standard cycle time*. A machine running at a higher speed than standard may require more maintenance and generate more scrap, while a machine running slower than standard can cause substantial loss of product.

The situation is not as simple as stated, however, and operators who may be responsible for the operation of a considerable number of machines may deliberately adjust the cycle time of a machine to cope with a particular situation. Variations in the quality of raw material, for example, may influence the choice of cycle time. In other instances, such as drilling operations, the operator may not be aware of increasing cycle time until it has become excessive. A 10% change in a machine which has a relatively long cycle time is not easy to recognize without a specific measurement, but the effect of such a condition persisting over a considerable period may be expensive.

Machine stoppage may also be a problem in certain operations. Long-term breakdown will always be recognized, but frequent short stoppages may not be so obvious. An automatic stamping machine, for example, may have a cycle time of a few seconds. It may be stopped periodically by an operator to clear jams, and it may even be considered that periodic jamming is inevitable. If a situation arises, however, where the frequency of jamming increases, an appreciable loss of production may occur before an operator recognizes the trend.

In a busy machine shop it is difficult for an operator to recognize whether machines are operating at their optimum and even when the operator does, there is little time to perform more than basic adjustments. So far as supervision is concerned, the problems are even more acute. A supervisor responsible for the manning and operation of possibly 100 machines has little chance of maintaining a close control over their efficient running without some mechanical assistance.

Computer Monitoring of Cyclic Machines

It has long been recognized that a single contact, signaling the end of a machine cycle, provides a considerable amount of information about that machine. If it is assumed that each cycle produces a good part, the total number of contact closures represents a product count. If the contacts are not operating periodically, the machine may be assumed to be out of service. Also, if the time between successive contact closures does not coincide with the standard cycle time, then the machine is working outside its specification.

In addition to cyclic contacts, it is possible to provide a machine with a simple data entry unit from which the operator may enter supplementary information. The simplest of these is the telephone dial, which may be used to enter ten single-digit codes, each representing a machine condition. These codes may represent such things as operator sign-on/off and reasons for the machine being out of service.

A number of systems have been designed around these basic inputs using multichannel chart recorders. One type uses pens which are normally raised from the paper and are lowered in response to each cycle switch closure. Operation of a telephone dial causes a number to be printed in the spaces between the end-of-cycle marks. With these systems, the measurement of individual cycle time for a machine is not possible with any accuracy, because of restricted chart speed. This was overcome by using an alternative measurement that defines machine efficiency as the number of productive machine cycles in a given time divided by the possible number of machine cycles in that time. The real difficulty experienced with these systems however, is one of record analysis. Their analysis, and the preparation of summary reports is a time-consuming manual operation. This creates a situation where results are often not available until several days after the event.

While recorder-based systems are limited in their application, replacement of recorders by small computers completely transforms their scope. The most important feature of the computer system in this context is its ability to operate in *real time*. This means that the computer response to an event is sufficiently close in time to be considered concurrent with the event. For example, the computer can measure the cycle time of a machine. It can then compare it with a stored standard cycle time, and turn on a warning light to indicate an out-of-limits situation, within a fraction of a second after the end of the cycle. So far as the operator is concerned, this is instantaneous. In the practical situation, it is usual to compare several machine cycles, before warning of an out-of-limits condition. The collection and analysis of dialed input is also in real time, and both types may be arranged and displayed in a way that provides a supervisor with a knowledge of the current state of the machine shop. Finally, stored information may be used to provide either the supervisor or management with historical information in ways most meaningful for their requirement.

The computers required for many of these installations are small, often falling into the category described as microcomputers. In many instances, however, a larger computer is used to allow the inclusion of many additional functions. For example, it is possible to store the number of parts required to complete an order, the required completion date, and the machines on which the parts will be produced. The computer can then be programmed to count the parts as they are produced on each machine, sum them, and estimate a completion date. In an ideal world this is not a difficult manual operation, but in the machine shop environment it can be very difficult. In reality machines break down, operators fall sick, material supplies run out, machines cycle with times that are not standard, and an unpredictable amount of scrap is produced over the manufacturing period. The calculation of completion date has two values. It provides a warning if completion is going to be late, but equally important is the prediction of an early completion date. This allows either of two policies to be followed. It is possible to allow the order to complete prematurely, in which case tools must be prepared early to meet the requirements of the next order. Alternatively it is possible to remove one, or more, machines from service without impacting the original completion date, and free them for other work.

It is not difficult to imagine many more applications that may be included in such a system to assist the operation of a machine shop, but one is assuming increasing importance. Many firms use one computer for their order scheduling and one for accounting operations. One of the outputs of this computer is often a printed scheduling list delivered to the machine shop by hand. Now many installations connect the two computers by a communications line, and data are transferred between the two as generated. This eliminates delays and errors associated with a manual system and is especially important in installations where the two computers may be separated by many miles.

Shop-Floor Data Collection

There are basically two sources of data entry from the shop floor. There are the data entered automatically from plant and machinery that originate from sensors; and there are the data entered by operators and other personnel. The majority of this total input of data is digital in form, and even where analog data are involved, such data are often converted before transmission to the computer.

Many plant sensors are switches of one type or another, and in monitoring operations usually signal the completion of an operational sequence. In the previous section, for example, a contact simply indicated the end of a machine cycle. Any variation in time between contact closures indicated that something was wrong, and the fact was signaled by turning on a warning lamp at the machine. It then became the operator's responsibility to find out what was wrong and rectify the situation. Where a computer controls the operation of a machine, it becomes necessary for the computer to monitor individual machine operations. If any operation fails, the computer is programmed to attempt to recover control before signaling the failure to an attendant. These recovery routines, as they are described, constitute a large part of any control program. The estimated proportion of these routines is often quoted as between 60 and 80% of the total.

Switch contacts are sensed by a computer input adapter described as digital input. These may include interface circuits to allow direct connection of contacts to the input but in many cases digital input circuits are simple transistor switches that must be provided with an external circuit. One of these circuits is illustrated in Figure 16.8 and is based on a device called an *optoisolator*. The isolator incorporates a light-emitting diode and a phototransistor.

The operation of these devices is simple and effective. When current flows through the diode, it glows. The light it produces is focused on the light-sensitive area of a phototransistor causing it to switch. Because the diode and transistor are physically separated, the transistor will survive the application of several thou-

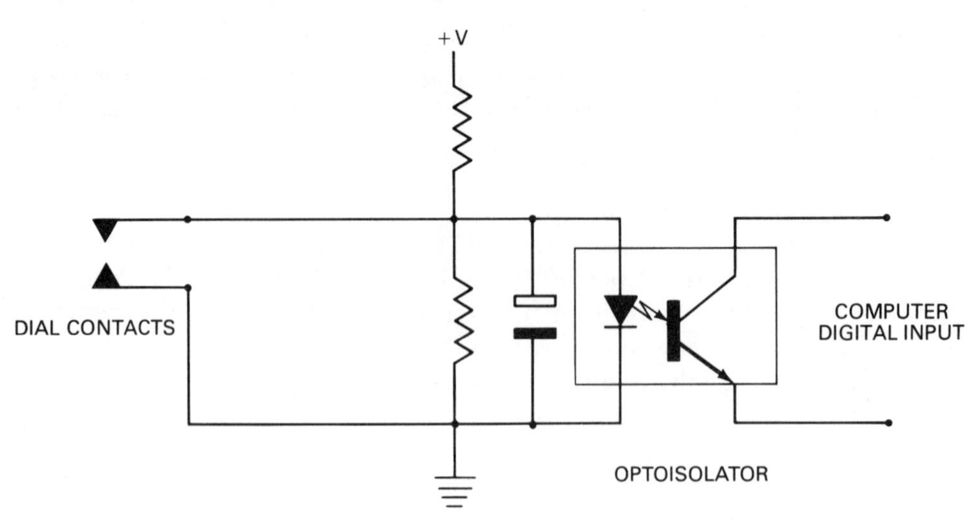

Figure 16.8
Optoisolator Circuit

sand volts to the input before failure. Also, the diode requires currents in excess of 10 mA before sufficient light is produced to switch the transistor. This makes the circuit highly immune to low-power interference voltages.

Where data are entered by an operator such data may originate from manually operated switches. These may be single toggle switches indicating on/off conditions or they may be banks of digital switches used to enter numeric information. These digital switches are commercially available as individual units that may be assembled into multidigit switches. Each switch is a ten-way device built into an enclosure with a window displaying a digit between 0 and 9. One side of the enclosure is a wheel centered on the switch and with protrusions resembling large teeth. When the switch is recessed into an instrument panel, the protrusions allow the switch to be stepped through its ten steps. Because of the way they are operated, they are sometimes described as "thumbwheel switches." Figure 16.9 illustrates one bank of such switches wired to four digital input lines of a computer in a BCD format. To economize on computer digital input points, the banks of switches are often connected to one set of input lines and they are switched into circuit sequentially by the computer. Sometimes an additional contact is provided to allow the operator to signal that the switch has been set, and sometimes a computer response light is provided.

Where it is necessary for an operator to enter substantial quantities of information, the video display terminal provides an attractive input terminal, and rugged designs are available to withstand workshop environments. Where appropriate, they can be used to enter complex sequences of information, and provide an operator with extensive prompts. Many semiautomatic machines now employ microcomputers in their control systems, and video displays are used as a standard means of providing the operator with machine-status information.

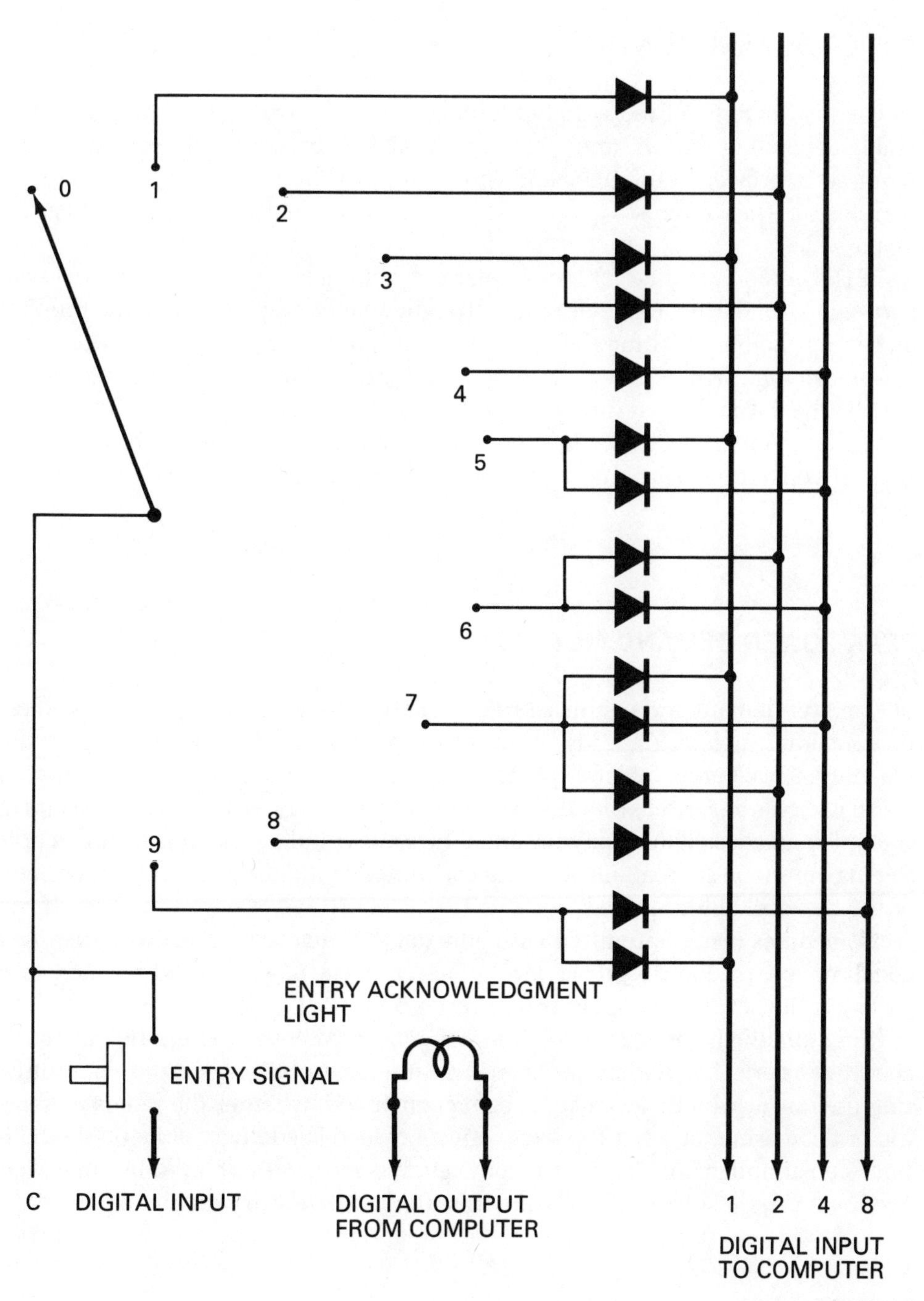

Figure 16.9
BCD Input Switch

POWER DEMAND CONTROL

In the vast majority of industrial enterprises, electricity usage fluctuates considerably depending on the time of day and year. Electricity supply companies are required to provide equipment and cables to meet the maximum load and it is, therefore, in their interests to persuade customers to reduce their peak load as far as possible.

It is common practice for a customer to be charged for electricity with a two-part billing rate. The customer pays a straight charge proportional to the number of kilowatt-hours consumed and, in addition, pays a surcharge based on *peak power demand.* This surcharge may be determined in a number of ways but usually the day is divided into half-hour periods and a *peak power demand meter* is used to monitor electricity in each of these periods. A surcharge based on the highest consumption in any demand period may be levied for an entire billing period.

PEAK POWER DEMAND METERS

Peak power demand meters measure electricity in much the same way as domestic consumer meters. An eddy current disk rotates at a rate determined by the current and voltage applied to energizing poles. In some meters the rotation of the disk drives a pointer around a dial, the pointer being reset to zero at the end of every demand period. A slave pointer on the same spindle is not reset, and records the maximum position attained by the consumption indicator from the beginning of the demand billing period.

Sometimes meters are fitted with adjustable contacts to raise an alarm when consumption reaches a preset level. The contacts may be used to *shed* load automatically, or the task may be the responsibility of a plant engineer.

Such methods of load shedding are not predictive. When the alarm is sounded, a serious situation has been encountered, but there is no way of estimating the rate at which electricity is being consumed, whether the rate is increasing or decreasing, or what the likely effects of load shedding will be on the half-hour consumption. In these situations, engineers usually "play safe" and over-shed, with the result that an operation suffers unnecessary disruption.

PREDICTIVE CONTROL

If some type of predictive control system is to be designed, it is necessary to obtain two control signals from the demand meter to drive the controller. One signal must be proportional to consumption, and the other must provide a synchronizing signal to indicate the end of the demand period. These signals are obtained by modifying a standard peak power demand meter. The rotation of the disk is used to provide a pulse train with a frequency proportional to disk rotation speed, and a switch on the reset mechanism provides a synchronizing signal. Figure 16.10 illustrates a typical demand meter set up for computer monitoring; and Figures 16.11 and 16.12 illustrate typical interfaces that may be used to connect the meter sensors to the digital input.

COMPUTER CONTROL OF PEAK POWER DEMAND

If load shedding is to be performed with minimum disruption to a production operation, it is necessary to select suitable loads to shed in advance. It is also necessary to decide how frequently and how long they may be switched, and to choose the order in which they should be selected.

The difference between the maximum possible consumption in a half-hour period and the total savings possible due to load shedding in that period provides a

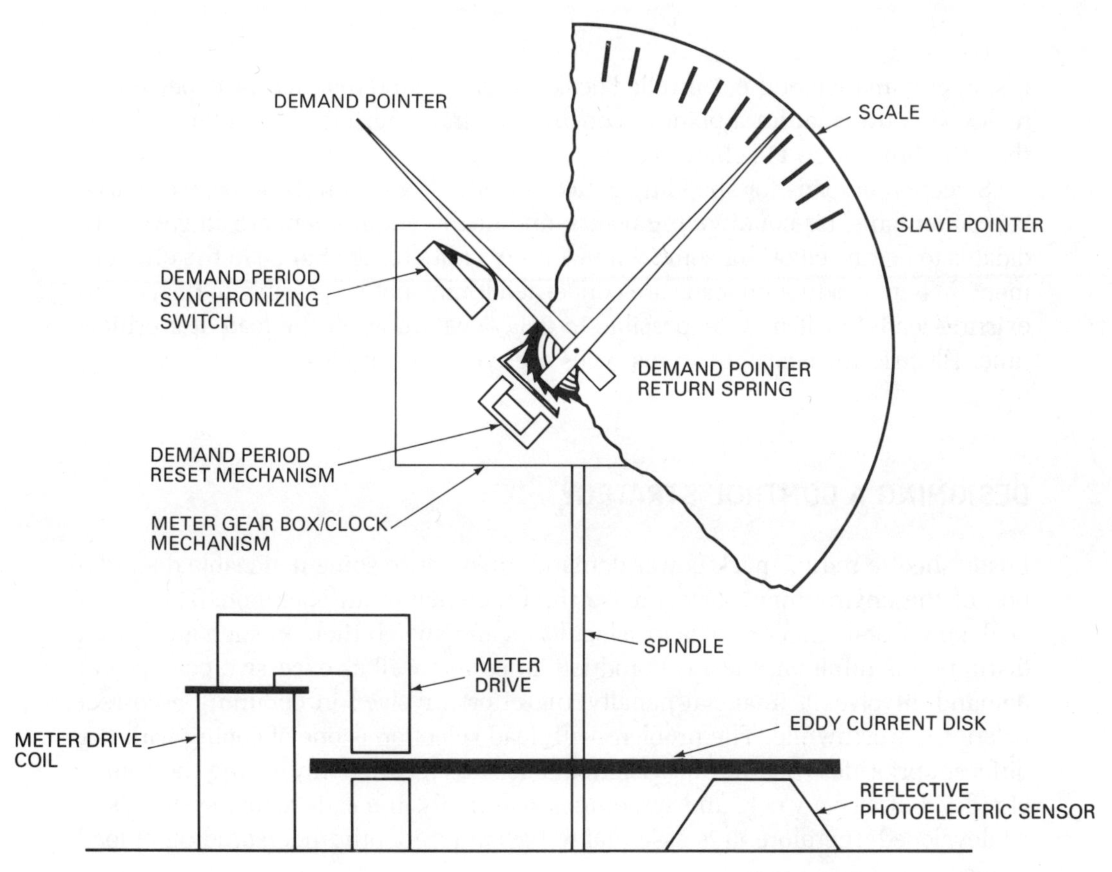

Figure 16.10
Maximum Power Demand Meter with Computer Output

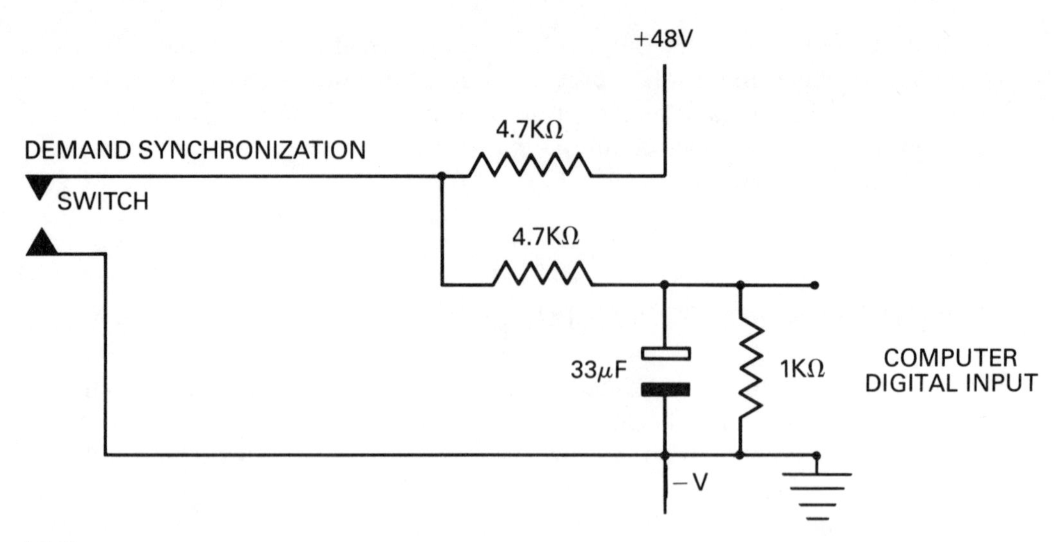

Figure 16.11
Synchronization Switch Computer Interface

first approximation of the controlled peak power demand that may be expected. In reality, it is unusual for a plant to consume its maximum load, so a target lower than the limit often is achievable.

Selection of loads for shedding is not as difficult as might be expected. Large ventilation fans, air-conditioning plants, and heating equipment are all good candidates to be turned off for some time without a noticeable change in the environment. Where production loads are concerned it may not be possible to turn off an existing load, but it may be possible to delay switching on the load at a critical time. Electric furnaces of various types are typical examples.

DESIGNING A CONTROL STRATEGY

Loads shed to reduce peak power demand must cause some noticeable degradation of the environment, or decrease the efficiency of an operation. The act of peak power demand control is to select loads and switch them in such a way that disruption is minimal. Because production loss, as well as excessive peak power demand, involves a financial penalty, the effort involved in obtaining a correct balance is worthwhile. The problem with load selection is one of coping with the different priorities that may be assigned to load usage, not only during the course of a day, but also a week, and sometimes seasonally. If a satisfactory system is to be developed, therefore, it is essential to have an in-depth understanding of load usage.

There are a number of ways in which optimum load switching can be achieved; one selects a strategy based on a "day type" where each day of the year

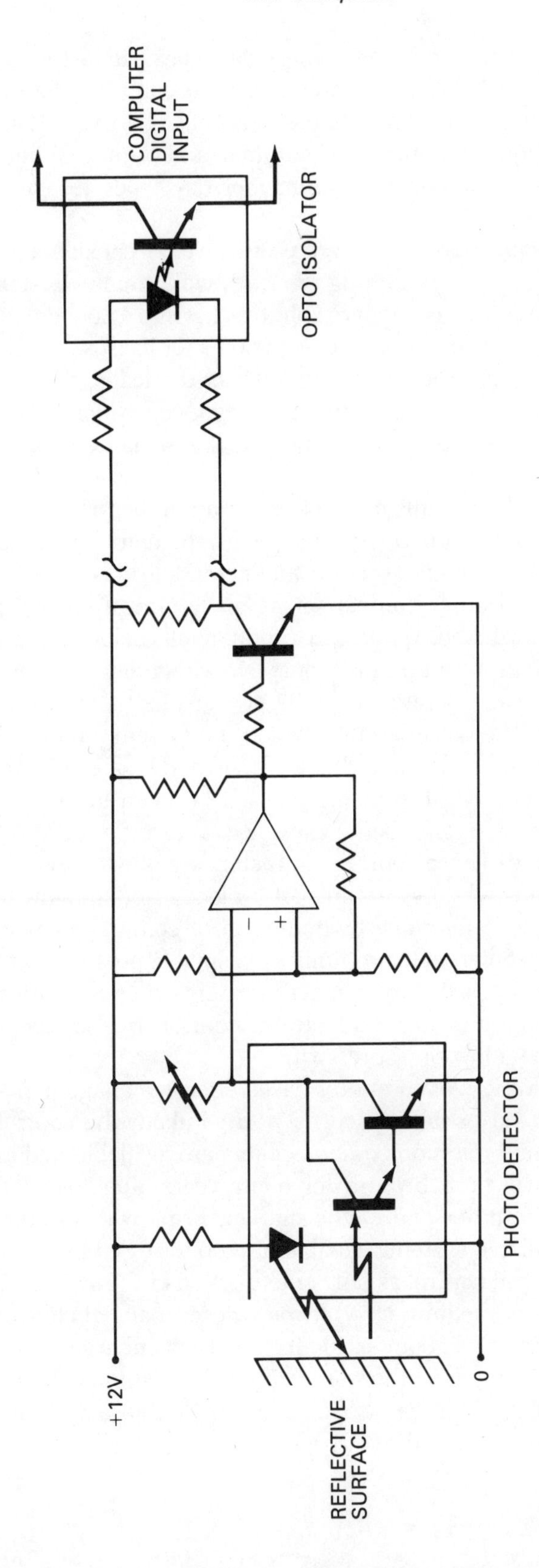

Figure 16.12
Photoelectric Sensor Computer
Interface

is allocated a day type number. Many of these day types will be the same. Monday through Friday, for example, of any week may be common, Saturday may be another, and Sunday yet another. Where seasonal changes exist, Monday to Friday may have different numbers in summer and winter. While it is in the programmer's interests to keep the number of day types to a minimum, the concept is very flexible.

The detailed specification of a day type involves a considerable amount of effort. It is unlikely that load switching priorities will remain constant over a 24-hour period. Consequently it is usual to split the day into a convenient number of time sectors and specify load switching sequences for each sector. Each switching sequence must specify the switching order of the loads, the length of time, and how many times they may be switched. In many operations plant and equipment are only used for part of the day, and it is also usual to make provision for specifying turn-on/turn-off times.

Peak power demand programs may operate in a number of ways. The simplest monitor energy consumption periodically through the demand period and if consumption exceeds the specified target for that point, a load is switched off. At the next sampling period another comparison is made, and if consumption is still high, another load is switched. When consumption falls below the target, a load is switched on again. To prevent an unreasonable number of load switching actions it is usual to specify two operational limits of consumption between which no switching will occur. The target around which the system attempts to control is often less than an allowable demand that is considered reasonable. If at any point during the demand period all available loads have been switched off and the allowable demand is in danger of being exceeded, a warning is given that manual intervention is required. These control parameters are illustrated in Figure 16.13.

There are occasions when switching off a piece of equipment to satisfy demand control could be dangerous. Switching off a compressor when the compressed air supply has fallen to its minimum acceptable pressure is one example. In these instances, it is possible to connect override switches to the computer, to signal that an equipment is not available for control. In the case of the compressor, this would be a low-pressure switch.

The number and variety of computer-based energy management systems are growing steadily. For very small concerns worried about the control of lighting and heating loads small microprocessor systems are available, tailored to a specific operation that will probably run for many years after installation without being changed. At the other end of the market, firms with multimillion dollar energy bills have available systems capable of monitoring in excess of 1000 sensors, with power demand control as only one function of the system. Between the two extremes are many medium-sized firms whose financial controllers are becoming increasingly worried about escalating electricity charges and whose operations could benefit from one of the minicomputer-based systems. It is generally considered that a 10% saving is a reasonable target to achieve, but this is a rule-of-thumb estimate and often a conservative value.

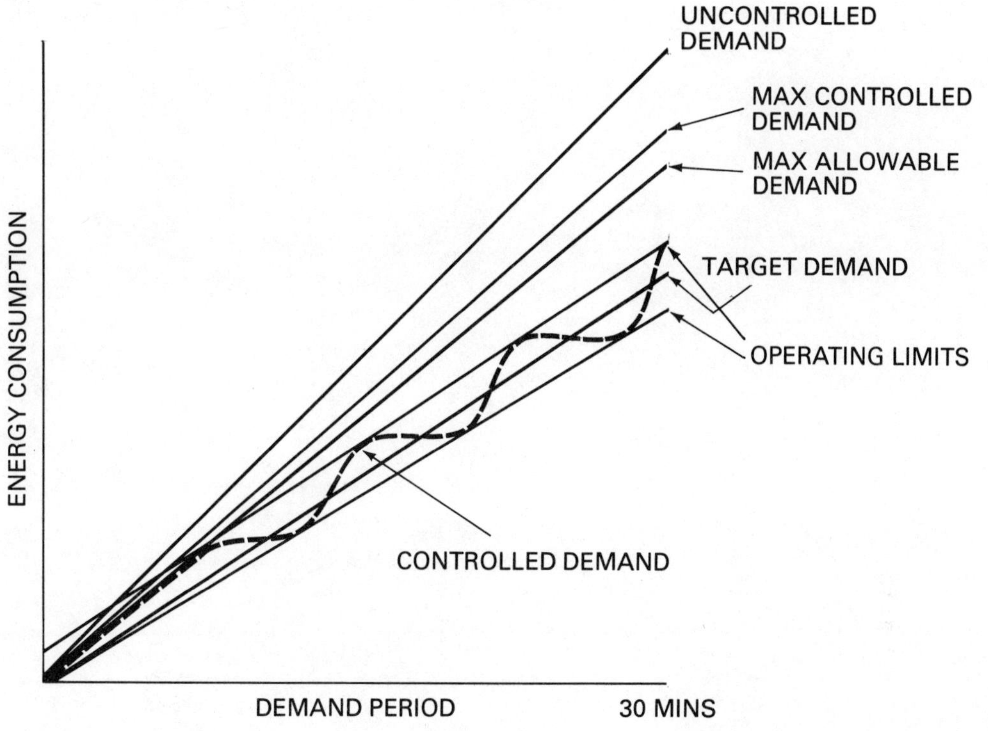

Figure 16.13
Graphical Representation of Maximum Power Demand Control Parameters

CHAPTER 17

ELECTRICAL INTERFACING

ELECTRICAL INTERFERENCE AND CABLING

The recording equipment of an instrumentation system is connected to a sensor by a cable, which may under certain conditions pick up electrical interference or noise. Noise on input lines can cause a reduction in measurement accuracy in analog circuits and can trigger false digital input signals. In output circuits it can cause chattering of control elements and spurious operation of digital equipment.

The effects of noise may be alleviated by the introduction of filters and by careful design of the recording system interface, but these are palliatives. The only sure way of preventing the undesirable effects of noise is to prevent it from getting into circuits in the first place. Much can be done in this direction by good cabling practice and careful selection of cable, but in a typical industrial plant the installation will inevitably be a compromise between what is technically possible and what is economically feasible. Even when a cable is installed and working satisfactorily, there is no guarantee that this state will continue indefinitely. Subsequent damage to the cable or modification to plant electrical wiring can create the problems that the original installation was designed to avoid.

Broadly speaking there are two types of noise that may be picked up by cables. Electrostatic noise is usually caused by capacitance coupling between signal cables and noise sources. It may be at the ac supply frequency, but in certain instances may result from high-frequency generators. Electromagnetic noise is introduced where signal cables run through areas of high magnetic fields of varying intensity.

Electrostatic Noise

Consider the case illustrated in Figure 17.1 in which two signal cables run parallel to two power cables, one of which is at ground potential. One signal cable is coupled to one power cable by capacitance C_1 and to ground by capacitance C_4. These capacitors will act as a voltage divider, raising the potential of the signal cable to a value of:

$$\frac{VC_1}{C_1 + C_4}$$

Similarly, the second signal cable will be raised to a value:

$$\frac{VC_2}{C_2 + C_5}$$

If the values of capacitance are different, a resultant voltage

$$V\left(\frac{C_1}{C_1 + C_4} - \frac{C_2}{C_2 + C_5}\right)$$

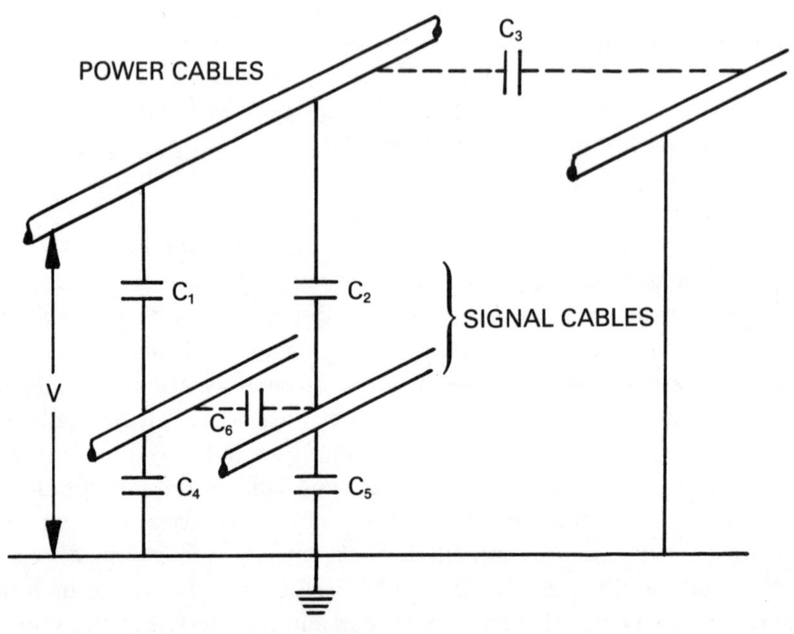

Figure 17.1
Capacitance Coupling Between Power and Signal Cables

will appear across the signal lines which is indistinguishable from the signal and may be of the same order as or greater than low-level analog signals. There are two methods of reducing this difference. If C_4 is made equal to C_5 and C_1 equal to C_2 the expression becomes zero. One method uses twisted-pair cable which causes equal voltages to be induced on each line and which are canceled at the equipment terminals.

An alternative, or additional way of reducing this type of noise, is to provide an electrostatic shield around the conductors. A shield may be braided copper wire mesh or it may be a copper or aluminum tape wrapped around the conductors. The more expensive cables may employ two shields of foil that are insulated from each other. Foil-wrapped cables often have a spiral-wrapped copper wire in contact with the foil, known as a *drain wire*, that provides a low-resistance path to ground for the noise currents.

It is difficult to obtain a true estimate of the effectiveness of electrostatic shields. Their effectiveness depend on the frequencies produced by the noise generator, the value of the termination resistance, and even the cable size and insulation thickness. The figures quoted in Table 17.1 provide a guide to the order of reduction that can be expected in a practical situation.

Table 17.1 Noise Reduction Due to Electrostatic Shielding

Cable type	Noise reduction ratio
Twisted-pair cable	70 to 1
Twisted-pair cable with braided shield	150 to 1
Parallel-pair cable with spiral-wrapped wire shield	300 to 1
Twisted-pair cable double shielded with aluminum tape and drain wire	5000 to 1

The principle of the electrostatic shield is simply that of coupling the noise source to the grounded shield instead of to the signal conductors as shown in Figure 17.2.

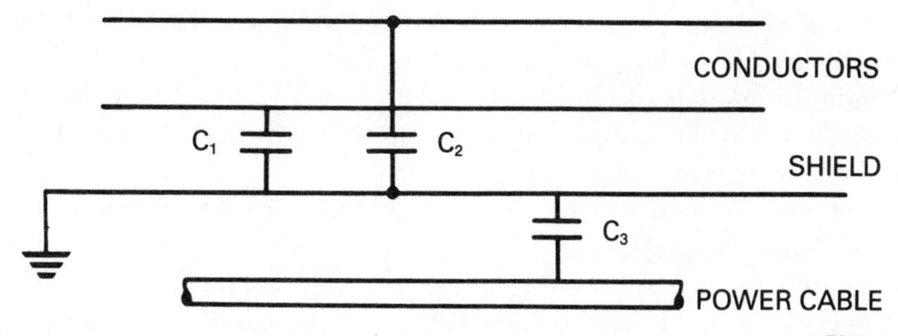

Figure 17.2
Electrostatic Shielding

Grounding Practice

When electrostatic shielding is used it must be connected to ground with as low a resistance path as possible. Sometimes it is possible to ground both ends of the shield without any noticeable effect and, in freak situations, this may be necessary to eliminate interference from the signal lines. In the vast majority of cases, however, this is poor practice. The ground potentials at each end of the cable may be at different levels and grounding each end may cause heavy currents to flow in the shield, creating a situation far worse than if no shield were used. Where cables are grounded in more than one place they are said to provide a ground *loop*.

Normal procedure is to ground the shield at one end only, to provide a continuity path for noise without the complications of a ground loop. The choice of grounding the cable at the sensor end or the system end depends on the situation. Where the electrical transducing section of a sensor is isolated from ground the shield may often be successfully grounded at the system end. In general, a better ground exists at the system and this practice has a practical advantage. Noise often appears on an installation quite unexpectedly when the outer covering of one cable is damaged and a ground loop is produced. It may be necessary to *lift* the grounds to locate the faulty cable and this is more easily accomplished when all the grounds are located in one place. In some installations, however, and where the electrical transducer is grounded, it is necessary to ground the shield at the sensor end to provide adequate noise reduction. In theory this will provide maximum reduction and is often offered as the recommended practice. Since it is not possible to state hard and fast rules about grounding, the only reliable procedure is to test each installation and eliminate problems as they arise.

Electromagnetic Noise

If a conductor moves in a magnetic field, a voltage is produced across its length which is proportional to $\beta \nu l$, where β is the flux density of the field, ν is the velocity of the conductor at right angles to the field, and l is its length. It is easy to imagine the converse situation applying where the conductor is stationary and the field intensity changes. This is the mechanism of electromagnetic interference.

Since nonferrous metals have no influence on magnetic fields, electrostatic shields are of no value in rejecting electromagnetic noise. The use of magnetic shields is restricted by cost, as high-permeability materials are very expensive. Steel conduit is probably as good a magnetic shield as is available for practical purposes. This provides a rejection ratio of about 20 to 1.

Twisted pairs provide better rejection than conduit, the reduction ratio being a function of the number of twists per foot. Figure 17.3 illustrates the effect of twisting on noise reduction.

The more twists per foot of cable length, the more expensive the cable, which implies that it is desirable to design the cable with the minimum twist compatible with its noise rejection capability. From Figure 17.3 it can be seen that the optimum is about 24 turns per meter (eight turns per foot).

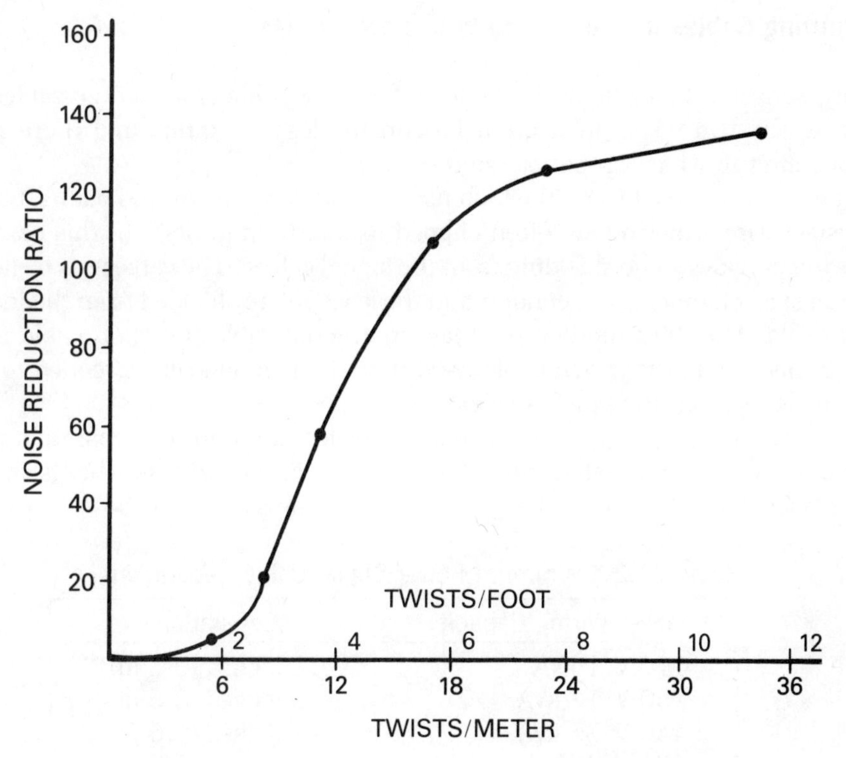

Figure 17.3
Relationship Between Noise and Twist of Signal Cables

Crosstalk

Crosstalk is the name given to the interference experienced on one pair of cables from signals applied to another pair and may be troublesome with multiconductor cables of the *planetary-layered* variety. Twisted-pair cables are a considerable improvement, and where crosstalk must be kept to a minimum, shielded twisted-pairs eliminate the problem.

Common-Mode and Series-Mode Interference

The types of interference so far discussed in this chapter, result from disturbances external to the sensor circuit. The connection of a sensor under certain circumstances can also cause interference voltages to appear at the equipment terminals. Where voltages referenced to ground appear on each terminal, the interference is described as common-mode interference. Where two common-mode voltages are not equal, they give rise to a voltage difference across the equipment terminals described as series-mode interference. These types are discussed in more detail in Chapter 6.

Routing Cables in the Proximity of Power Lines

Ideally, sensor cables should be kept as far as possible from power cables. In practice, the two types must often be run in close proximity and there is an obvious limit to what can be tolerated.

There is a variety of ways in which power cables are run around plants. Cables with steel wire armoring are often clipped to walls and girders. In this case the armoring provides some shielding from the signal cables. The cable may be laid in covered steel channels or in conduit and these again are shielded from the instrument cable. The third method is to lay unarmored cables on open cable trays. Where this occurs the signal cable itself must be magnetically screened to provide shielding from the power cables.

Even when the two types of cable are shielded, a minimum separation must be maintained where possible, and Table 17.2 provides a guide to what is considered reasonable.

Table 17.2 Minimum Power/Signal Cable Separation

Power Wiring Capacity	Separation
110 V 1 kVA	12 inches (0.3 m)
250 V 12 kVA	20 inches (0.5 m)
440 V 75 kVA	24 inches (0.6 m)
5 kV 5 MVA	60 inches (1.5 m)

These are minimum distances and should be increased if possible. Where signal and power cables cross, the two should be at right angles. It is not always possible to attain the ideal and in circumstances where there is no alternative but to reduce parallel separation, signal cables should be enclosed in conduit and the run made as short as possible.

CABLE TYPES

There are a number of types of instrumentation cable in use; the more common are illustrated in Figure 17.4. Of these the relative merits of four types are summarized in Table 17.3. These are rated in the order preferred, satisfactory, and adequate. "No" indicates that they are unsatisfactory.

Type 1: Unshielded Twisted Pairs—These are the normal telephone-type cables which may have solid or standard conductors. Usually the insulated conductors are protected by an outer plastic sheath.

Type 2: Shielded Twisted Pairs—This is a cable in which individual twisted pairs are shielded, either with copper braid or with aluminum tape. Each shield is insulated from adjacent ones by a plastic tape.

Type 3: Coaxial Cable—Coaxial cable has a single central conductor with a cylindrical insulator, and is shielded by a copper braid. The braid is protected and insulated by an overall plastic sheath.

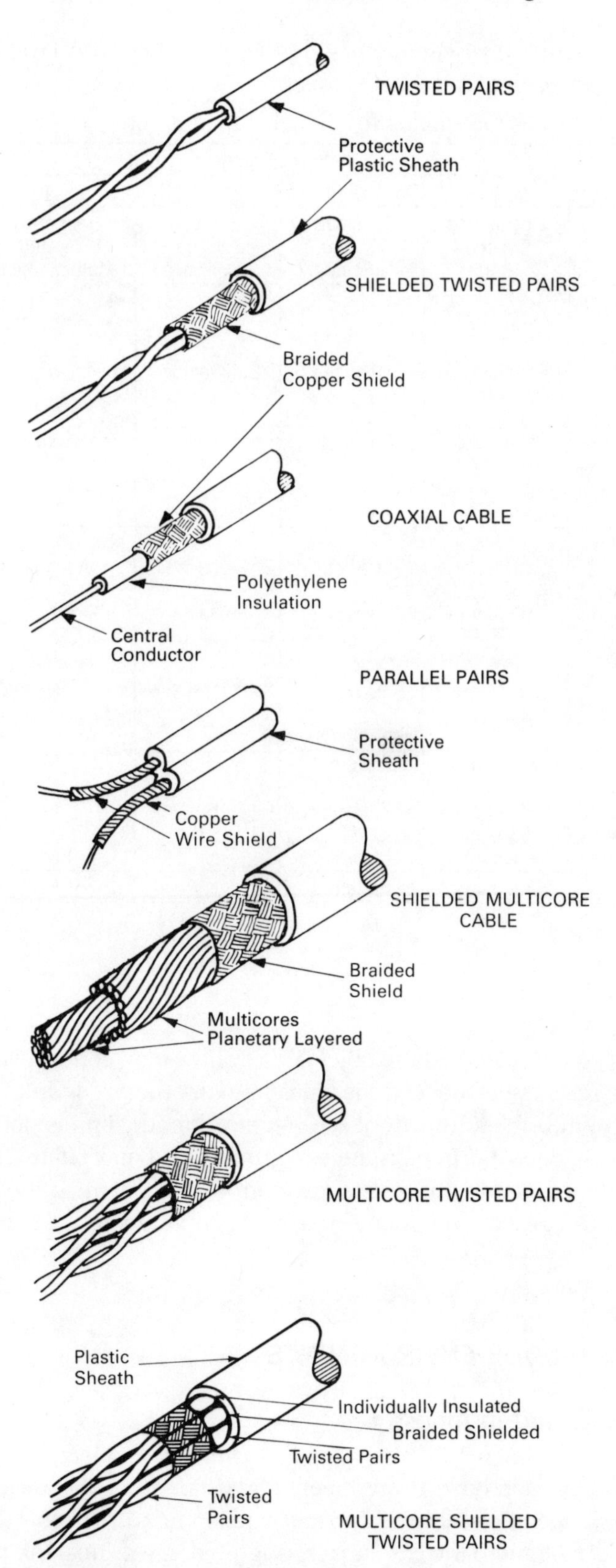

Figure 17.4
Typical Instrumentation Cables

Type 4: Planetary-Layered Multiconductor—These are concentric layers of conductors with an overall screen and plastic sheath.

Table 17.3 Preferred Cables for Signal Transmission

	Wiring Category			
	1	2	3	4
Function	Unshielded Twisted Pairs	Shielded Twisted Pairs	Coaxial	Planetary-layered Multiconductor
Digital input/output; logic levels (0–5 V)				
Data rate 300 kHz; cable length 200 ft (60 m)	Satisfactory	Satisfactory	Satisfactory	No
Data rate 10 kHz; cable length < 3000 feet (1000 m)	Adequate	Adequate	Preferred	No
Data rate < 3 kHz; cable length > 3000 feet (1000 m)	Adequate	Adequate	Preferred	No
Digital output (open collector switching)	Satisfactory	Satisfactory	Preferred	Adequate
Data rates and cable length as above (relay contacts)	Satisfactory	Satisfactory	Satisfactory	Satisfactory
Analog-input high-level relay multiplexers (0–50 V typical)	Satisfactory	Satisfactory	No	No
Analog-input low-level relay multiplexers (0–100 mV typical)	Adequate	Preferred	No	No
Analog-input high-level solid state	Satisfactory	Satisfactory	No	No
Analog-input low-level solid state	Adequate	Preferred	No	No
Analog output	Adequate	Preferred	No	No

Protection from Water

The insulation used to protect electrostatic shields on cables may contain voids and pinholes. While the resistance between the shield and dry metal conduit may be in the region of 50 or 100 megohms over a 100-meter length, immersion in water can cause a dramatic decrease in resistance to ground, and may fall to a few hundred ohms. Normally the conductor resistance to ground is not affected, but failure of the outer sheath can be a notorious source of ground loops.

INTERFACING IN HAZARDOUS ENVIRONMENTS

Explosion and Fire Hazard in Industry

The dangers of fire and explosion have always been a preoccupation of engineers involved with coal mining operations. Fire may occur when a mixture of coal dust and air, or naturally occurring methane gas and air, is ignited. If a critical mixture of these ingredients is ignited, under certain conditions combustion spreads through the mixture at high speed and an explosion occurs. From the time that coal was first mined, ignition of these mixtures has been responsible for the destruction of numerous mines and the death of an untold number of miners, and even now disasters occur.

During the nineteenth century, development of safe equipment and strict enforcement of safety regulations considerably reduced the incidence of disaster, but the introduction of electric equipment at the beginning of the twentieth century added a new dimension to the problem. Also about this time new industries, particularly petrochemical, were evolving in which the hazards of fire and explosion were as acute as those experienced in the mines. Development of the safe equipment currently available stems from a combination of activities. Extensive laboratory investigation coupled with the findings of exhaustive inquiries into accidents has led to the design of equipment with a very high standard of safety. Even so, no electric equipment can be regarded as absolutely safe. Combinations of the most unlikely failures, or human error, or negligence, can create hazards where none previously existed. In these circumstances, equipment design and government legislation can only go so far in providing standards of safety. Ultimately, responsibility rests with the engineer. The quality of an installation and the engineer's vigilance in maintaining its standards are often the only real barrier between life and death.

Electric equipment capable of producing sparks is always a potential source of ignition in a flammable atmosphere. Where equipment is designed so that the thermal energy in the spark is insufficient to ignite the gas/air mixture, the equipment is described as intrinsically safe. Where equipment produces sparks capable of igniting a flammable atmosphere, then the source of ignition must be isolated from the atmosphere. One method uses an enclosure constructed such that if ignition does occur, the flame cannot propagate outside the enclosure and spread to the surrounding atmosphere. These are described as explosionproof enclosures. Another method uses purging systems to provide an atmosphere inside the enclosure where ignition cannot occur.

Ignition of Inflammable Gases

The theory of ignition and uncontrolled combustion is very complex and may not be described in a few pages. Basically, if a jet of flammable gas in contact with a surplus of air is ignited, the resulting flame consists of a number of layers. At the center of the jet is a concentration of cool pure gas. Outside this layer is a critical mixture of gas and air that produces a chemical reaction that is strongly *exothermic* (i.e., it produces a considerable amount of heat described as combustion). Beyond the combustion layer, a rapidly expanding mixture of combustion products and air at high temperature is generated. The chemical reaction is sustained by new air being brought into contact with the gas by convection currents and maintaining the critical mixture. The speed of the chemical reaction, and consequently the intensity of combustion, is determined essentially by how quickly, and how effectively, the critical mixture can be created. If a jet of gas is allowed to burn in convecting air at the end of a pipe, mixing is inefficient and a large low-temperature flame results. If gas and air are premixed, a more concentrated hotter flame is produced at the end of the pipe. The operation of a Bunsen burner is the best known demonstration of this phenomenon.

Suppose a source of ignition is introduced into an atmosphere consisting of a critical mixture of flammable gas and air. The mixture in contact with the source will ignite and this will, in turn, ignite the mixture in contact with the burning gas. The outcome of this is a combustion wavefront that spreads through the atmosphere, leaving behind it hot products of combustion. The combustion wave propagates at high speed, and the consequent expansion of the hot gases produced is described as an explosion.

From this point, the subject of ignition becomes extremely complex. The critical gas/air ratio is only critical insofar as the chemical reaction is and hence combustion is complete at this ratio. Also at this ratio the energy required to ignite the mixture is minimal. However, the combustion wave will still propagate with a limited range of mixtures outside the critical, but the thermal energy required for ignition also increases. Propagation of the combustion wave is also effected by the type of gas, its temperature and its pressure. Where ignition is caused by arcs during the opening of electrical contacts, the situation is further complicated. For example, the current required to ignite a mixture, using tin-plated copper contacts and a noninductive circuit, is considerably more than when using cadmium-plated contacts. Where an inductive circuit is concerned, this discrepency disappears.

Classification of Hazardous Areas

The classification of hazardous areas varies throughout the world and is the subject of local definition. In the United States, the fundamental classification originates from the National Electrical Code (NEC) prepared by the NEC Committee of the National Fire Protection Association. Article 500 of this code broadly defines hazardous areas into three classes:

- Class I constitutes those locations where flammable gases or vapors are present in air, in quantities sufficient to produce an ignitable or explosive mixture.
- Class II constitutes locations where combustible dusts may be present in sufficient quantity to cause a fire or explosion hazard.
- Class III constitutes locations where easily ignitable fibers and flyings are present, but not normally suspended in air in sufficient quantities to produce ignitable mixtures.

Hazardous materials are divided into groups to provide a more specific subclassification of the hazard. Class I materials are divided into four groups, A through D, the materials of each group representing a hazard of the same general character. Similarly, class II materials are divided into groups E through G.

The following is typical class I subgrouping.

Group:	A	B	C	D
Atmosphere:	Acetylene	Hydrogen	Carbon monoxide	Acetone

The following is typical class II subgrouping.

Group:	E	F	G
Material:	Aluminum dust	Carbon black	Flour

Finally, areas are designated according to the probability of hazardous material being present in ignitable concentrations.

- Division 1 designates a location where hazardous concentrations will exist either continuously or periodically during normal operation; or maintenance of the plant. Also included in this division are those locations where equipment or process failure might simultaneously release hazardous concentrations of material and also cause failure of electric equipment (e.g., metal dusts).
- Division 2 designates a location which is presumed to be hazardous only in abnormal conditions. These include areas where flammable liquids or gases are processed but normally confined to closed containers, or where an area is maintained in non hazardous conditions by forced ventilation.

The cost of providing safety in hazardous environments is high and consequently there is considerable financial incentive when designing a plant to provide the conditions that will result in the lowest classification practicable. However, classification is a complex process. The types of material being handled and their quantities obviously feature prominently in any consideration, but not exclusively. Geographic location of a project, prevailing wind, the construction of plant and buildings, and even the safety record of an organization, can influence a decision.

Safety Considerations for Electric Equipment

When considering equipment for use in hazardous locations, each circuit is carefully analyzed as to its safe operation. In general, it is not difficult to recognize potentially dangerous electric equipment and isolate it from the environment in which it operates. The problem is rather one of recognizing possible sources of failure and determining their effect on the safety of the circuit. A thermocouple, for example, is incapable of generating sufficient energy to ignite a flammable mixture. If, however, the thermocouple is connected to electronic equipment, it is possible that a fault developing on the input of the equipment may cause large currents to flow through the thermocouple. In this event the thermocouple will become a potential source of ignition.

There are a number of ways this circuit can be prevented from becoming dangerous if equipment failure occurs. It is possible, for example, to enclose the thermocouple in a sheath and it is possible to insert an electrical barrier between the thermocouple and the equipment. Barrier circuits are designed to prevent the current flowing through the barrier from rising to a dangerous value. Either or both of these protective measures may be introduced.

Considering the construction of a sheathed thermocouple, in a typical design, the sheath is integral with a housing that contains a terminal block. This housing, often described as the thermocouple head, may be an explosionproof container with a threaded conduit entry for cables. Before an explosion can occur in a failure condition, it is necessary for the sheath to rupture in such a way that an explosive mixture builds up around the thermocouple. Simultaneously, a heavy current must flow through the thermocouple, causing the wires to fuse. This is an unlikely coincidence, and if a barrier is installed, the possibility of coincident failure of all three is almost negligible. Even so, this circuit is not necessarily safe. There is also the possibility that as a result of a rupture of the sheath, flammable material will perculate along the conduit, eventually escaping at some considerable distance from the source. It is even possible that the point of escape into the atmosphere may be in an area considered safe, and where flameproofing is not considered necessary. To prevent this type of hazard from occurring, it is normal to seal the ends of the conduit.

Many instrumentation systems involve the use of large numbers of switches to indicate the state of a controlled system. Sometimes these switches are enclosed in explosionproof cases, but often advantage is taken of proximity devices to provide the switching function. The level switch illustrated in Figure 11.10 is an example. The float of this switch moves a magnet located in the hazardous area, while the switch, operated by the repulsion of a complementary magnet, is located in a safe area. Separation of the two areas is maintained by a nonmagnetic diaphragm. Another type of switch in common use uses fiber optics where light between a transmitting and receiving optical bundle is interrupted by a shield at the moment of switching.

Barrier Protection

The use of electrical barriers as a method of preventing the rise of current to a dangerous level has already been mentioned. One type uses Zener diodes to restrict the rise of voltage in a faulty circuit, and to short-circuit the current until a fuse is blown. Zener diodes are diodes that behave like normal diodes until a reverse voltage is raised above a specified value. At this point the resistance of the diode falls to a low value allowing a high current to flow, thereby preventing any further rise in the reverse voltage. There is an obvious limit of power that the diode can dissipate and circuit parameters must be designed to prevent failure.

A basic Zener barrier circuit is shown in Figure 17.5. The supply AB is connected through a fuse to a Zener diode Z_1. One Zener and a series resistance R_1 in normal operation will prevent the voltage and current rising in the hazardous area beyond the safe value. In practice, redundancy is added to the circuit in the form of another Zener diode, Z_2 to guard against failure of Z_1.

There are several variations of the basic circuit. Where currents are small, the fuse may be replaced by a high-value resistance. In another variant where a differential supply is used, the circuit may be duplicated for each line of the supply as shown in Figure 17.6.

Zener barriers are available as commercial units and their application is well supported technically by firms that have considerable experience in their installation.

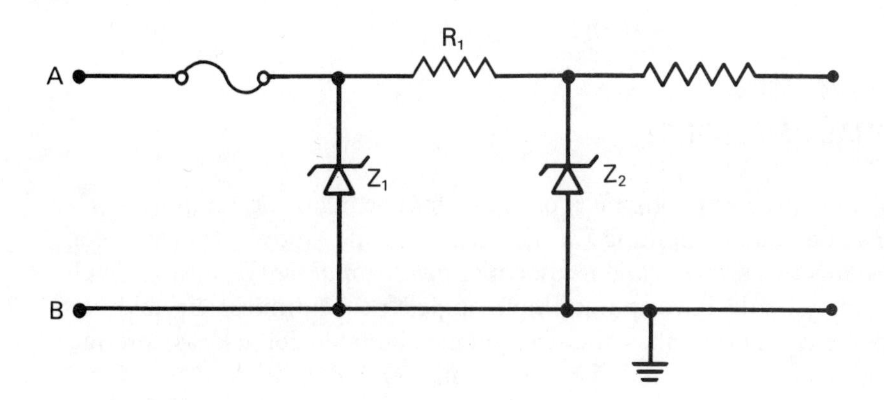

Figure 17.5
The Zener Barrier

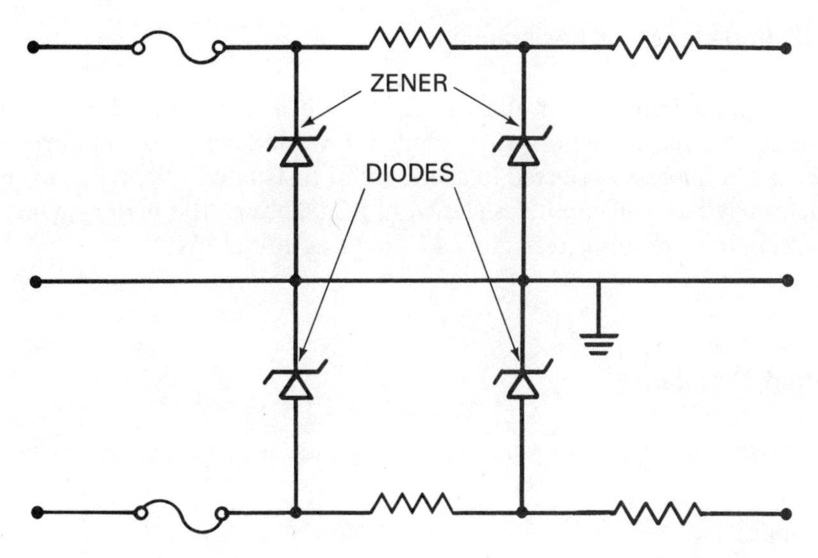

Figure 17.6
Barrier Protection for Differential Supplies

INTERFACING WITH INTEGRATED CIRCUITS

There is now a wide variety of integrated circuits available that can be used for interfacing sensors to an instrumentation system. They have the advantage of being inexpensive, and the engineer need not be an electronics specialist to develop simple interface equipment.

Operational amplifiers are the most common, but it is possible to obtain time-out circuits, comparators, bistables, and many other logic circuits that are adaptable to a wide variety of applications. These circuits are well advertised and well documented and this chapter is intended to provide only an introduction to their application.

OPERATIONAL AMPLIFIERS

Operational amplifiers were originally designed for use in analog computers. The mathematics of analog computing was developed for the study of transient control response in servo systems and is known as *operational mathematics*. Ideally, these amplifiers should have infinite input impedance, infinite gain, and zero output impedance. This makes them eminently suitable for use as interface amplifiers between a sensor and an instrumentation system input.

To understand operational amplifiers; it is necessary to understand the basic parameters associated with them.

Input Impedance or Resistance

This is the impedance to an alternating current at some specified frequency. Because operational amplifiers are intended to work down to direct current, impedance is often loosely referred to as the input resistance. While less expensive amplifiers may have an input resistance of 2 megohms, the better quality FET amplifiers may have input resistances as high as 10^{11} ohms.

Output Resistance

Ideally the output resistance value is zero. In fact, it is a finite value calculated from:

$$\frac{V_{out}(\max)}{I_{out}(\max)}$$

and may be as low as 0.1 ohm.

Gain

The gain of an amplifier is expressed as V_{out}/V_{in}; where V_{out} is the voltage measured at the output terminal and V_{in} is the voltage across the input. The gain of an operational amplifier is usually very high and may be in excess of 100,000 to 1. This is known as the open-circuit gain.

Often gain is expressed as a power ratio in decibels (dB). The decibel is a power ratio obtained from the formula:

$$dB = 20 \log_{10} \frac{V_{out}}{V_{in}}$$

To convert decibels to voltage ratios. Table 17.4 can be used to provide a close approximation:

Table 17.4 Decibel Conversion Table

dB	Gain
0	1
3	$\sqrt{2}$
6	2
9	$2\sqrt{2}$
12	4
15	$4\sqrt{2}$

Frequency Response

If the input of an amplifier is supplied with a sinusoidal input of constant amplitude but of increasing frequency, the output will remain essentially constant up to a certain value of frequency and then will decrease. The frequency over which the amplifier maintains a constant amplitude output is known as the *flat* part of its frequency response curve. This is the frequency over which the output is essentially proportional to the input. It is common practice, however, to extend the frequency range of the amplifier to the point where the response curve falls to 3 dB of its level value and accept the incurred inaccuracy at the high-frequency end of the response.

Feedback

A proportion of the output voltage is fed back to the negative input terminal of the amplifier. This is known as *negative feedback*. The greater the feedback, the less the closed-loop gain of the amplifier, or the gain with feedback, but the greater the frequency response. Negative feedback may, therefore, be used to adjust the overall gain of the amplifier or to extend its frequency response to suit a particular application.

APPLICATION OF OPERATIONAL AMPLIFIERS

Figures 17.7 through 17.13 illustrate some of the ways in which operational amplifiers may be connected and indicates how their circuit parameters are calculated. These circuits are well documented in manufacturer's literature, which should be referred to for detailed design data.

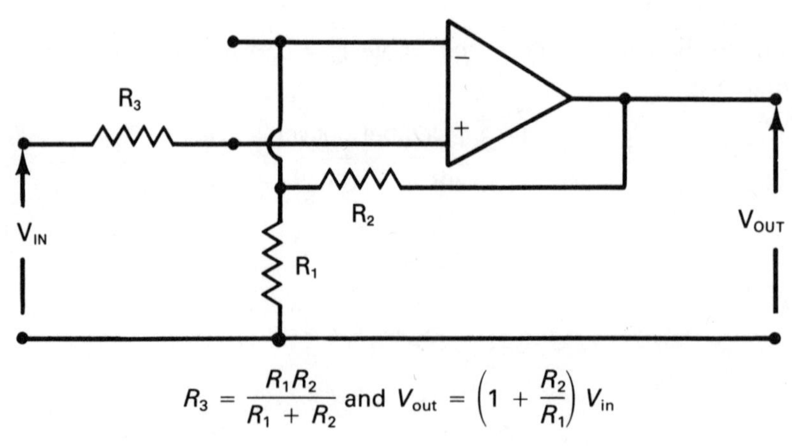

$$R_3 = \frac{R_1 R_2}{R_1 + R_2} \text{ and } V_{out} = \left(1 + \frac{R_2}{R_1}\right) V_{in}$$

Figure 17.7
Noninverting Amplifiers

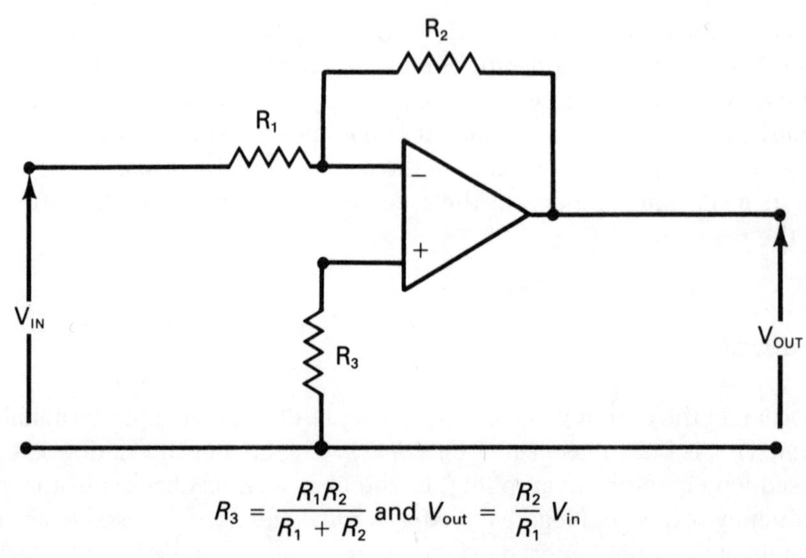

$$R_3 = \frac{R_1 R_2}{R_1 + R_2} \text{ and } V_{out} = \frac{R_2}{R_1} V_{in}$$

Figure 17.8
Inverting Amplifier

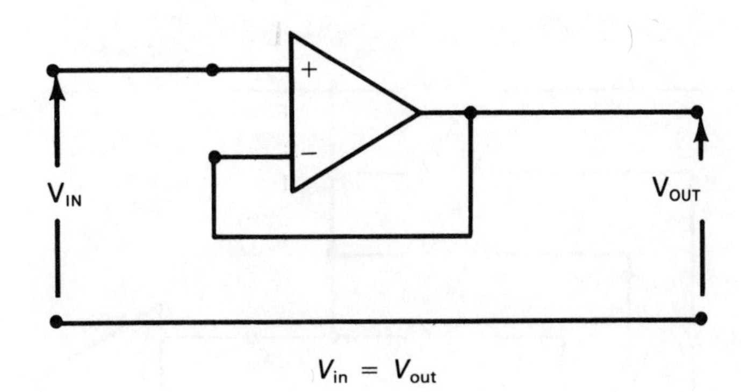

$V_{in} = V_{out}$

Figure 17.9
Follower

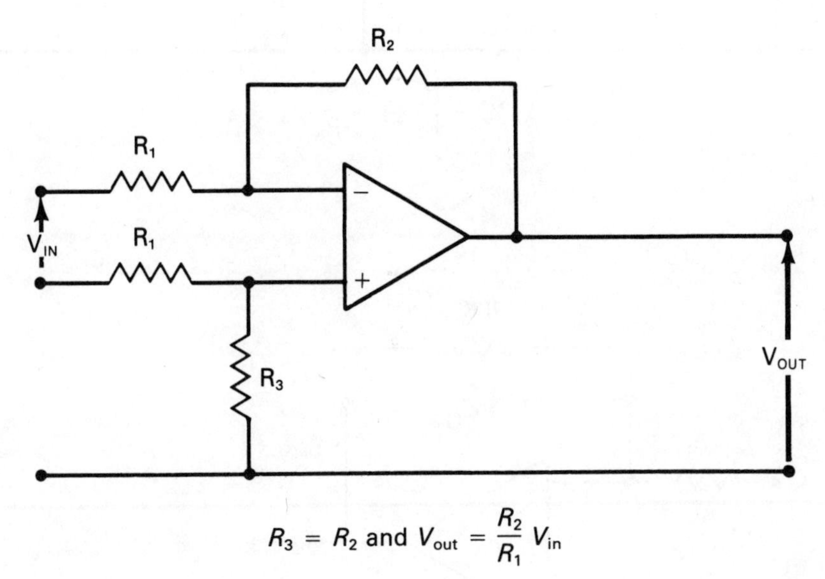

$R_3 = R_2$ and $V_{out} = \frac{R_2}{R_1} V_{in}$

Figure 17.10
Differential Amplifier

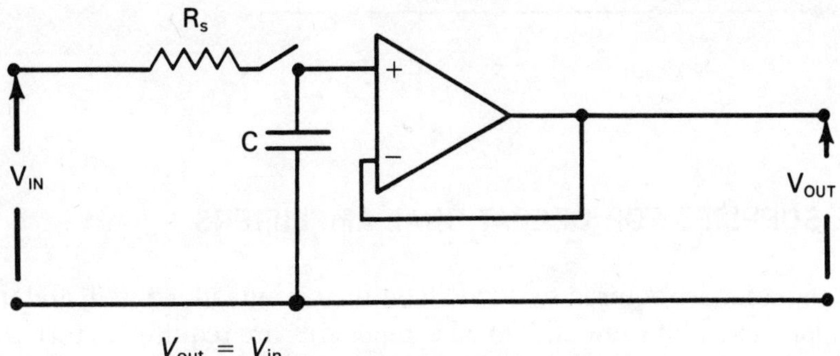

$V_{out} = V_{in}$
Charge time $= 7R_sC$ R_{in}: Input resistance
Discharge time $= 7R_{in}C$

Figure 17.11
Sample and Hold Amplifier

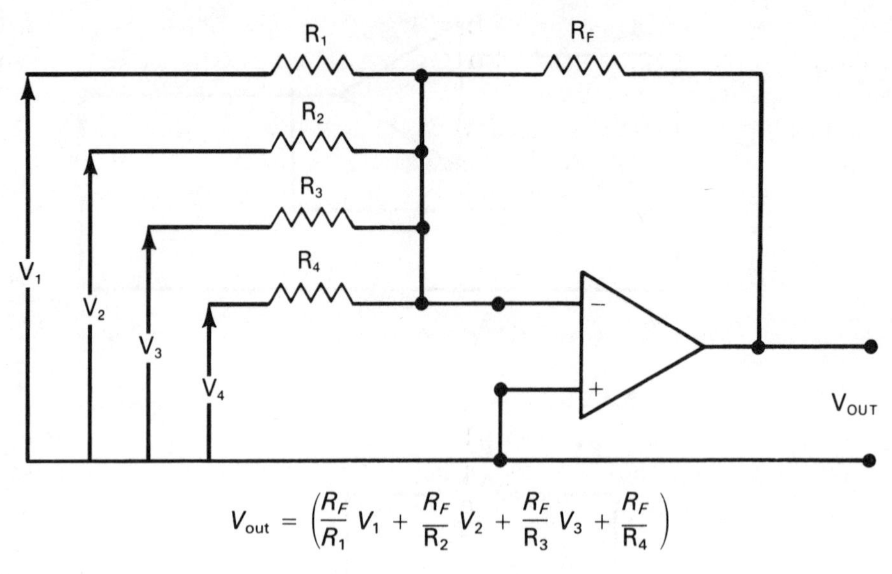

$$V_{out} = \left(\frac{R_F}{R_1} V_1 + \frac{R_F}{R_2} V_2 + \frac{R_F}{R_3} V_3 + \frac{R_F}{R_4}\right)$$

Figure 17.12
Summing Amplifier

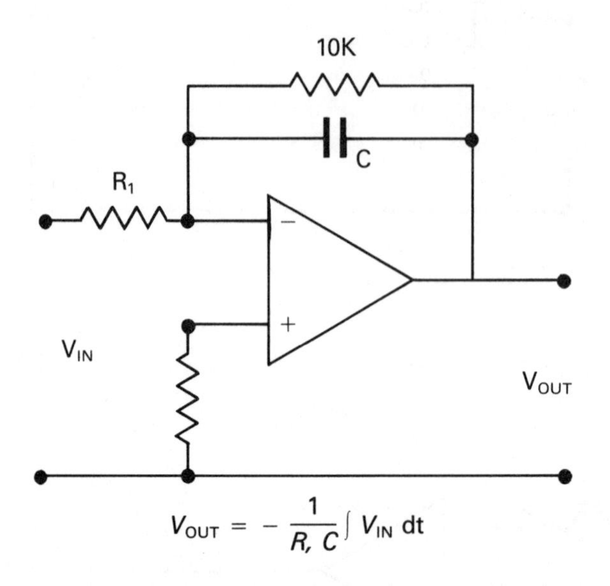

$$V_{OUT} = -\frac{1}{R, C}\int V_{IN}\,dt$$

Figure 17.13
Integrator

POWER SUPPLIES FOR OPERATIONAL AMPLIFIERS

Little explanation is required to understand the circuits illustrated, and it can be readily appreciated how few additional components are required to turn the basic amplifier into a functional unit. In addition to the components shown, however, these circuits also require power supplies. Usually these are dual supplies supplying power that is both positive and negative to a ground reference. Where high performance is not required, batteries are often used as power supplies, and where a number of amplifiers are to be driven from a central source, a simple unstabilized supply is often used with Zener diodes providing a degree of stabilization as illustrated in Figure 17.14a. Where high performance is required, the unstabilized supplies may be provided with voltage regulators. Regulators are also integrated circuits that add very little cost or complexity to a basic unstabilized supply to provide a high degree of regulation. This is illustrated in Figure 17.14b.

INTEGRATED TIME-OUT CIRCUITS

Integrated time-out circuits may be used in two ways. They may be used as a *one-shot timer* or as *monostable*, which produces a change of output at some predetermined time after an input has been applied. Before a second output is obtained, the input must be again initiated. In the alternate method of operation the circuit switches itself on and off at predetermined regular time intervals. In this configuration it is described as a *multivibrator* or *astable*.

The wiring of timers and their component values depend on their design and the manufacturer's literature should be consulted. The type shown in Figures 17.15 and 17.16 is a typical type of device common to a number of manufacturers.

When using timers at low frequency, large values of C_1 and R_1 are required. Care must be taken in selecting the capacitor so that minimum leakage is experienced. In addition, timers may become unstable due to noise pickup in long leads. To work efficiently, therefore, good wiring practice must be observed.

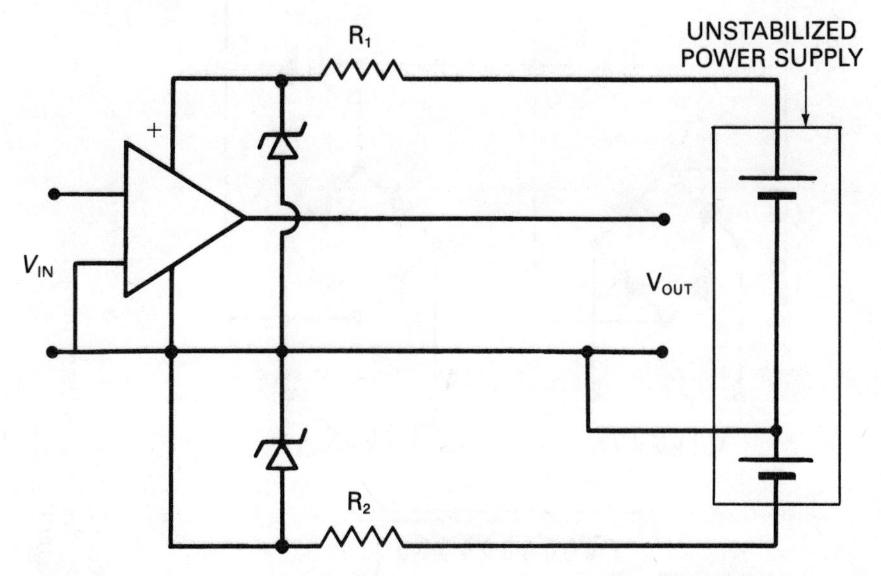

Figure 17.14a
Amplifier Power Supply Stabilization

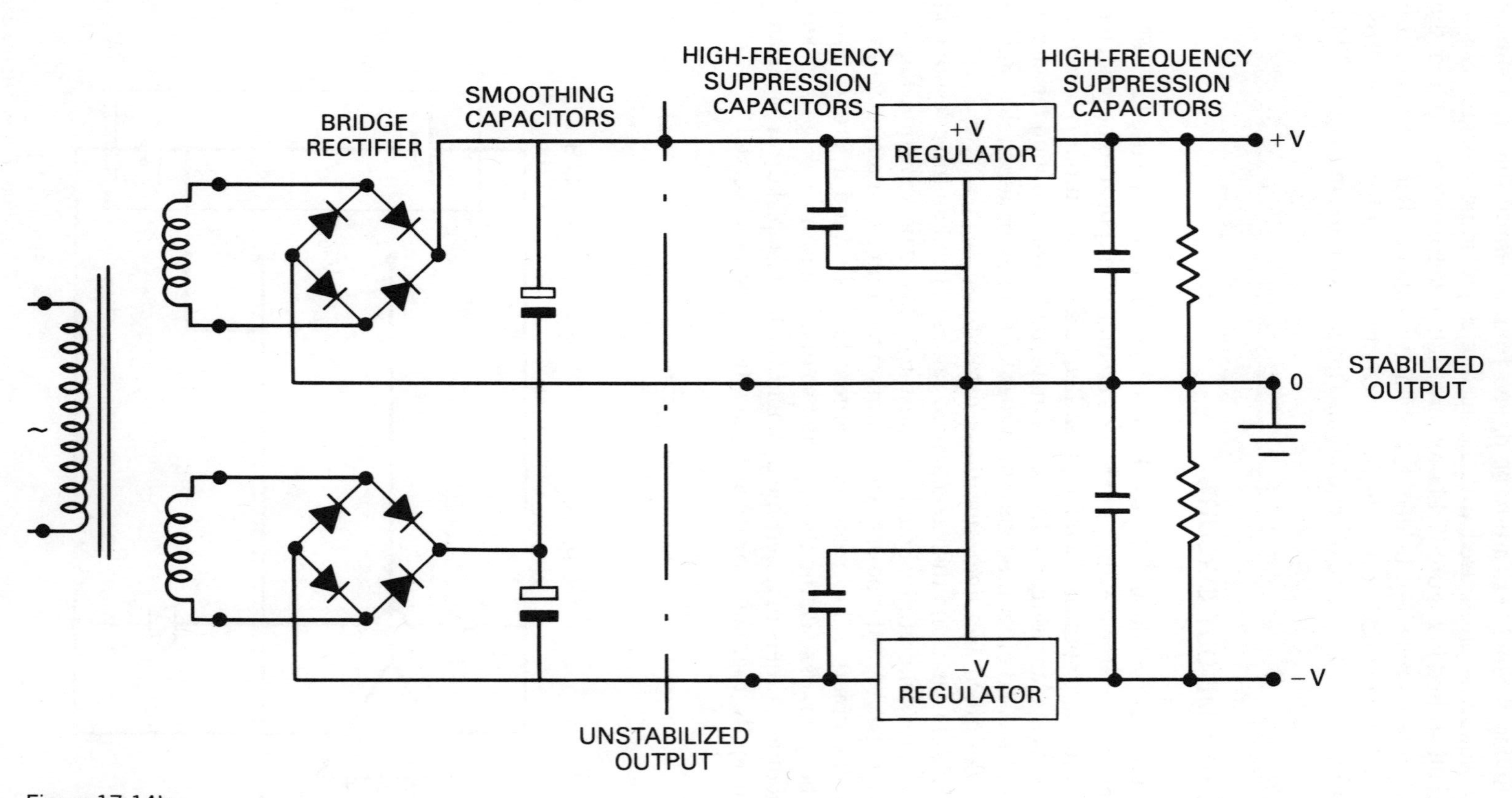

Figure 17.14b
Stabilized Supply Using Integrated
Circuit Regulators

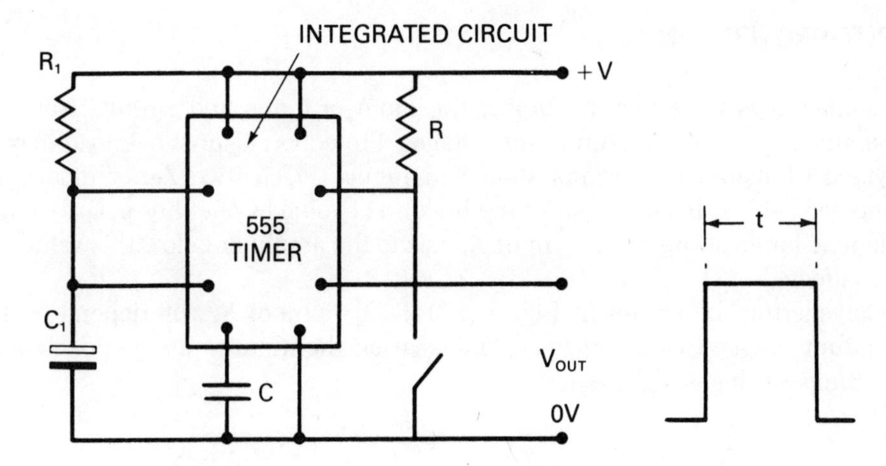

$T = R_1C_1$
where T is in seconds, R_1 in ohms, and C_1 in farads.

Figure 17.15
Monostable Operation of Integrated Circuit Timer (courtesy of RS Components Ltd.)

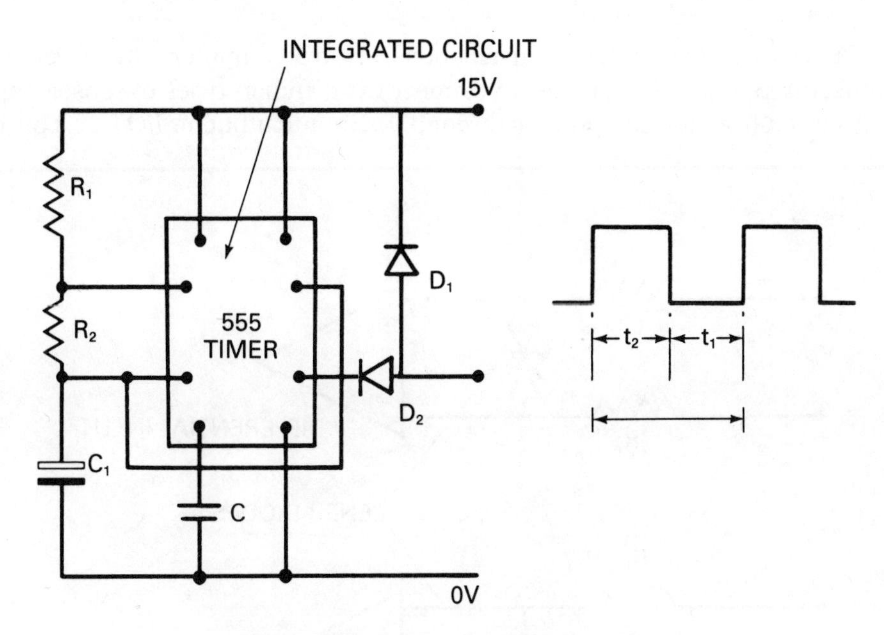

t_1 (load unenergized) $= 0.685\ (R_1 + R_2)\ C_1$
t_2 (load energized) $= 0.685\ R_2C_1$
Total period $T = t_1 + t_2$

Figure 17.16
Multivibrator Operation of Integrated Circuit Timer (courtesy of RS Components Ltd.)

INPUT PROTECTION

Sometimes it is necessary to protect the input of integrated circuits from the accidental application of high input voltages. Protection against overload may be provided by using the circuits shown in Figure 17.17. Two Zener diodes are connected back-to-back to restrict the buildup of voltages of either polarity to the avalanche limits of the diodes. Input signals to the amplifier below this value will be unaffected.

Considering the circuit in Figure 17.17, the value of R_1 will depend on the maximum voltage that is expected to appear on the input in an overload condition. Suppose this is V_{in}, then:

$$R_1 = \frac{V_{in} - V_z}{I_z}$$

where V_z is the Zener avalanche voltage and I_z is the maximum allowable Zener current.

V_z should be chosen at some value between the maximum signal voltage and the maximum voltage the input can accept without damage.

INTERFACING TO COMPUTERS

Many modern instrumentation installations involve computers and frequently manufacturers provide interfaces to connect the common types of sensor. Input to and output from the computer is through the Input/output, or *I/O bus*. This is a

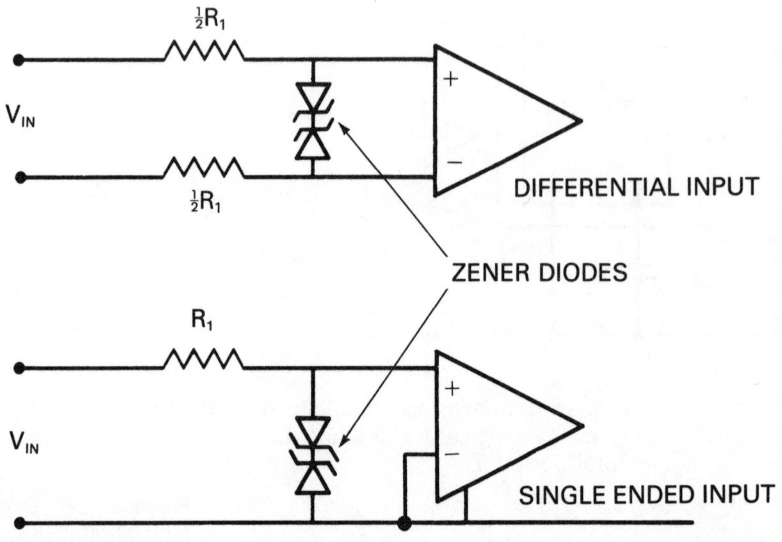

Figure 17.17
Input Overload Protection

set of parallel input and output circuits with appropriate control lines. I/O bus design depends on the architecture of the computer and each bus is usually unique to a computer family. Connection between attached equipment and the I/O bus is normally through an adapter card, provided with its own microprocessor. These microprocessors are used to organize data and control signals between external equipment and the main processor. This allows much routine work to be transferred from the main processor to the microprocessor, thus allowing the main processor to be utilized much more efficiently.

The use of adapter cards makes possible the attachment of almost any type of equipment requiring electric drive signals or providing electrical output. In this chapter the more common interfaces will be considered.

Most manufacturers of computers provide parallel digital input/digital output adapters to connect other electronic equipment. With suitable networks and external power supplies these adapters may be used to sense remote contacts or switch low-power intermediate relays.

Analog input is often provided by a special type of data logger supplied by the computer manufacturer. This data logger connects to the computer through its own adapter card. In many instances, however, manufacturers of data loggers provide their equipment with interfaces compatible with those fitted to most computers. This arrangement often allows the logger to be sited remotely from the computer with a considerable saving in cable costs using another important type of adapter described as the *serial* or *communications line* adapter. Such equipment may operate in a number of ways; the most common use *binary synchronous communication* (BSC) or *asynchronous communication* (Asynch) methods.

Binary synchronous communication operates with a *clock* at the transmitting station that pulses a binary code onto a line at a fixed rate. A clock at the receiving station samples the line at fixed intervals, maintaining its synchronism by adjusting itself to line transitions.

Asynchronous transmission systems have been in use for many years for low-speed telegraph systems. They use what are known as start bits and stop bits to indicate to the receiving equipment the beginning and end of a character. Transmission is at a fixed rate and it is assumed that the receiving equipment will maintain synchronism over the length of a character. These systems for many years have been used extensively for computer input and output. Asynchronous transmission codes have been standardized into the well-known ASCII code (American Standard Code for Information Interchange). Many types of equipment are fitted with asynchronous interfaces and most computers are provided with compatible adapters.

When large numbers of widely scattered sensors and controls are to be connected to a computer, serial transmission systems are available where sensors and controls are connected to conveniently placed concentrators. A number of these concentrators may be connected to the computer by a common transmission line. This technique is described as *multidropping*. Each concentrator has its own unique address code and responds only to messages from the computer prefixed by that code.

CAMAC INTERFACE

One of the major problems of computing is the difficulty experienced in interfacing instruments and equipment to the many different computer I/O buses. One system addressing this problem is the CAMAC interface. In 1969, the ESONE Committee of the European Atomic Energy Community published a specification entitled "CAMAC Modular Instrumentation System for Data Handling." This requires that all conforming computer peripherals be designed with an interface compatible with the CAMAC interface. The CAMAC bus and the computer bus are linked by a custom-designed controller that converts one type of interface to the other. Such an arrangement allows replacement of one computer in an installation by another with only the minimum change.

The CAMAC specification requires that all devices be connected to what are described as instrumentation crates. Each crate containing its own controller is an enclosure accepting up to 25 compatible devices into a bus structure called a *dataway*. This dataway provides 24 data lines and 32 addresses and control lines. The controller is responsible for the provision of buffering, control, and data transfer operations between individual devices and the computer. There are basically three methods of connection requiring the provision of a different type of controller for each installation. These are illustrated in Figures 17.18, 17.19, and 17.20.

The direct-connect method provides controllers that interface each crate directly to a computer I/O bus. Data transmission rate may be up to 24 megabits per second with a maximum transmission length of about 300 feet (100 m). The second type of controller is a "branch highway" controller; it allows up to seven crates to be connected serially by a 66-pair cable.

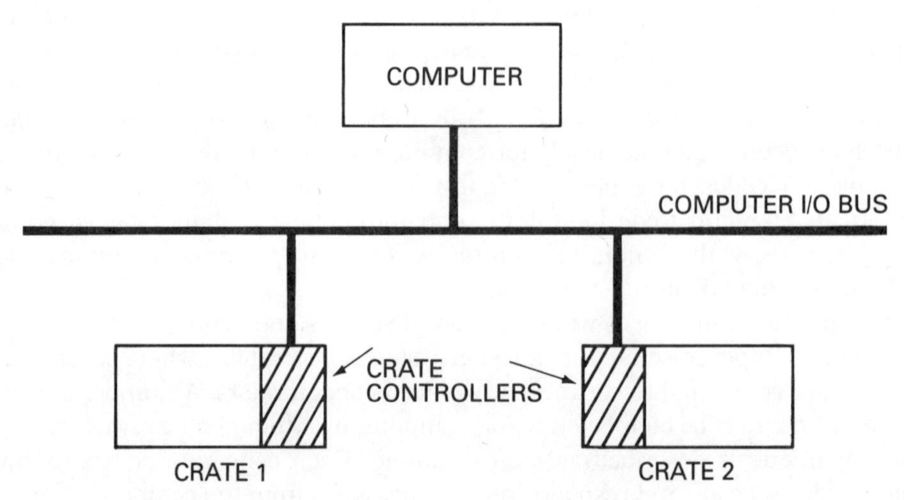

Figure 17.18
CAMAC Interface to I/O Bus

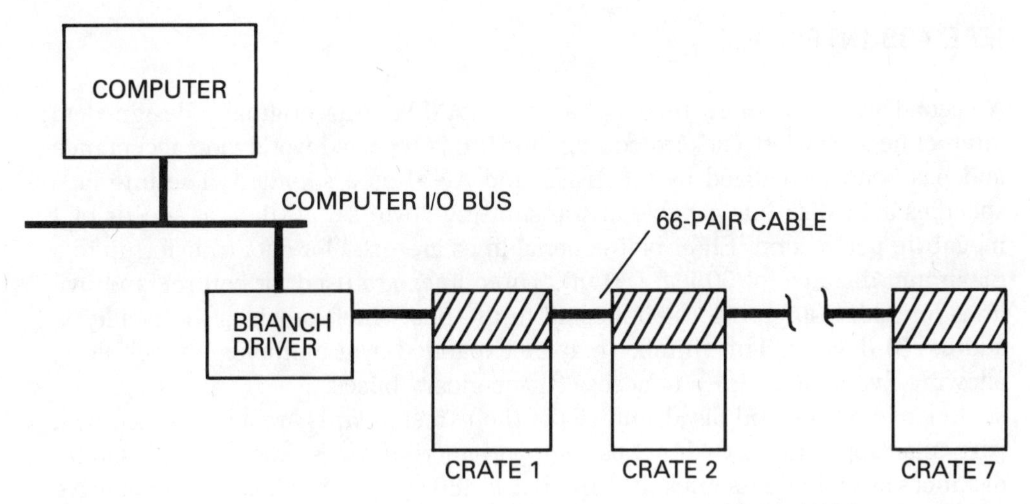

Figure 17.19
CAMAC Branch Highway Connection

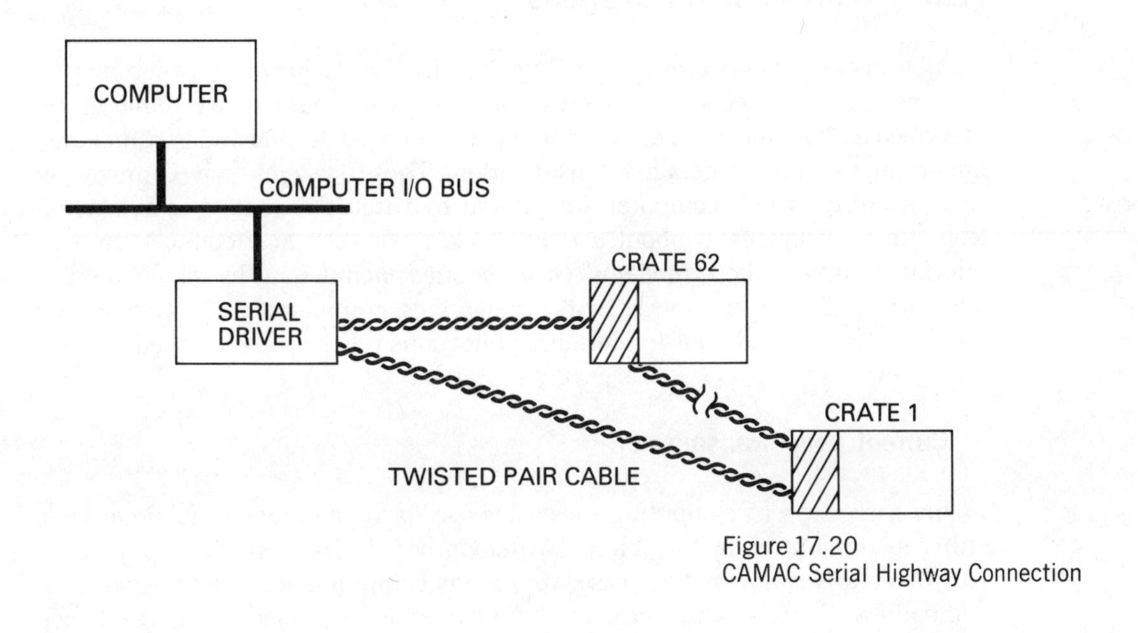

Figure 17.20
CAMAC Serial Highway Connection

The serial transmission systems operate on a twisted pair or coaxial cable loop. Up to 62 crates may be attached and data rates of up to 5 megabits per second may be achieved over unlimited distances. This system, as well as allowing much greater transmission distances, also provides considerable economy in cable installation cost.

IEEE 488 INTERFACE

A second interface more recent than the CAMAC was originally designed to interconnect Hewlett-Packard equipment. It has received worldwide acceptance and has been formalized by the IEEE and ANSI as a standard. The interface specifies a 16-wire bus capable of transmitting "byte serial" data at speeds of 1 megabyte per second. Eight bidirectional lines are used for data transfer up to a maximum distance of 60 feet (20 m). Three lines are used for control, and five lines for addressing devices connected to the bus which provides the facility to address 31 devices. This number may be expanded by a multiplexing technique allowing five address lines to access 31 secondary buses.

Relative merits and disadvantages of the two standards are accentuated by a particular application. CAMAC can support more devices, and transmit longer distances at higher data rates than the IEEE/488 system, but it is more expensive and complicated. Instruments that do not conform to either specification are more easily interfaced to the IEEE/488 system than the CAMAC.

SERIAL TRANSMISSION SYSTEMS

Serial transmission systems are rated by their transmission speed, which may be millions of bits per second, or as low as 50 bits per second. One particular group described as "low speed" operates at up to 9600 bits per second and is sufficiently important to warrant detailed consideration. These systems may connect peripherals directly to a computer attachment by what is described as a "current loop" up to distances of about a mile (1.5 km), or they may connect through interfacing units called "modems" (*mod*ulators/*dem*odulators) by telephone-type lines over unlimited distances. By far the greatest number of these systems use the 7-bit "ASCII" code but for certain applications other codes are used.

Current Loop Transmission

In the early days of computing most keyboard/printer input/output units were either manufactured by the Teletypewriter Corporation or were directly compatible. While this is now not the case, very many computers are still provided with compatible adapters. One form of this adapter is described as "current loop adapter" and may or may not provide power for the transmission loop. Figure 17.21a illustrates the situation where current is supplied by an external power source; Figure 17.21b illustrates the situation where power for each transmission line is provided by the adapter.

Current loop operation requires a loop current between 20 and 60 mA. Sometimes computer adapters are designed to provide only the lower limit of current and then the line resistance may be sufficiently great to prevent the device from operating satisfactorily. In this instance external supplies with a variable output

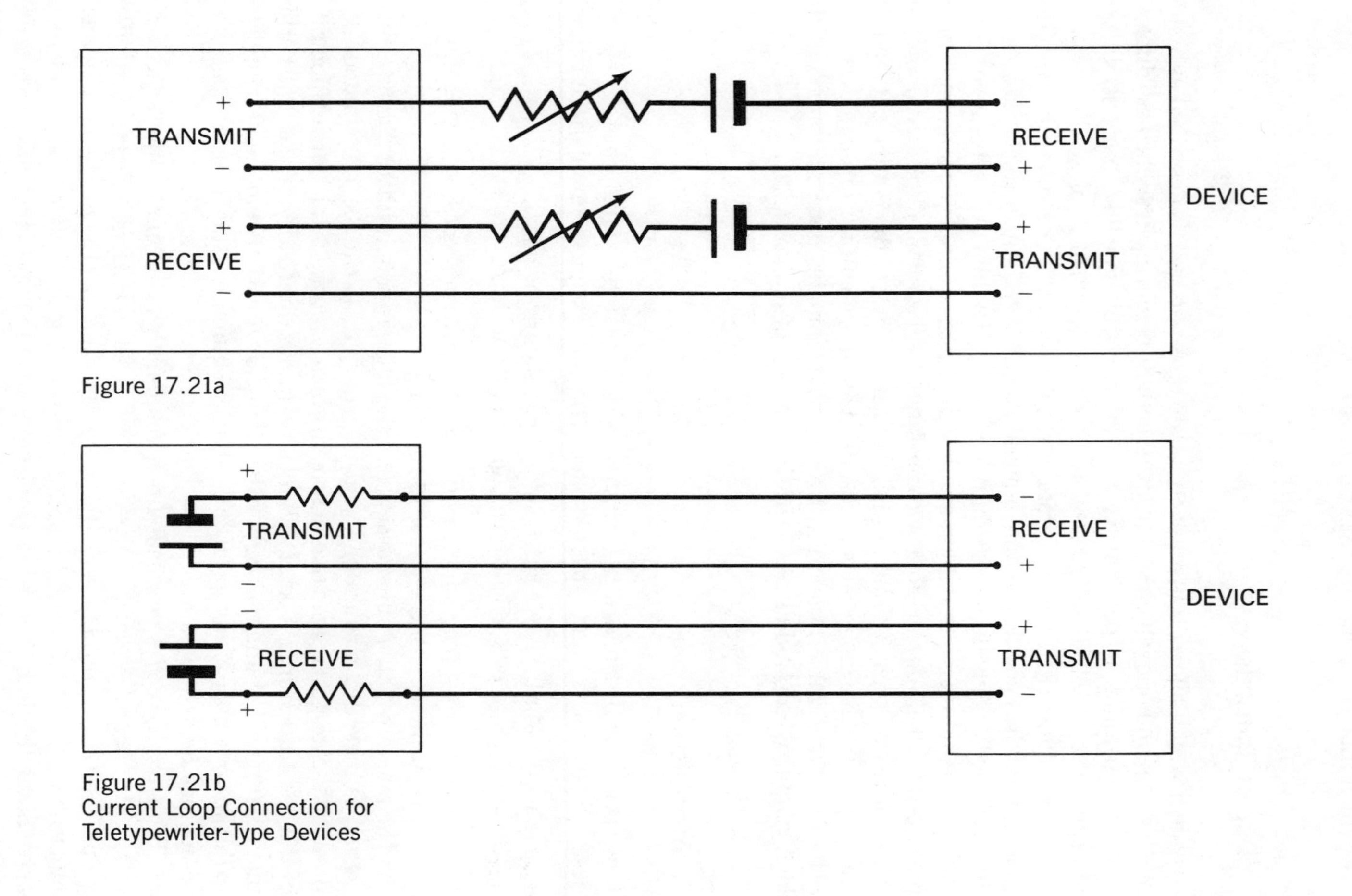

Figure 17.21a

Figure 17.21b
Current Loop Connection for
Teletypewriter-Type Devices

allow the operating current to be adjusted to take line resistance into account. Such adjustment, however, will be limited by the maximum common-mode voltage which may be applied to the equipment.

Duplex Transmission

A system which allows transmission in one direction only is described as a simplex system and one which transmits in both directions is described as "duplex." Two-way operation is possible on a single transmission line, but only in one direction at a time. Such a system is described as "half duplex." The systems shown in Figure 17.21 have separate transmit and receive circuits, and with this arrangement it is possible to transmit and receive simultaneously. These systems are described as "full duplex." A third method of operation uses separate transmit and receive circuits with equipment designed for half-duplex operation. This is described as "half-duplex, four-wire operation." Where half-duplex equipment is used with two wires, a finite time is required to switch the equipment from its transmit to receive mode. This is described as equipment *turnaround* delay. The use of separate transmit and receive circuits eliminates delay and much of the logic circuitry required to cope with the change of mode.

Transmission Code

Low-speed serial data are transmitted as a "two-state" signal code. These are described as "mark" and "space" states. The convention is summarized below:

Binary 1	Mark	Negative voltage or no current flow
Binary 0	Space	Positive voltage or current flow

While this convention is normal, it is not standard and occasionally equipment operating in the opposite sense is encountered.

The rate at which a transmission line can change state without degradation of signal defines the maximum speed at which data can be transmitted. The number of changes in state that may be made in a second is the "modulation rate" of the line with a unit of *baud*. Modulation rate is a measurement of line performance. It does not automatically follow that transmission equipment connected to the line operates at this speed, even though equipment transmission speeds are sometimes quoted in bauds.

Transmission codes are made up of combinations of "mark" and "space" elements. In many systems, where a number of marks or spaces occur consecutively, there is no change in the state of the line that the receiving equipment must decode. This is done by specifying the length of time a signal element appears on the line, and then employing a sampling system in the receiving equipment which examines the state of the line at these intervals. Figure 17.22 illustrates the form of such a five-element serial code.

Just as the receiver must sample precisely at a fixed rate, so the transmission

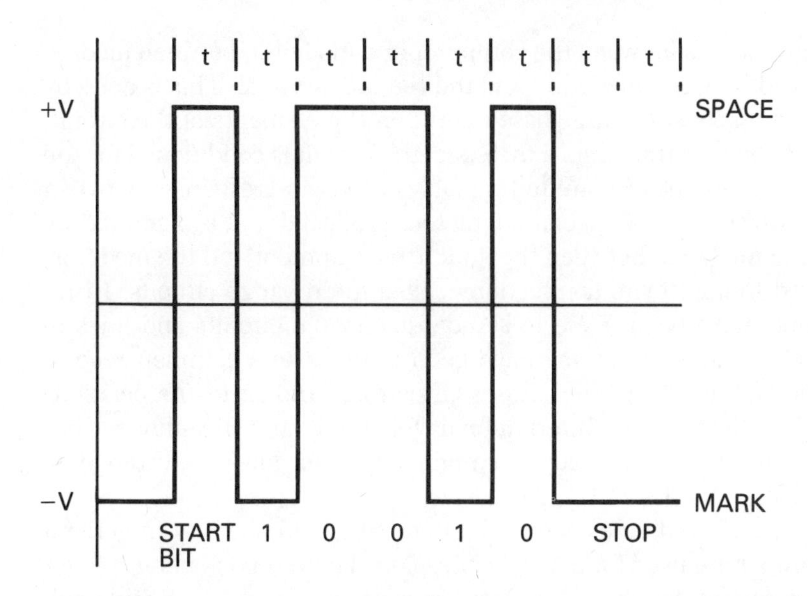

Figure 17.22
Five-Element Serial Code

equipment must provide signal elements of precise length. Asynchronous transmission systems rely on the ability of the separate clocks to maintain synchronization for the length of time required to transmit the code representing a complete character. Normally a quiescent line is maintained in its negative, or mark, state and each character transmitted is prefixed by a *start bit* that drives the line into a positive state. This positive transition starts the receiver clock which then samples the line a predefined number of times at predefined time intervals.

A further element added to the end of the character code is known as the *stop bit.* Stop bits force the line into a negative state for a period of one or two signal elements to allow the receiver time to complete its function and prepare for the next start bit.

TRANSMISSION OF SIGNALS OVER TELEPHONE LINES

When transmitting signals long distances a current loop system is not satisfactory and it then becomes necessary to interface the computer and the remote equipment by a modem at each end of the line as illustrated in Figure 17.23.

Modems may operate on one of two principles. The transmitted character code may be frequency modulated with mark and space signals represented by different frequencies. These frequencies are in the speech spectrum. A typical 200-baud system may have mark signals of 980 Hz and spaces of 1180 Hz when transmitting from modem 1 and signals of 1650 Hz–1850 Hz, respectively, from the other. With the alternative method of transmission, marks and spaces are represented by changes of phase in a carrier frequency. Typical operating frequencies from each end for a 600-baud line would be 1500 Hz and 1800 Hz for the two modems. If a two-way communication is to be maintained, it is necessary to organize the flow of data between the computer and modem 1, between modem 1 and modem 2, and between modem 2 and the remote terminal. This is done by control signals and responses being passed between the elements of the circuit, so that when the computer transmits a message, the circuit is conditioned for the remote terminal to accept the transmitted signal. Controls and responses between the modems are in the form of special characters produced and transmitted by logic circuits in the modems. Between the attached equipment and the modems, control is achieved by multiwire connections called interchange circuits. In recent years attempts have been made to standardize signal circuits and lines to control transmission. Line definitions have been reduced to a common recommendation by the EIA in the United States (Electronic Industries Association) and in Europe by the CCITT (International Telegraph and Telephone Consultative Committee). The respective recommendations for low-speed transmission are EIA RS232.C and CCITT V 24.

The complete interface defines about 37 interchange circuits, but in general only a limited number are used for data transmission. There is no general rule for connecting two equipments with a restricted version of the interface and each must be treated individually. Table 17.5 provides a list of the more important interchange circuits with their equivalent reference numbers:The precise definitions of these circuits are listed in the appropriate CCITT and EIA documents. For convenience they are summarized following.

Table 17.5 Telecommunication Standard Interchange Circuits

CCITT Circuit	EIA Circuit	Circuit Name
101	AA	Frame Ground
103	BA	Transmitted Data
104	BB	Received Data
105	CA	Request to Send
106	CB	Clear to Send
107	CC	Data Set Ready
109	CF	Data Carrier Detector
108.1		Connect Data Set to Line
108.2	CD	Data Terminal Ready
125	CE	Ring Indicator
102	AB	Signal Ground

101 This circuit provides continuity of ground between the data terminal and any external equipment for electrical safety. It is not normally connected to circuit 102.

103 Serial data are transmitted from the data terminal on this line. Voltage levels between +3 and +25 V represent spaces and −3 to −25 V represent marks. When no data are being transmitted, this line is held at the mark level.

104 Serial data are received from the remote equipment on this line. A mark level is −3 to −25 V, and a space +3 to +25 V.

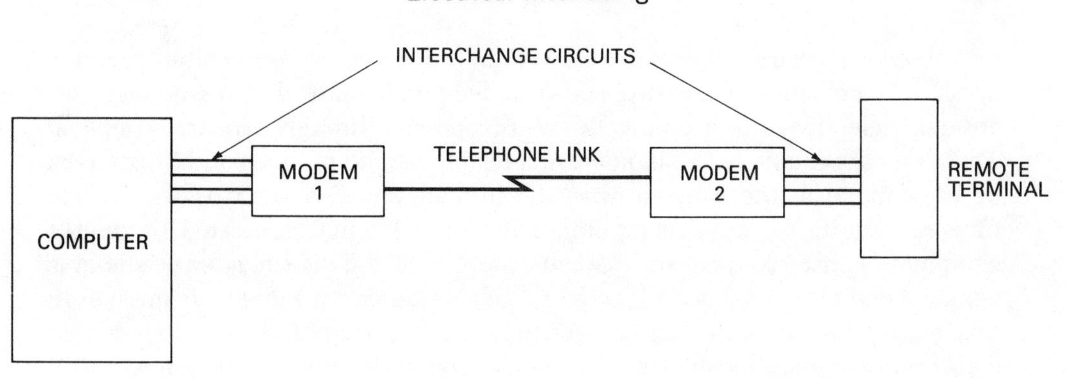

Figure 17.23
Remote Connection Using Modems

105 When a positive voltage is applied, this circuit is turned ON and indicates that the equipment is ready to transmit data.

106 When the remote equipment is ready to receive data, the circuit is turned ON. It is a reply to the request to send signal 105.

107 Data set ready is turned ON when the remote equipment is switched on.

108 This line may have two definitions. When two equipments are connected directly 108.2 is applicable and will connect to 107 of the remote equipment. An ON condition indicates that the local equipment is in a ready state.

The circuits just described are interpreted as being connections between directly connected equipment. The equipment local to an operator is often described as the data terminal and the connected equipment as a data set. Where the connection is a telephone link, the modem local to the terminal is often regarded as the data set. Interchange circuits between the terminal and the modem serve the functions described but transmission and decoding of transmitted control signals are functions of the modems.

The use of a modem often includes the provision of additional interchange circuits, two of which are listed below:

108.1 "Connect Data Set to Line" is the CCITT alternative definition of this circuit. When turned ON the circuit causes immediate connection to the transmission line. If the circuit is turned off prematurely, the line does not disconnect until transmission of data is complete. There is no EIA equivalent of this line.

109 Data Carrier Detector is sometimes described as Data Channel Received Line Signal Detector. It is a signal which when ON indicates to the terminal that received signals are within defined limits.

AUTOMATIC CALLING CIRCUITS

By far the most secure way of transmitting data from a distant station to a computer is by a permanently connected line. Where data have to be transmitted over the public network, the cost of leasing a line can be expensive, especially if transmissions are infrequent and only of short duration. In these situations it is possible to transmit the data over the switched public network. One method uses ordinary telephones and connects the equipment through *acoustic couplers*. These devices couple to the handset of a telephone and produce audio signals to be transmitted in the same way as the human voice. When it is required to transmit data to a central computer, a modulated signal generated by remote equipment is used to drive the acoustic coupler. At the central site the signal is demodulated by a modem connected to the computer. Two-way transmission in this way is possible, but in many cases the system is used to "dump" prerecorded data from a remote location to the central computer. Reverse transmission in these instances is confined to control signals. Such systems are useful where remote equipment is portable and in constant movement between sites.

Where more permanent arrangements can be made, dial-up modems may connect remote equipment to telephone lines. An operator at the central station dials the remote terminal in the same way as any ordinary call is made. Connection to an *auto answer* modem is automatic and will remain intact until the call is terminated by the operator. In some installations manual dialing can be eliminated and the computer programmed to perform the task at specified times. This is described as an "auto call" and "auto answer" facility and is provided by special modems.

The interface for auto call circuits is specified by CCITT in their $V^{25}/_{200}$ Series recommendation and by EIA in the specification RS-366. These are listed in Table 17.6.

Table 17.6 Telecoms Auto Call Interchange Circuits

CCITT	EIA	Circuit Name
212	FGD	Frame Ground
211	DPR	Digit Present
205	ACR	Abandon Call and Retry
202	CRQ	Call Request
210	PND	Present Next Digit
213	PWI	Power Indication
201	AB	Signal Ground
204	COS	Call Original Status
206	NB1	Digit 1
207	NB2	Digit 2
208	NB4	Digit 4
209	NB8	Digit 8
203	DLO	Data Line Occupied

In summary, these circuits may be defined as follows:

202 When this line is turned ON, the auto call modem connects to a line. The circuit must be turned OFF between calls.

203 When turned ON, this circuit indicates to the terminal that the data line is in use.

204 When a connection is made, the remote station transmits a signal back to the calling equipment, which turns this circuit ON.

205 This line is turned ON when preset times between successive events in a calling procedure have been exceeded. It instructs the terminal to abandon the call.

206–209 These four lines are set by the terminal as binary representations of the number to be dialed and are converted by the modem into dialing pulses. Each number is held on the lines as long as circuit 211 is ON.

210 An ON state indicates to the terminal that the automatic calling equipment is ready to accept the next combination of binary digits on circuits 206–209.

211 When turned ON by the terminal, the next digit is available for conversion on lines 206–209.

213 This is turned ON when the automatic calling equipment is switched on.

DIGITAL INPUT

Digital input and output are important aspects of sensor-based system operation and warrant attention in some detail. Many computer digital input circuits are designed to operate at what are often referred to as logic levels, but which should be more correctly described as Transistor Transistor Logic levels (TTL). These are defined as:

Binary 1 or "Up State" +2.4 to +5 V
Binary 0 or "Down State" 0 volts to +0.5 V

Any voltage changes between +0.5 and +2.4 V may cause unpredictable operation of the circuit and must therefore be avoided. Currents flowing in TTL circuits are often very low and are dependent on the characteristics of the circuits. The impedance of these circuits may be such as to restrict current flow to 1 or 2 mA. Within the confines of electronic equipment, or where equipment is placed in close proximity to the computer, TTL levels are adequate for reliable operation. Where voltage levels are sensed over considerable distances, however, these circuits are sensitive to triggering by electrical noise. It is a feature of TTL switches that they will trigger from low-energy microsecond pulses, and in environments where fluctuating electrical fields are present the chance of spurious operation is high. Noise "spikes," as they are often called, may be generated by contacts carrying heavy currents being switched, or rapid changes of current occurring in inductive circuits. Generation of noise spikes may be unpredictable and their source very difficult to locate. Their effect on computer operation may, however, be serious, causing faulty program operation or, in the extreme case, damage to the input circuits. The worst effects of noise may be eliminated by either modifying the input circuit or by introducing a filter.

It is possible to reduce the impedance of an input circuit by connecting a low-value resistor across it. If a switching circuit is terminated by a 250-ohm resistor, a current of about 10 mA will be required to raise the input voltage to the minimum switching level. A low-energy noise spike will be incapable of producing this current and the transient voltage will never rise to a point where it can switch the circuit. The successful operation of this technique requires that the switching source can produce sufficient current to switch the circuit.

Where switching speeds are not important, it is possible to introduce a measure of filtering to reduce the effect of noise. A filter increases the time for the pulse to reach the circuit triggering level, attenuates the peak value of the signal, and also slows down the discharge time as illustrated in Figure 17.24. Where signal pulses are of adequate amplitude and relatively long duration the circuit will respond satisfactorily while noise pulses will be suppressed by the time constant of the filter.

Sensing Small Digital Voltage Changes

It is not always possible to choose the voltage levels which are to be sensed by digital circuits. Where these are lower than the level recognized as a change by a TTL circuit, preamplification is necessary. Since it is not desirable to transmit low-level signals any distance it is preferable to amplify at the sensor end. One method employs a voltage comparator illustrated in Figure 17.25 and available as an integrated circuit.

Comparators operate from two voltage sources, the signal and a reference. The one illustrated operates on the principle that when no input signal is applied, the output is +5 V. If an input is greater than the reference voltage applied, the comparator switches its output to V_0. These circuits are capable of producing currents of the order of 50 mA and will, therefore, provide appreciable noise rejection when operating into a low-impedance circuit. When referenced by a voltage marginally lower than the input signal, spurious input to this level will not cause the circuit to switch. Comparators, therefore, are useful devices to amplify millivolt signals to a level where they will operate TTL switches.

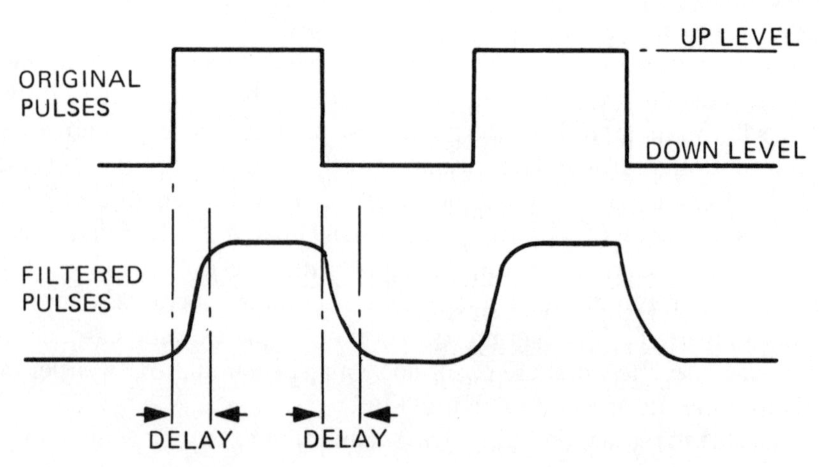

Figure 17.24
Effect of Filtering on a Switched Signal

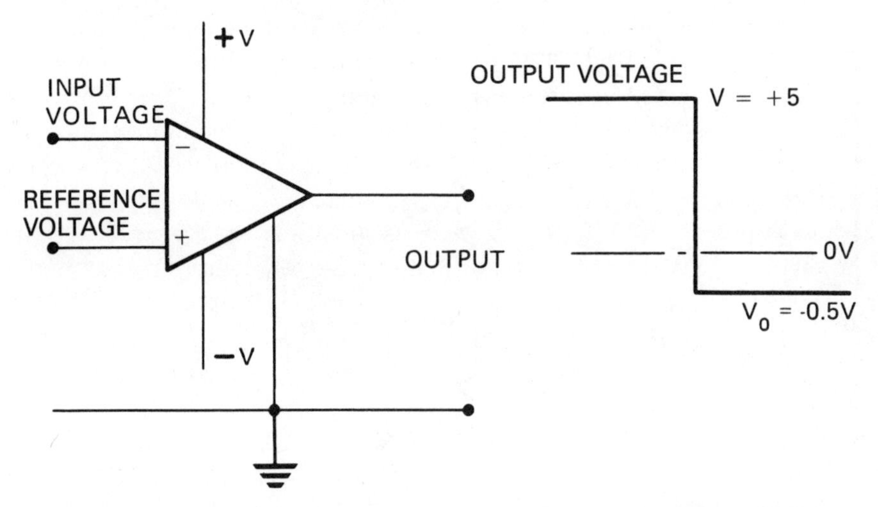

Figure 17.25
Amplification of Digital Signals

Sensing High-Voltage Changes

High-voltage inputs which may be of the order of 250 V will require attenuating to make them acceptable to logic-level input circuits. In these instances there is a danger that during a fault condition the input will still represent a shock hazard to maintenance personnel who may not be aware of the dangers.

A shock hazard is usually considered to exist if a voltage on an exposed terminal is more than 42.4-V peak open circuit and is capable of delivering more than 5 mA through a 1.5-kΩ resistor.

When designing alternators for high voltages it must be remembered that although the output voltage may be of the order of 5 or 6 V, failure of one of the resistors could raise the value to the full line voltage. Not only is this likely to produce a shock hazard, but it is also possible that input circuits may be destroyed.

Suppose a 250-V supply is to be attenuated to 6 V to operate an input circuit, and suppose it must be fully protected not only for personal safety, but to protect the input circuit voltage from rising above 7 V. It is not possible to design a simple attenuator where the 42-V 5-mA limits can be guaranteed in a fault condition and it is therefore necessary to develop a more complex circuit.

Consider a basic attenuator illustrated in Figure 17.26. If this circuit is analyzed, it can be seen that the most serious failure condition is where R_1 becomes short-circuited. This fault will raise the voltage at point A to the input voltage and create a shock hazard. In addition it is probable that the input circuit will be destroyed.

One method of preventing the voltage rising to its line value uses a Zener diode as shown in Figure 17.27. With this circuit the Zener diode holds the equipment input to its rated voltage while the resistance R_3 prevents the current

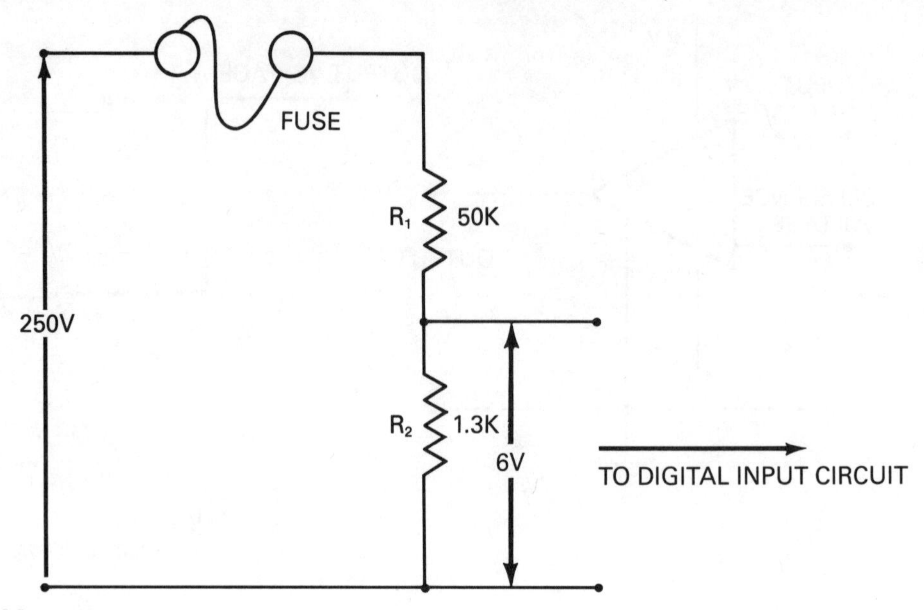

Figure 17.26
Basic Attenuator

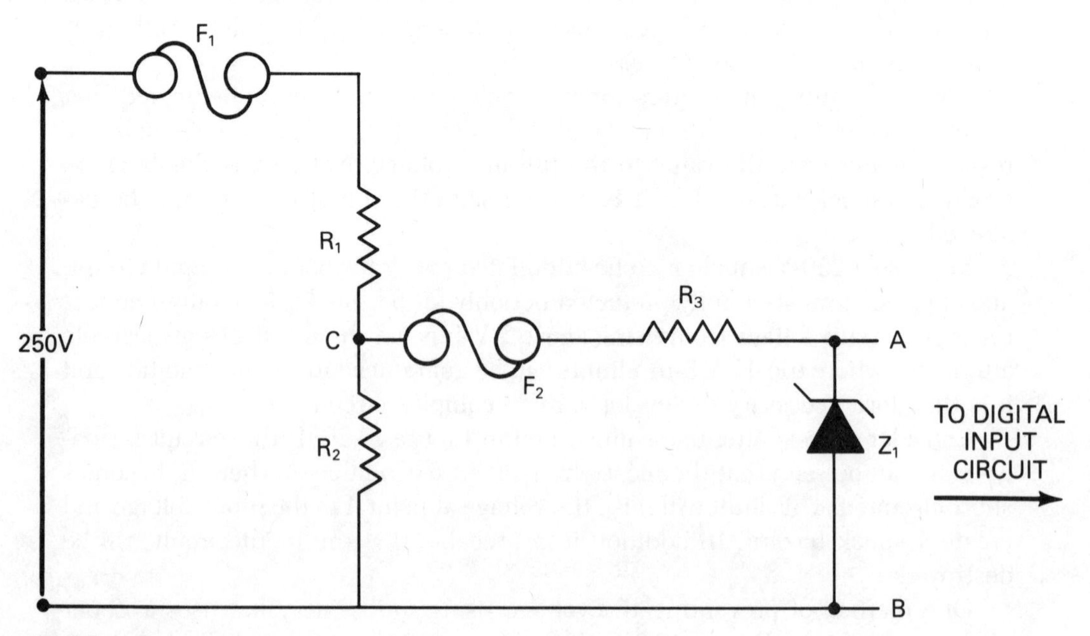

Figure 17.27
Zener-Protected Attenuator

rising to a value that will cause the Zener to fail. If the components are chosen correctly, the current flowing through the Zener may be made high enough to blow a fuse $F2$.

Suppose Z1 is a 7-V Zener capable of carrying 100 mA and suppose $F2$ is rated at 20 mA. If the circuit is to fail safe, in the fault condition a design current in excess of 20 mA must flow through R_3. The voltage at junction C with R_1 short-circuited will be 250 V and at A will be 7 V. If the current is allowed to rise through R_3 to 90 mA the value of R_3 will be:

$$\frac{250\text{ V} - 7\text{ V}}{90} = 2.7\text{ k}\Omega$$

If the alternative situation exists, where R_2 is open circuit, then R_1 will still be in series with R_3 and the current will be reduced to a value:

$$i = \frac{250}{R_1 + R_3} \qquad (1)$$

For the fuse $F1$ to blow this current must be greater than 20 mA (say 40 mA) and expression (1) becomes

$$\frac{250}{R_1 + 2.7} = 40\text{ mA}$$

or

$$R_1 = 3.6\text{ k}\Omega$$

If, with the circuit intact, the voltage across R_2 is designed to be 6 V the Zener will not conduct and the only current flowing in CA will be due to the resistance of the equipment input, which is usually high.

The voltage drop across R_1 in this condition will be 250 − 6 V = 244 V. The current flowing in R_1 will be 244/3.6 K= 68 mA. The resistance of R_2 will be 6 V/68 mA = 0.088 kΩ = 88 ohms.

In the event of a short circuit occurring across R_1 and R_2, F_1 will blow. The value of this fuse must be in excess of the operating current through R_1 and R_2, i.e., 68 mA (say 200 mA)

The circuit illustrated in Figure 17.28 is fully protected and only a combination of failures, including the failure of Z_1, will cause destruction of the input circuit. If this is of concern, a further resistance followed by a second Zener can be inserted in line A. It will be necessary, however, for the safety of maintenance personnel to enclose the whole alternator in a grounded metal case with suitable hazard notices attached. It is preferable that these cases be mounted in a rack separate from the electronic equipment so that only low voltage levels are connected to input circuits in these racks.

Contact Sensing with Digital Input

Digital input circuits designed to operate with TTL signal levels are usually connections between the bases of transistors and ground. It is common practice

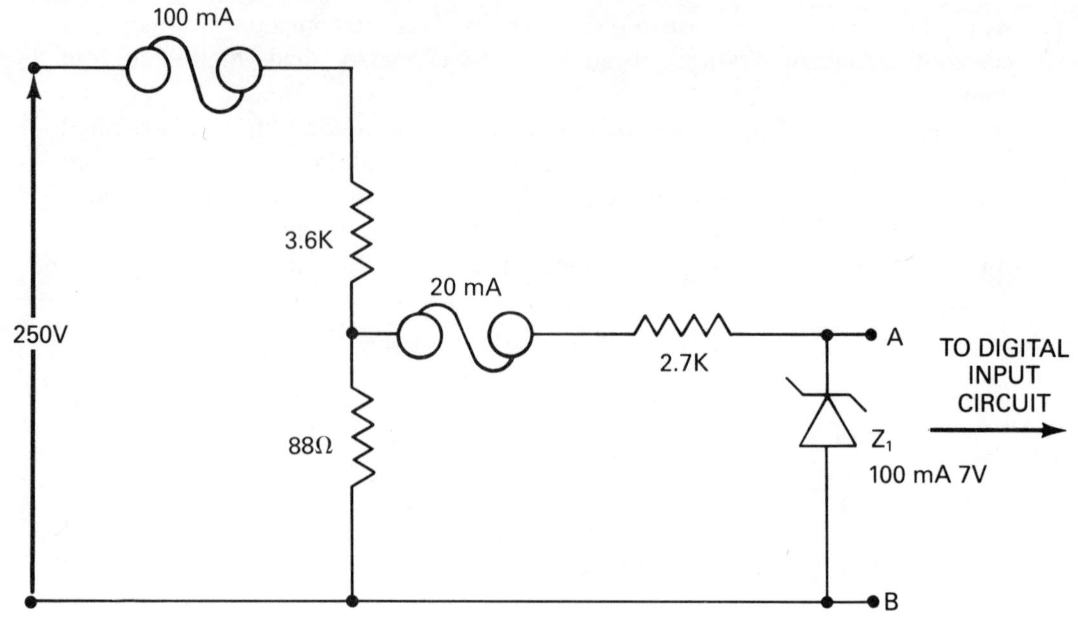

Figure 17.28
Zener Protected Attenuator Design

to connect a *pull-up* voltage to the bases to prevent spurious operation of any circuits not in use. This is an internal voltage applied through a resistor and raises the base to about +5 V. In principle, switch contacts connected across the base will change the state of this input when they are operated. In practice, to connect long lines to the input and leave them *floating* when the switch is open will probably cause spurious operation by noise pickup. Contact bounce is also a problem. The erratic contact resistance experienced during contact closure can lead to gross errors in counting applications. Finally the pull-up voltage is only intended to deliver small currents and may not be sufficient to break down deposits which accumulate on base metal switch contacts. When sensing contacts, therefore, it is usual to use a network excited from an external supply and provided with some filtering. Figure 17.29 illustrates the simplest of these circuits.

This circuit consists of a simple attenuator R_1 and R_2 energized by a suitable power supply. The voltage at A is designed to match the specification of the digital input with the capacitor and R_2 acting as a simple filter to suppress electrical interference. To ensure satisfactory operation of this circuit, the value of R_1 should be chosen to provide a current flow through the contacts of at least 10 mA. This circuit is adequate for short cable runs and low-resistance contacts but the one illustrated in Figure 17.30 is better for long cable runs.

The circuit employs a 48-V sensing voltage which is sufficient to break down contact resistance resulting from deposition on the switch contacts. A current flow through the closed switch of approximately 10 mA is maintained by the 4.7K series resistance R_1. With the contact open, the voltage appearing at the digital input is effectively the voltage appearing across the 1K resistor, which is part of a dividing circuit made up of R_1, R_2 and R_3. This is approximately 5 V. The addition of a capacitor across R_3 turns this circuit into a low-pass filter with a time constant of about 30 ms, which is adequate to smooth the effects of contact bounce in most switches.

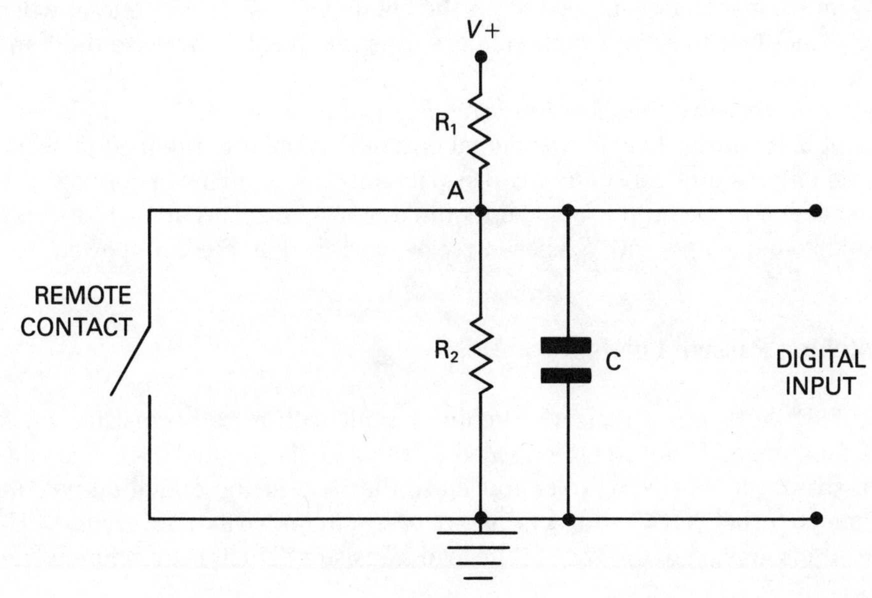

Figure 17.29
Basic Filter Circuit

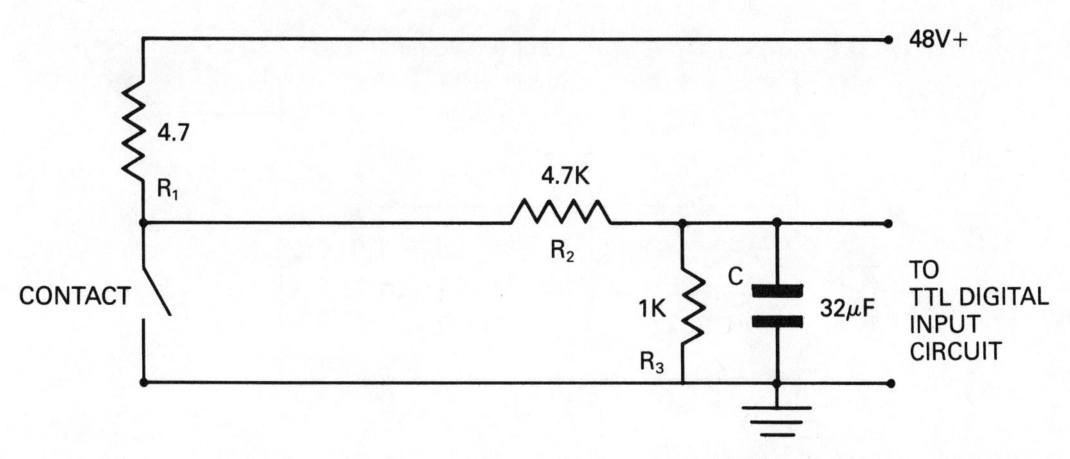

Figure 17.30
Circuit to Sense Base Metal Contacts

DIGITAL OUTPUT

Most general-purpose computers are provided with digital output facilities which are effectively transistor switches. These switches may be simple transistors which, when operated, ground an output line, or the transistor may operate a relay.

In the case of a relay output, the relay coil is switched by a transistor circuit, and provides a convenient output for switching ac or inductive loads. Relay contacts are often rated as high as 250 V but whether or not it is safe to connect high voltages to the contacts will depend on the computer design. In general where relays are required to switch high voltages, it is much safer to house them in a separate rack.

Digital output switches are often rated and priced by the power they are capable of dissipating. Low-power digital output circuits are intended to switch electronic circuits and consequently dissipate only a few milliwatts of power. In other applications the output may be required to switch relays at currents up to 0.5 A and voltages up to 50 V. These may be described as medium power.

Switching Passive Loads

Figure 17.31 illustrates a typical switching circuit with a load transistor whose base is conditioned by a voltage provided by the digital output. This is described as a passive load. With zero volts applied to the base of the digital output, the transistor is turned OFF and acts as an open-circuit switch to ground. The voltage at *A* is applied to the base of the load transistor, which is therefore turned

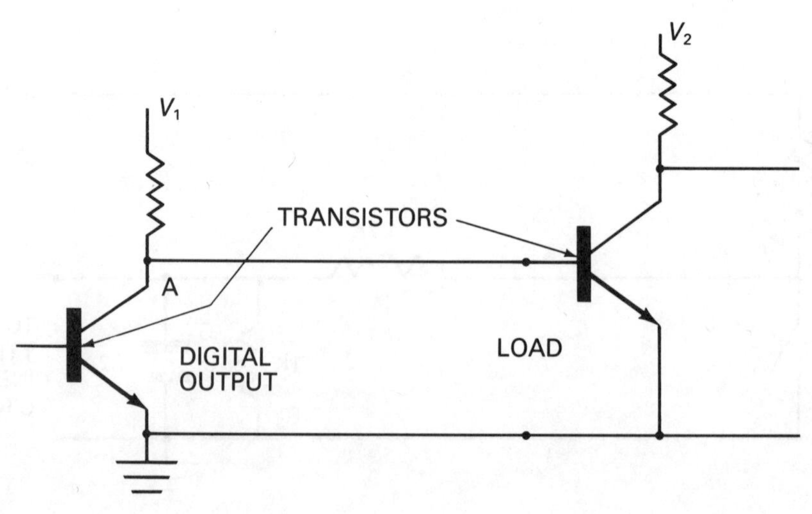

Figure 17.31
Passive Load Switching

ON. When a positive voltage is applied to its base, the digital output is turned ON and the voltage appearing at *A* is short-circuited to ground. This removes the signal from the base of the load transistor and turns it OFF.

Switching Active Loads

Figure 17.32 illustrates a circuit where the load transistor is provided with its own pull-up voltage and is described as an active load. The operation of the digital output when switching active loads is very similar to that when switching passive ones. On this occasion, however, the digital output must conduct to ground the current through the resistors R_1 and R_2 to reduce the voltage at B to zero. This is described as "sinking" the current.

Open Collectors

Some digital output switches are not provided with the voltage on their collectors and operate as a switch with one side grounded. These are described as transistor switches with "open collectors." Where the digital output is used to switch an active load the switching action is the same as that described above. Where a load is passive, an external pull-up voltage must be used.

Open collector switches are usually capable of sinking comparatively high currents and are used for switching relays and lights from an external supply as illustrated in Figure 17.33. These switches operate in much the same way as a mechanical switch with a series diode. Current only flows one way through the switch, and supply polarity must therefore be observed.

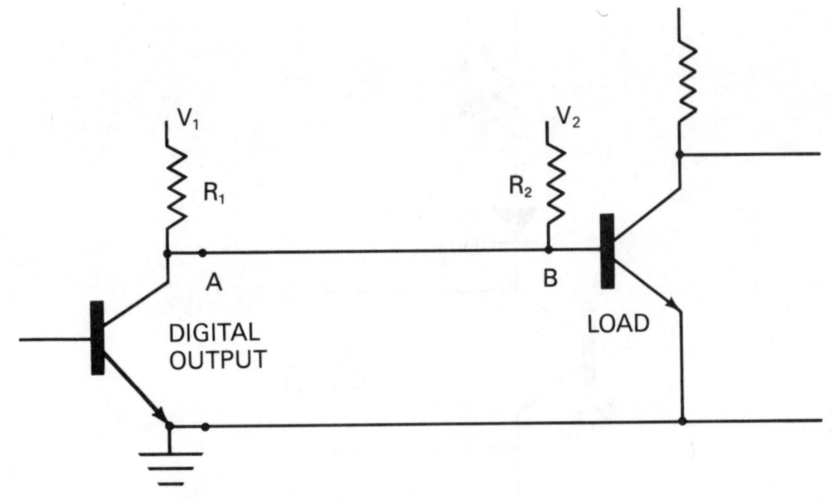

Figure 17.32
Switching Active Loads

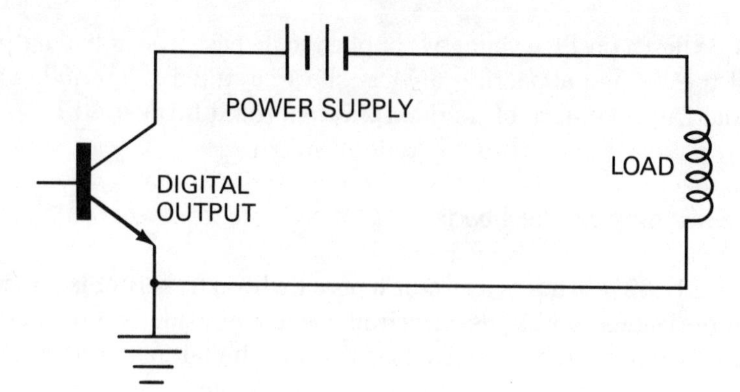

Figure 17.33
Open Collector Operation

Composite Switches

A transistor switch that may be used either to switch TTL loads or be operated as an open collector is a much more convenient device than those described. The problem with designing such a switch is the range of voltages that it may be required to switch in the open collector mode. Any load voltages higher than the internal supply will discharge back through that supply, and if the difference becomes too great, damage will occur. One method of protecting power supplies uses a blocking diode as illustrated in Figure 17.34.

When driving passive or active loads an internal supply provides the collector voltage. When operating in open collector mode, the diode isolates this supply from reverse current flow. Composite switches are usually designed with power transistors capable of sinking 500 mA or more, and operating with voltages up to 50 V.

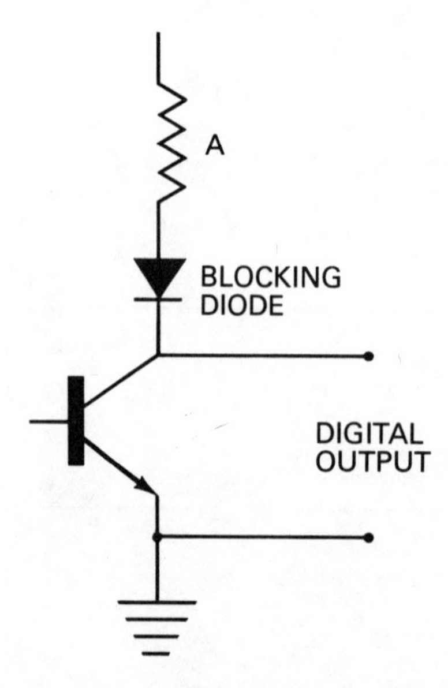

Figure 17.34
Composite Transistor Switch

ANALOG INPUT

Most manufacturers of general-purpose computers provide facilities that allow analog measurements to be made in much the same way as with a data logger, described in Chapter 14. It is often convenient, however, to make analog measurements with a data logger at a site remote from the computer and to connect the two by a serial transmission link.

Most modern data loggers are provided with interfaces compatible with computers. Parallel BCD is often available which may be connected to the digital input feature of the computer, or the data logger may be provided with a serial asynchronous interface. Even older data loggers with punched tape or printer output can often be converted to interface to the digital input/digital output feature.

ANALOG OUTPUT

Analog output is most frequently provided by digital-to-analog (D to A) converters. These may be a manufacturer-supplied feature, or they may be purchased devices connected to the computer digital output feature. Where large numbers of analog output signals are required, D to A converters may multiplex into a number of sample and hold amplifiers. In some cases when only a few output signals are required it may be reasonable to provide separate D to A converters for each output.

An inexpensive form of analog output where none is readily available is the motor-driven potentiometer. A small dc motor driving a two-ganged potentiometer through a gear box is easily assembled and relatively troublefree. Figure 17.35 illustrates a typical device.

Two computer digital output points are required to provide the output. One switches supplies to the motor armature, and the other controls the direction of rotation by switching the direction of field current. A “feedback” signal to the computer is required to allow correct positioning of the potentiometer and this is provided by connecting one of the potentiometers to the computer analog input feature.

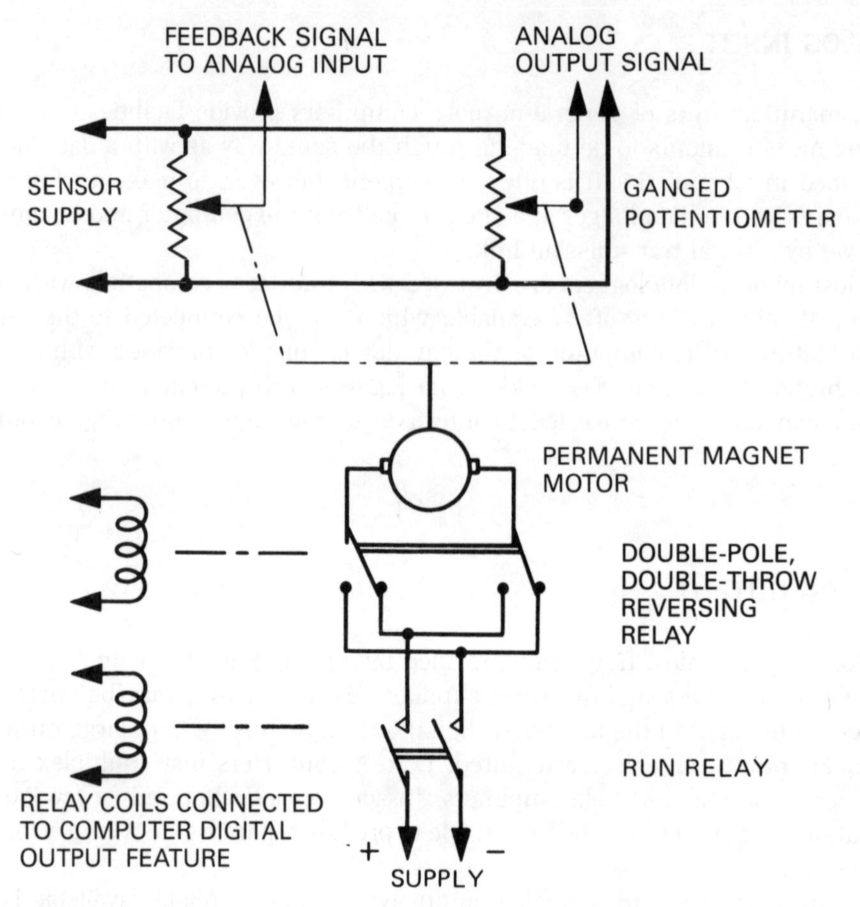

Figure 17.35
Motor-Driven Potentiometer

APPENDIX

INSTRUMENT SOCIETY OF AMERICA STANDARDS AND RECOMMENDED PRACTICES

The Instrument Society of America standards and recommended practices listed on the following pages outline accepted instrumentation procedures used throughout the process measurement and control field.

B.1 ISA-RP2.1, "Manometer Tables"
This recommended practice presents abbreviations and fundamental conversion factors commonly used in manometry, recommended definitions of pressure in terms of a column of mercury and water, and for a large number of liquids, tables of various pressures indicated by, or equivalent to, heights of columns at various temperatures.

B.2 ISA-RP3.2, "Flange Mounted Sharp Edged Orifice Plates for Flow Measurement"
ISA-RP3.2 is intended to cover any orifice plates designed for bolting between flanges, with flat or raised face, whose bolting dimensions are in accordance with ANSI B16 series standards.

B.3 ANSI/ISA-S5.1-1975 (1981), "Instrumentation Symbols and Identification"
This standard establishes a uniform means of designating instruments and instrumentation systems used for measurement and control. The differing established procedural needs of various organizations are recognized (where not inconsistent with the objectives of the standard) by providing alternative symbolism methods. A number of options are provided for adding information or simplifying the symbolism, if desired.

B.4 ANSI/ISA-S5.2-1976 (R1981), "Binary Logic Diagrams for Process Operations"
This standard provides symbols, both basic and nonbasic, for binary operating functions. Intended to symbolize the binary operating functions of a system in a manner that can be applied to any class of hardware, whether electronic, electrical, fluidic, pneumatic, hydraulic, mechanical, manual, optical, or other.

B.5 ISA-S5.3-1982, "Graphic Symbols for Distributed Control/Shared Display Instrumentation, Logic and Computer Systems"
The purpose of this standard is to establish documentation for that class of instrumentation consisting of computers, programmable controllers, minicomputers, and microprocessor-based systems that have shared control, shared display, or other interface features. Symbols are provided for interfacing field instrumentation, control room instrumentation, and other hardware to the above. Terminology is defined in the broadest generic form to describe the various categories of these devices.

B.6 ANSI/ISA-S5.4-1976 (R1981), "Instrument Loop Diagrams"
This standard provides a method and practice for the preparation and use of instrument loop diagrams in the design, construction, checkout, startup, operation, maintenance, and reconstruction of instrument systems in industrial plants.

B.7 ISA-RP7.1-1956, "Pneumatic Control Circuit Pressure Test"
This recommended practice is intended to provide a satisfactory procedure for the testing of pneumatic control circuits for leaks, together with reasonable criteria for acceptance of work done and suitable aids for performance.

B.8 ANSI/ISA-S7.3-1975 (R1981), "Quality Standard for Instrument Air"
This standard establishes instrument air quality values for maximum allowable moisture content, maximum entrained particle size, and maximum allowable oil content, and establishes an awareness of possible source of corrosive or toxic contamination—all of which affect the satisfactory operation of instrumentation.

B.9 ANSI/ISA-S7.4-1981, "Air Pressures for Pneumatic Controllers, Transmitters and Transmission Systems"
The purpose of this standard is to establish standard operating pressure ranges for pneumatic intelligence transmission systems, and standard air supply pressures (with limit values) for operations of pneumatic controllers and transmitters, pneumatic intelligence transmission systems, current-to-pressure transducers, and similar devices.

B.10 ISA-RP12.1-1960, "Electrical Instruments in Hazardous Atmospheres"
This recommended practice provides general guidance for the safe installation of electrical instruments using appropriate means to prevent ignition of flammable gases and vapors.

B.11 ISA-S12.4-1970, "Instrument Purging for Reduction of Hazardous Are Classification"
This standard covers a technique for reducing the hazard classification by the continuous addition of air or an inert gas within a general-purpose enclosure. It refers only to hazards created by gases or vapors and is concerned only with those system design criteria related to electrical ignition of a hazardous gas or vapor.

B.14 ISA-S12.11-1973, "Electrical Instruments in Hazardous Dust Locations"
This standard summarized the requirements for safe and economical installation of electrical instruments in locations made hazardous by a cloud or blanket of dust. (See ISA S12.10 for classification of such areas.) It is in conformance with (and attempts to expand and clarify) the National Electrical Code and the Canadian Electrical Code.

B.15 ISA-RP16.1,2,3-1959, "Terminology, Dimensions and Safety Practices for Indicating Variable Area Meters (Rotameters, Glass Tube, Metal Tube, Extension Type Glass Tube)"
The purpose of this recommended practice is to (a) establish uniformity of connection dimensions to permit interchangeability of one manufacturer's meters with another manufacturer's meters of the same size; (b) provide a common ground of understanding the terminology, use, and component parts and accuracies of these meters; and (c) provide a reference for the safe working pressures of these meters.

B.16 ISA-RP16.4-1960, "Nomenclature and Terminology for Extension Type Variable Area Meters (Rotameters)"
This recommended practice defines the nomenclature and terminology of various types of extensions applicable to 5-inch (125-mm) glass and metal tube variable-area meters (rotameters) covered in ISA RP16.1,2,3.

B.17 ISA-RP16.5-1961, "Installation, Operation, Maintenance Instructions for Glass Tube Variable Area Meters (Rotameters)"
This recommended practice covers the general considerations important to the installation, operation, and maintenance of glass tube variable-area meters (rotameters) to obtain the most reliable results.

B.18 ISA-RP16.6-1961, "Methods and Equipment for Calibration of Variable Area Meters (Rotameters)"
This recommended practice describes the methods and equipment used for calibrating the glass and metal metering tube area meters (rotameters) covered in ISA-RP16.1,2,3.

B.19 ANSI/ISA-S18.1-1981, "Annunciator Sequences and Specifications"
This standard covers electrical annunciators that call attention to abnormal process conditions by the use of individual illuminated visual displays and audible devices. Sequence designations provided can be used to describe basic annunciator sequences and also many sequence variations.

B.20 ISA-S20-1981, "Specification Forms for Process Measurement and Control Instruments, Primary Elements and Control Valves"
This standard provides forms (checklist) to promote uniformity in instrument specification, both in content and form, by listing and providing space for all principal descriptive options, supplying terminology, and facilitating quoting, purchasing, receiving, accounting, and ordering procedures.

B.21 ISA-RP25.1-1957, "Materials for Instruments in Radiation Service"
This recommended practice is intended to serve as a guide to the selection of materials for use in intense radiation fields such as those encountered in and around nuclear reactors.

B.22 ISA-S26 (ANSI MC4.1-1975), "Dynamic Response Testing of Process Control Instrumentation"
Incorporating four revised ISA recommended practices, the standard establishes the basis for dynamic response testing of measurement and control equipment with pneumatic output and electrical output, and for closed-loop actuators for externally actuated control valves and other final control elements. Pulse testing techniques as well as methods for sine wave, step, and pulse-type signals are included.

B.23 ANSI/ISA-RP31.1-1977, "Specification, Installation, and Calibration of Turbine Flowmeters"
This recommended practice establishes minimum ordering information; recommended acceptance and qualification test methods, including calibration techniques; uniform terminology and drawing symbols; and recommended installation techniques for volumetric turbine flow transducers having an electrical output.

B.24 ANSI/ISA-S37.1-1975 (R1982), "Electrical Transducer Nomenclature and Terminology"
This standard establishes uniform nomenclature for transducers and uniform simplified terminology for transducer characteristics.

B.25 ISA-RP37.2-1982, "Guide for Specifications and Tests for Piezoelectric Acceleration Transducers for Aerospace Testing"
This recommended practice covers piezoelectric acceleration transducers, primarily those used in aerospace test instrumentation. Terminology used in this document folows ISA RP37.1, except that additional terms considered applicable to piezoelectric vibration transducers are defined.

B.26 ANSI/ISA-S37.3-1975 (R1982), "Specifications and Test for Strain Gage Pressure Transducers"
This standard establishes uniform minimum specifications for design and performance characteristics; uniform acceptance and qualification test methods, including calibration techniques; uniform presentation of minimum test data; and a drawing symbol for use in electrical schematics for strain gage pressure transducers.

B.27 ANSI/ISA-S37.5-1975 (1982), "Specifications and Test for Strain Gage Linear Acceleration Transducers"
This standard establishes uniform minimum specifications for design and performance characteristics; uniform acceptance and qualifications test methods including calibration techniques; uniform presentation of minimum test data; and a drawing symbol for use in electrical schematics for strain gage pressure transducers.

B.28 ANSI/ISA-S37.6-1976 (R1982), "Specifications and Tests of Potentiometric Pressure Transducers"
This standard establishes uniform minimum specifications for design and performance characteristics; uniform acceptance and qualification test methods, including calibration techniques; uniform presentation of minimum text data; and a drawing symbol for use in electrical schematic for potentiometric pressure transducers.

B.29 ANSI/ISA-S37.8-1977 (R1982), "Specifications and Tests for Strain Gage Force Transducers"
This standard outlines uniform general specifications, acceptance and qualification methods, and methods for data presentation, and includes a drawing symbol used in electrical schematics for tension, compression, and combination tension/compression transducers.

B.30 ANSI/ISA-S37.10-1975 (R1982), "Specifications and Tests for Piezoelectric Pressure and Sound-Pressure Transducers"
This standard establishes uniform specifications for describing design and performance characteristics, acceptance and qualification test methods and calibration techniques, and procedures for presenting test data for piezoelectric (including ferroelectric) pressure and sound-pressure transducers.

B.31 ANSI/ISA-S37.12-1977 (R1982), "Specifications and Tests for Potentiometric Displacement Transducers"
This document covers potentiometric displacement transducers, primarily those used in measuring systems. The specifications are not intended to cover transducers used in hazardous locations as specified in the National Electrocal Code nor are all requirements covered for transducers used in nuclear power plants.

B.32 ISA-RP42.1-1982, "Nomenclature for Instrument Tubing Fittings (Threaded) (Revised)"
This recommended practice defines nomenclature for tubing fittings most commonly used in instrumentation. It is not intended as a substitute for manufacturer's catalog numbers, nor does it apply to special fittings. It is intended to apply to a mechanical fitting rather than a sweat fitting.

B.33 ANSI/ISA-S50.1-1982, "Compatibility of Analog Signals for Electronic Industrial Process Instruments"
This standard applies to analog dc signals used in process control and monitoring systems to transmit information between subsystems or separated elements of systems.

B.34 ANSI/ISA-S51.1-1979, "Process Instrumentation Terminology"
This standard is intended to include many specialized terms used in the industrial process industries, to describe the use, performance, operating influences, hardware, and product qualification of the instrumentation and instrument systems for measurement, control, or both. This document is a guideline for vendor/user understanding when referring to product specifications, performance, and operating conditions. It is primarily intended to cover the field of analog measurement and control concepts, and makes no effort to develop terminology in the field of digital measurement and control.

B.35 ISA-RP52.1-1975, "Recommended Environments for Standards Laboratories"
Recommendations for three levels of standardization are presented—from the more general National Bureau of Standards, through commercial, industrial, and government laboratories. Requirements for nine environmental factors are discussed.

B.36 ANSI/ISA-RP55.1-1975 (R1983), "Hardware Testing of Digital Process Computers"

This recommended practice establishes a basis for evaluating functional hardware performance of digital process computers. It covers general recommendations applicable to all hardware performance testing, specific tests for pertinent subsystems, and system parameters. This document includes a brief glossary of terms used.

B.37 ISA-RP6.8-1978, "Electrical Guide for Control Centers"

This recommended practice assists the design enginer in establishing the electrical requirements of a control center; it is also intended to comply with the provisions of the NEC. Special considerations which may apply to particular devices or circuits are not taken into account in this recommended practice.

B.38 ISA-RP60.9-1981, "Piping Guide for Control Centers"

This recommended practice assists the designer or engineer in the definition of piping requirements for pneumatic signals and supplies in control centers. Because of the special nature of each control center, specific rules are not practical and accepted guidelines should take precedence. Piping external to the control center (field piping) is beyond the scope of this document.

B.39 ANSI/ISA-S61.1-1977, "Industrial Computer System FORTRAN Procedures for Executive Functions, Process Input-Output, and Bit Manipulations"

This document presents external procedure references for use in industrial computer systems. These external procedure references permit interface of programs with executive routines, process input and output functions, allow manipulation of bit strings, and provide access to time and data information. Intended for use with programs written in FORTRAN, conforming to ISO Programming Language FORTRAN R1539-1972 (ANSI X3.9-1966).

B.40 ANSI/ISA-S61.2-1978, "Industrial Computer System FORTRAN Procedures for File Access and the Control of File Contention"

This standard presents external procedure references for use in industrial computer control systems. These external procedure references provide means for accessing files, and also provide means for resolving problems of file access contention in a multiprogramming/multiprocessing environment. In such an environment it is expected that concurrent programs will attempts to access the same file at the same time; therefore, the external procedure references defined in this standard provide the information necessary for the processor to resolve such simultaneous access in an orderly manner. The method for resolution of access control is left to the processor.

B.41 ANSI/ISA-S67.01-1981, "Transducer and Transmitter Installation for Nuclear Safety Applications"

This standard covers the installation of transducers for nuclear-safety-related applications, excepting those for measurands of liquid metals. It establishes requirements and recommendations for the installation of transducers and auxiliary equipment for nuclear power plant applications outside of the main reactor vessel. This standard considers mounting structures and materials, seismic design, separation, protection from mechanical damage, ambient variations (environmental) signal connections, process connections, and service and test provisions.

B.42 ANSI/ISA-S67.02-1980, "Nuclear-Safety-Related Instrument Sensing Line Piping and Tubing Standards for Use in Nuclear Power Plants"
This standard covers design, protection, and installation of nuclear-safety-related instrument sensing lines for light-water-cooled nuclear power plants. The standard covers the pressure boundary requirements for piping, capillary tubing, and tubing lines up to and including 1-inch (25.4-mm) outside diameter or ¾-inch nominal pipe. The boundaries of this standard span from the process tap to the upstream side of the instrument panel, bulk head fitting, or instrument shutoff valve. It is also applicable to systems (except liquid metal) in other types of nuclear power plants.

B.43 ISA-S67.03-1982, "Standard for Light Water Reactor Coolant Pressure Boundary Leak Detection"
This standard covers identification and quantitative measurement of reactor coolant system leakage in light-water-cooled power reactors. Leak detection for gas- and liquid-metal-cooled reactors and for containment building structures surrounding the reactor coolant pressure boundary is not covered in this standard.

B.44 ISA-S67.04-1982, "Setpoints for Nuclear-Safety-Related Instrumentation Used in Nuclear Power Plants"
The purpose of this standard is to develop a basis for establishing setpoints for actions determined by the design basis for protection systems and to account for instrument errors and drift in the channel from the sensor through and including the bistable trip device.

This standard defines minimum requirements for assuring that setpoints are established and held within specified limits in nuclear-safety-related instruments in nuclear power plants.

B.45 ANSI/ISA-S75.01-1977, "Control Valve Sizing Equations"
This standard establishes equations for flow-through control valves throttling either compressible or incompressible fluids. The equations are not, however, intended for use with mixed phases. Use of these equations for mixed phases, wet or dry slurries, or for non-Newtonian liquids may result in inaccuracies.

This standard does not apply to fluid power components as defined in the National Fluid Power Association (NFPA) Standards.

B.46 ANSI/ISA-S75.02-1982, "Control Valve Capacity Test Procedure"
This test standard utilizes the mathematical equations outlined in ANSI/ISA S75.01, "Control Valve Sizing Equations," in providing a test procedure for obtaining the following: Valve Sizing Coefficient, C-V; Liquid Pressure Recovery Factor, F-L; Reynolds Number Factor, F-$; Liquid Critical Pressure Ratio Factor, F-F; Piping Geometry Factor, F-P; and Pressure Drop Ratio Factors, X-T and X-TP.

This standard is intended for control valves used in flow control of process fluids and is not intended to apply to fluid power components as defined in the National Fluid Power Association Standard NFPA T.3.5.28-1977.

B.47 ANSI/ISA-S75.03-1979, "Uniform Face-to-Face Dimensions for Flanged Globe Style Control Valve Bodies"
This standard applies to flanged control valves sizes ½ inch through 16 inches having top, top and bottom, part, or cage guiding. This standard aids users in their piping designs by providing ANSI Class 125 flat face and ANSI Classes 150, 250, 300, 600 raised face flanged control valve dimensions without giving special consideration to the manufacturer of equipment to be used.

B.48 ANSI/ISA-S75.04-1979, "Face-to-Face Dimensions of Flangeless Control Valves"

This standard applies to flangeless control valves in sizes from ¾ inch through 16 inches for ANSI Classes 150 through 600. This standard establishes face-to-face dimensions for flangeless control valves so that piping design will accommodate this category of control valves without giving special consideration to the manufacturer of equipment to be used.

This standard applies to flangeless ball control valves, utilizing a full ball or a segment of a ball, and other rotary-stem or sliding-stem flangeless control valves. It does not apply to weld-end valves, butterfly valves, or other rotary-stem valves which may be covered by other standards.

B.49 ISA-RP75.06-1981, "Control Valve Manifold Designs"

Control valves and bypass valves are sometimes manifolded in piping systems to allow manual manipulation of the flow through the systems in those situations in which (usually) the control valve is not in service. This recommended practice presents six control valve manifold types with space estimates for various sizes. Each of these six types consists of a straight-through globe control valve, isolation upstream and downstream block valves, and bypass piping with a manual activated valve.

B.50 ANSI/MC96.1-1982, "American National Standard for Temperature Measurement Thermocouples"

This standard covers coding of themocouples and extension wire; coding of insulated duplex thermocouple extension wires; terminology, limits of error and wire sizes for thermocouples and thermocouple extension wires; temperature EMF tables for thermocouples; plus appendices that cover fabrication, checking procedures, selection, and installation.

NOTE TO READER

Although the information provided in this Appendix was current when this edition went to press, the Instrument Society of America revises, reaffirms, or withdraws all ISA standards on a five-year cycle and regularly publishes new standards. A listing of current ISA standards, as well as complete purchasing information, will be provided upon request from:

Instrument Society of America
67 Alexander Drive
P.O. Box 12277
Research Triangle Park, NC 27709
(919) 549-8411
Telex 802-540

For further reference regarding Thermocouple Data, see tables ANSI MC96-1-1975, sponsored by ISA.

Process Instrumentation Terminology Extracts are published by ISA. See publication ANSI/ISA S51.1 1979.

Tables of Fundamental Constants and Conversion Factors are published by the National Bureau of Standards. See Misc. Publ. No. M 233.

Index

E

F

G

N

O

P

R

U

V

W

Y

Z